D0857081

Foreword

This publication, *Ground Water and Vadose Zone Monitoring,* contains papers presented at the symposium on Standards Development for Ground Water and Vadose Zone Monitoring Investigations, which was held on 27–29 Jan. 1988 in Albuquerque, New Mexico. The symposium was sponsored by ASTM Subcommittee D18.21 on Ground Water Monitoring, a subcommittee of ASTM Committee D-18 on Soil and Rock, and was developed in cooperation with the U.S. Environmental Protection Agency's Environmental Monitoring Systems Laboratory and the U.S. Geological Survey's Office of Water Data Coordination. David M. Nielsen, of Blasland, Bouck & Lee, presided as chairman of the symposium and also served as editor of this publication. In addition, A. Ivan Johnson, of A. Ivan Johnson, Inc., also served as editor of this publication.

Contents

Overview

The decade of the 1980s has been a period of explosive growth for the field of ground-water and vadose-zone monitoring and a time of both great achievement and confusion for those involved in conducting investigations of ground-water contamination. Passage of the Resource Conservation and Recovery Act (RCRA) by Congress in 1976 and subsequent promulgation of the first of the regulations authorized under RCRA by the U.S. Environmental Protection Agency (EPA) in May 1980 provided the primary impetus for the growth of the field. RCRA, which is EPA's main tool for managing hazardous waste from generation through disposal, includes provisions for establishing ground-water or vadose-zone monitoring systems, or both, at all of this country's hazardous waste treatment, storage, and disposal facilities, which number in the hundreds of thousands. Recent provisions of RCRA specify similar monitoring systems for each of the country's solid-waste facilities (i.e., sanitary landfills), which number in the thousands. Still other provisions of recent Amendments to RCRA (the Hazardous and Solid Waste Amendments of 1986) call for the installation of ground-water or vadose-zone monitoring systems, or both, at underground storage tank locations, which number in the millions across the country.

Passage of the Comprehensive Environmental Response, Compensation and Liability Act (CERCLA), better known as Superfund, by Congress in December 1980 addressed the national threat caused by so-called "uncontrolled" hazardous waste sites, which probably number in the tens of thousands. Cleanup of these sites requires the installation of monitoring devices to investigate the extent of environmental contamination and to monitor the progress of the cleanup. Ground-water and vadose-zone monitoring is also done under other environmental regulatory programs, and for a variety of nonregulatory purposes, creating a tremendous demand for knowledge in this relatively young field.

Like most fields that experience such tremendous surges in growth, the ground-water and vadose-zone monitoring field, if it can truly be called that, saw prolonged periods of disorder and disorganization. In the early 1980s, those persons involved in the newly created field of ground-water monitoring were cautioned that they had to learn from the mistakes that scientists conducting surface-water monitoring programs had made in the 1970s. But the cautions went largely unheeded. Many ground-water quality monitoring investigations were conducted strictly to meet the letter of the law, and many data of poor quality were produced. No real procedural guidelines or standards were developed for those conducting ground-water monitoring investigations to follow. At the same time, the technology for monitoring ground water and the vadose zone was evolving at such a rapid rate that it was difficult for practitioners to keep up. Clearly there was a need to step back and take a long, hard look at the direction in which the field was headed.

Many questions had to be answered. How could we address the millions of sites that now fell under government regulation? Were enough trained and experienced specialists available to do the work and to evaluate the work that would be done? Could we do the work that we were being asked to do with existing methods and technologies?

It was soon realized that if practitioners of ground-water monitoring were to establish credibility for the investigations they were conducting, the state of the art or science had

1

to be improved through the development of useful, practical guidelines and standards. But the idea of developing standards for a mostly inexact science (hydrogeology), in which one commonly has to deal with unexpected conditions (the subsurface) in an exceedingly inhomogeneous environment, was by no means without controversy or without its detractors. Yet, a number of technical fields in which there were similarly difficult and seemingly insurmountable obstacles to standards development have now adopted the "standards approach." The chemical industry or profession, the energy industry, the medical profession, the computer industry, the biotechnology industry, the petroleum industry, and other industries have succeeded in improving the state of their art or science through the development and use of voluntary consensus standards. Could ground-water monitoring follow suit?

The answer to this question was initially explored by ASTM through the conduct of a symposium on Field Methods for Ground-Water Contamination Studies and Their Standardization, sponsored by ASTM Committee D-18 on Soil and Rock and ASTM Committee D-19 on Water, in Cocoa Beach, Florida, in February 1986. The papers from that symposium have been published as *Ground-Water Contamination: Field Methods, ASTM STP 963*. Following this symposium, ASTM Subcommittee D18.21 on Ground-Water Monitoring, a subcommittee of ASTM Committee D-18, was formed to begin the task of identifying where standards were needed and how they could be developed. Subcommittee D18.21 is charged with the responsibility of developing standards for methods and materials used in the conduct of ground-water and vadose-zone investigations. Sections within the subcommittee have been formed to address a variety of narrower subject areas, including (1) surface and borehole geophysics; (2) vadose-zone monitoring; (3) well-drilling and soil sampling; (4) determination of hydrogeologic parameters; (5) well design and construction; (6) well maintenance, rehabilitation, and abandonment; (7) ground-water sample collection and handling; (8) design and analysis of hydrogeologic data systems; (9) special problems of monitoring in karst terrains, and (10) ground-water modeling. With this organization in place it was then possible to start a concentrated effort to use the ASTM consensus process to develop standards needed to ensure the collection of high-quality data that are comparable, compatible, and usable, no matter where or by whom collected. In January 1989 the name of Subcommittee D18.21 was changed to Ground Water and Vadose Zone Investigations to indicate more properly its broad subsurface coverage and interest in all types of ground-water investigations, not just monitoring. Although hundreds of existing ASTM standards related to ground-water quantity and quality investigations already are available, and many others are in the draft stage of development by ASTM Committees D-18 on Soil and Rock, D-19 on Water, and D-34 on Waste Disposal,[1] there are many other standards that are needed.

But what are these "standards" that need such serious development for use in groundwater investigations? If one turns to the definition used by ASTM, a standard is defined as a "rule for an orderly approach to a specific activity, formulated and applied for the benefit and with the cooperation of all concerned," which is essentially what Subcommittee D18.21 is trying to develop as speedily as possible for each of the many operations that can be involved in ground-water investigations. However, the effort to develop standards is by no means meant to discourage new ideas or stifle innovation. Rather, it is an attempt to bring order to a science that is currently struggling to keep pace with sister disciplines.

The editors believe that there are two key points, illustrated by the ASTM definition, that make the standards-developing process work for other professions or industries and

---

[1] See the *Annual Book of ASTM Standards,* most recent edition, Vol. 04.08.

will make it work for ground-water science. The first is that an orderly approach be developed. Few people would argue against this being a desirable goal for any endeavor. The second is that the approach be developed and applied for the benefit and with the cooperation of all concerned. In order to make the standards development process work, the community of professionals in ground-water science must be enlisted—there are now many hard-working volunteers working many hours and days to develop those standards needed for ground-water investigations, but additional expert help is needed. In addition to having more specialists volunteer to work on the ground-water sections and task groups, potential future standards can be found among scientific methods papers presented by authors at symposia and, thus, the incentive for organizing symposia is provided.

To provide a bank of information on new methods that may lead to the development of needed new standards, Subcommittee D18.21 sponsored another symposium in Albuquerque, New Mexico, in January 1988, on Standards Development for Ground-Water and Vadose-Zone Monitoring Investigations. The papers contained in this Special Technical Publication were presented at that symposium and represent a collection of some of the information being used to develop standards for the rapidly growing and evolving field of ground-water and vadose-zone monitoring. The intent of the symposium was to foster interdisciplinary communication and to make available state-of-the-art technology to those scientists and engineers engaged in ground-water and vadose-zone monitoring. A side benefit, but an important one, is that some of the papers may be useful in developing acceptable standards.

The two-and-a-half-day symposium was sponsored by ASTM in cooperation with the U.S. Environmental Protection Agency's Environmental Monitoring Systems Laboratory in Las Vegas, Nevada, and the U.S. Geological Survey's Office of Water Data Coordination, in Reston, Virginia. Featured at the meeting were 40 invited presentations by some of the most noted authorities on the subjects discussed at the meeting. After three peer reviews and review by the editors, 22 papers were accepted for publication. The topics covered in this publication include: (1) vadose-zone monitoring; (2) drilling, design, development, and rehabilitation of monitoring wells; (3) aquifer hydraulic properties and water-level data collection; and (4) monitoring well purging and ground-water sampling.

David M. Nielsen
Blasland, Bouck & Lee, Westerville, OH 43081; symposium chairman and editor.

A. Ivan Johnson
A. Ivan Johnson, Inc., Arvada, CO 80003; editor.

Vadose Zone Monitoring

L. G. Wilson[1]

Methods for Sampling Fluids in the Vadose Zone

REFERENCE: Wilson, L. G., "**Methods for Sampling Fluids in the Vadose Zone,**" *Ground Water and Vadose Zone Monitoring, ASTM STP 1053,* D. M. Nielsen and A. I. Johnson, Eds., American Society for Testing and Materials, Philadelphia, 1990, pp.7–24.

ABSTRACT: This paper reviews available methods for sampling water-borne pollutants in the vadose zone. The "standard" method for sampling pore fluids in the vadose zone is core sampling of vadose zone solids, followed by extraction of pore fluids. The preferred method for solids sampling is the hollow-stem auger with core samplers. Core sampling does not lend itself to sampling the same location time and again. Membrane filter samplers and porous suction samplers are an alternative approach for sampling fluids in both saturated and unsaturated regions of the vadose zone. There are three basic porous suction sampler designs: (1) vacuum-operated suction samplers, (2) pressure-vacuum lysimeters, and (3) high-pressure-vacuum samplers.

Suction samplers are constructed mainly of ceramic and polytetrafluoroethylene (PTFE). The effective range of ceramic cups is 0 to 60 cbar of suction. The operating range of PTFE cups installed with silica flour is 0 to 7 cbar of suction. This paper reviews a method for extending the sampling range of suction samplers using an injection-recovery procedure. Factors affecting the operation of porous suction samplers include the physical properties of the vadose zone, hydraulic conditions, cup-wastewater interactions, and climatic conditions. Innovative procedures include the water extractor and the filter-tip system sampler. The free-drainage samplers include pans, blocks, and wick-type samplers. Techniques for sampling from perched ground-water zones include profile samplers, sampling from cascading wells, and sampling from dedicated wells.

KEY WORDS: ground water, vadose zone monitoring, soil core sampling, porous suction samplers, free-drainage samplers, perched ground-water sampling

Until recently, the major emphasis in monitoring programs at waste management sites was on ground-water sampling. This emphasis ignores the value of vadose zone monitoring techniques for the early detection of pollutant movement from a waste management unit. In regions with deep water tables, such as in the southwestern United States, the potential consequence of ignoring vadose zone monitoring is that the vadose zone and the ground-water system may become polluted before tangible evidence of leakage is evident in ground-water samples. Today, federal regulations for hazardous waste land treatment units (Subtitle C of the Resource Conservation and Recovery Act, Section 264.278 of 40 CFR, Part 264) mandate vadose zone monitoring. Some states (for example, California) also require vadose zone monitoring at waste management facilities.

Essentially vadose zone monitoring is a component of a comprehensive monitoring system at a given waste management facility. Such a system includes surface liquid and solids sampling, vadose zone monitoring, and ground-water monitoring. Similarly, there is a

---

[1] Hydrologist, Department of Hydrology and Water Resources, University of Arizona, Tucson, AZ 85721.

suite of methods available for vadose zone monitoring systems. Everett, Wilson, and McMillion [1] reviewed the available methods for premonitoring and active vadose zone monitoring, including both sampling and nonsampling methods. Both sets of methods are required since nonsampling techniques establish movement of liquids in the vadose zone (such as during a leak from an impoundment) and sampling techniques obtain liquid samples for laboratory analysis. For example, evidence of water content changes by neutron logging in access wells triggers the need to sample from pore-liquid samplers. Morrison [2], Everett, Wilson, and Hoylman [3], Rhoades and Oster [4], and Gardner [5] discuss available nonsampling (indirect) techniques for vadose zone monitoring in detail.

The primary purpose of this paper is to review techniques for extracting liquids from the vadose zone. Sampling techniques are divided into three groups: solids sampling followed by extraction of pore liquids, unsaturated pore-liquid sampling, and saturated pore-liquid sampling.

Solids Sampling for Pore Liquid Extraction

Solids sampling, followed by laboratory extraction of pore liquids, is the standard method for characterizing pollutant distribution in the vadose zone. Solids sampling also provides useful geological information, such as the lithological characteristics of the vadose zone layers, and the water content distribution. Comprehensive references on solids sampling include a book by Hvorslev [6], the text by Driscoll [7], an Environmental Protection Agency guidance document for ground-water monitoring at RCRA sites [8], and reviews by Riggs [9], Everett and Wilson [10], and Hackett [11]. There are two broad categories of solids sampling methods: hand-operated samplers and power-operated sampling rigs.

Hand-Operated Samplers

Hand-operated samplers are basic sampling units developed by agriculturalists for determining soil texture, soil water content, soil fertility, and soil salinity. Common units include tube-type samplers and auger samplers.

The Veihmeyer or King tube is a commonly used tube sampler for obtaining a long continuous sample near the land surface (see Fig. 1). This sampler consists of a hardened cutting point and head, threaded into a body tube. The upper end of the body is connected to a head containing two opposing, protruding lugs. A drop hammer is provided. The operation of the sampler is illustrated in Fig. 1. Gentle tapping on the head removes the sample. These samplers are commercially available in lengths from 1.22 m (4 ft) to 1.83 m (6 ft) [12]. The inside diameter of commercially available samplers is 1.9 cm (¾ in.).

Other commercial tube-type samplers are available for obtaining larger, 8.9-cm (3½-in.)-diameter, continuous soil cores [12]. Brass retaining cylinders, slipped into the sampler barrel, provide undisturbed samples for extraction of pore liquids. Samples are extracted from inside the drive barrel by means of a pusher rod.

Chong, Khan, and Green [13] described a soil core sampler for obtaining shallow samples using a hand-operated two-ton jack. Sharma and De Datta [14] reported a core sampler for obtaining undisturbed samples from the upper 10.2 cm (4 in.) of puddled soils.

Another common hand augering tool is the bucket auger. The basic components of bucket augers are shown on Fig. 2. Types of bucket augers include the common posthole auger, ditch augers, and regular or general purpose barrel augers [10]. Variations of the general purpose auger include sand and mud augers. These samplers are best suited for sampling shallow soil depths. However, using a tripod and pulley allows sampling to depths up to 24.4 m (80 ft) in some types of materials.

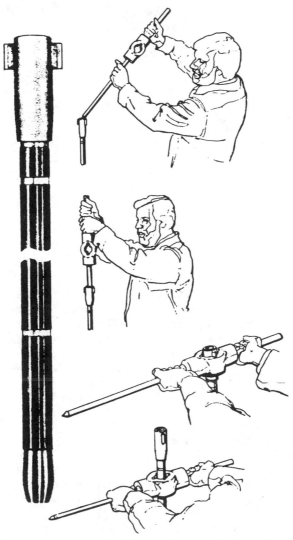

FIG. 1—*Tube-type sampler, showing the operational procedure* [12].

Power-Operated Sampling Rigs

Power-operated samplers are identical to the units used to sample below water tables. However, commonly used drill rigs, such as cable tool and rotary units, are not recommended because they generally require drilling fluid. Air drilling is undesirable when attempting to obtain pore liquids at field moisture content. Two suitable power-driven techniques are bucket augers and flight-type auger drill rigs.

Bucket augers are large-diameter, that is, 1.83-m (6-ft)-diameter cylindrical buckets with auger-type cutting blades on the bottom [7]. In practice, the bucket is rotated with depth in the vadose zone until the bucket is full. Sampling consists of extracting small-diameter core samples from the interior of the bucket after the full bucket is lowered to the ground.

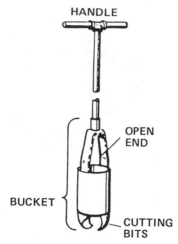

FIG. 2—*Bucket-type hand auger* [12].

This approach minimizes problems with cross-contamination of samples. Bucket augers are best suited for sampling from relatively stable formations. Common sampling depths are between 15.25 m (50 ft) and 45.75 m (150 ft) [7].

Solid and hollow-stem augers are also used for sampling. The most basic, power-driven flight auger is driven by a small air-cooled engine. Handles attached to the head assembly allow two operators to guide the continuous-flight auger into the soil. Additional flights are added as required. Samples are brought to the surface by the screw action of the auger.

Larger solid-stem augers are mounted on drill rigs. Attached to the lowermost, or leading, flight of augers is a cutter head, about 5 cm (2 in.) larger in diameter than the flights [7]. Auger diameters are available up to 61 cm (24 in.). Flight lengths are generally 1.52 m (5 ft). Typical drilling depths range from 15.25 m (50 ft) to 36.6 m (120 ft), depending on the texture of the vadose zone sediments [7].

The favored method for sampling in unconsolidated material is the hollow-stem, continuous-flight auger (see Fig. 3). This design simultaneously rotates and axially advances a hollow-stem auger column. The auger head contains replaceable carbide teeth that pulverize the formation deposits during rotation of the flight column. The solid flight column serves as a temporary casing while relatively undisturbed samples are obtained from within the hollow stem [7]. Recently, Hackett [11] published an excellent state-of-the-art review of this technique.

According to the U.S. Environmental Protection Agency (EPA) [8], wells with diameters up to 10.2 cm (4 in.) have been constructed with hollow-stem augers. In addition, attempts are under way to construct augers with inside diameters of about 25.4 cm (10 in.). The cutting diameter is somewhat greater than the flighting diameter because of the protruding carbide teeth. Individual flights are generally 1.52 m (5 ft) in length, although auger flights up to 3.05 m (10 ft) in length are available. The total completion depth is about 45.75 m (150 ft). Water is not added to the hole during augering to avoid diluting and contaminating pore fluids.

Tubular samplers provide "undisturbed" vadose zone sediments from inside a hollow-stem auger (see Fig. 3). Three popular tubular samplers are ring-lined barrel samplers, split tube samplers, and thin-wall "Shelby tube" samplers. These samplers are commercially available in a variety of diameters. Brass cylindrical rings inserted inside a barrel sampler

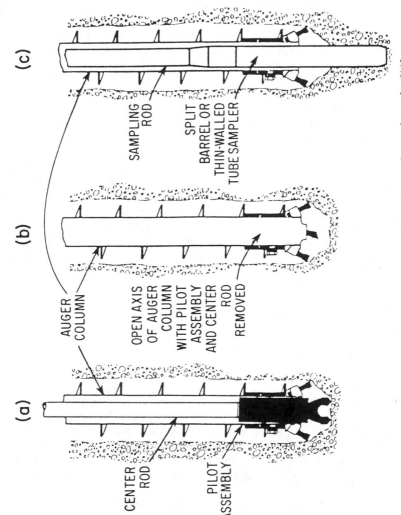

FIG. 3—*Hollow-stem auger, showing the method of sampling with sampling tubes* [11].

protect the samples during shipment to a laboratory. After retrieving the sampler, the sampling technician forces the cylinders out of the sampler, separates the cylinders with a spatula, covers the ends with Teflon sheets and rubber caps, and seals the caps onto the cylinders with electrician's tape. This technique is particularly important for reducing loss of volatile organics in the samples.

According to Hackett [11], continuous-sampling tube samplers are available to permit sampling as the auger column is rotated and axially advanced. Barrel samplers of 1.53 m (5 ft) are used. The sampler does not rotate with the augers [11]. According to Riggs [9], this system overcomes difficulties in sample recovery experienced with standard techniques. The auger teeth and flight continuously relieve overburden stresses and break the vacuum at the base of the sample at the end of sampling [9]. For additional information on samplers see Hvorslev [6], Riggs [9], Hackett [11], and Fenn et al. [15].

Advantages and Disadvantages of Solids Sampling

Solids sampling for extraction of pore fluids is valuable for characterizing depth-related changes in the composition of pore fluids. A major disadvantage is that core sampling is basically a destructive technique because additional samples cannot be obtained from the same location. This precludes taking samples from the same location at later times for determining trends. Also, as with all techniques for sampling pore liquids, extrapolating data from a single sampling hole to wider areas is risky because of the natural spatial variability of vadose zone hydraulic properties.

Cross contamination is an inherent danger when sampling with solid-flight and hollow-stem augers [11]. Contaminated solids are transported on the auger flights both upward and downward in the hole. This problem is largely circumvented when sampling by drive tube through a hollow-stem auger by discarding the uppermost cylinder. The sample in this cylinder contains solids originally at the base of the hole, possibly including sediment brought down during drilling.

Downward leakage of contaminated, perched ground water may occur through the borehole during drilling. Thus, if the principal water table is shallow, contaminants may be short-circuited to the water table.

Pore-Liquid Sampling from Unsaturated Regions of the Vadose Zone

Pore liquid samples cannot be obtained from open cavities in unsaturated regions of the vadose zone because the fluid is held at negative matric potentials. Thus, wells or sampling pans cannot be used to obtain liquid samples under these conditions. Methods for sampling under unsaturated conditions include cellulose-acetate hollow fibers, membrane filter samplers, and suction samplers, also called suction lysimeters. These units have an advantage over solids sampling in allowing sequential sampling from fixed locations in the vadose zone.

Jackson, Brinkley, and Bondietti [16], and Morrison [17] described the cellulose-acetate hollow-fiber sampler. These semipermeable fibers function as molecular sieves for the dialysis of aqueous solutions. Advantages that have been claimed for these samplers include flexibility, small diameter, minimal chemical interaction of solute with the tube matrix, and sample compositions similar to those obtained using ceramic cups [4]. So far these samplers are more suitable for laboratory studies than for field sampling.

Stevenson [18] presented the design of a suction sampler using a membrane filter mounted in a "Swinnex"-type filter holder (see Fig. 4). These filters are composed of polycarbonate or cellulose-acetate. The collector system draws vadose zone fluid by capillar-

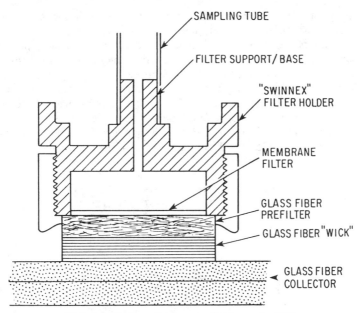

FIG. 4—*"Swinnex"-type pore fluid sampler* [16].

ity. The collected fluid then flows in the collector sheets toward the glass fiber wicks as a result of suction applied to the filter holder assembly [10]. The "wick and collector" assembly has a relatively large contact area with vadose zone sediments. A flexible tube, attached to the filter holder, permits applying a negative pressure to the system when delivering collected liquid to a sample bottle. So far extensive use of this technique has not been reported in the literature.

Porous Suction Samplers

Suction samplers (also called soil lysimeters) are the favored method for sampling pore fluids under suction in the vadose zone. Basically, these samplers comprise a porous cup mounted on the end of a hollow tube [generally of polyvinyl chloride (PVC)]. An alternative design incorporates a cylindrical porous section within the tubing rather than at the end of the tubing [2]. When placed into the vadose zone, the pores (and fluid) in the porous sampler form a continuum with the pores (and fluid) in the surrounding medium. The surrounding medium consists of either native material or silica flour. The porous samplers are made from various materials, including ceramic, Alundum, polytetrafluoroethylene (PTFE), and nylon [2,19]. Ceramic and Alundum cups are hydrophilic, whereas PTFE cups are hydrophobic [19].

There are three basic types of suction cup samplers: (1) the vacuum-operated sampler, (2) the pressure-vacuum-operated sampler, and (3) the pressure-vacuum sampler with check valves. Vacuum-operated samplers consist of a suction cup mounted on a body tube that projects slightly out of the soil surface. A negative air pressure, applied to the interior of the sampler through a small-diameter tube, draws pore fluid into the cup for delivery to a sample bottle at the land surface. The practical depth limitation on these samplers is about 7.62 m (25 ft). However, because of the expense of using long body tubes, these units are not generally used for depths greater than 1.83 m (6 ft).

Parizek and Lane [20] designed vacuum-pressure samplers for sampling at depths beyond the suction limit of vacuum samplers. As shown on Fig. 5, these samplers include a ceramic cup mounted on a body tube, a sample delivery line, and a pressure-vacuum line. When sampling, a negative pressure is applied to the inside of the sampler, drawing pore fluids into the cup. Subsequent application of pressure through the vacuum-pressure line forces the collected fluids through the discharge line to sample bottles. These samplers function most effectively at depths above 15.25 m (50 ft). Experience has shown that below 15.25 m (50 ft) application of pressure forces part of the collected sample out of the cup.

Pressure-vacuum samplers with check valves, designed by Wood [21], are used for sampling depths greater than 15.25 m (50 ft). Figure 6 shows the components of this unit. The

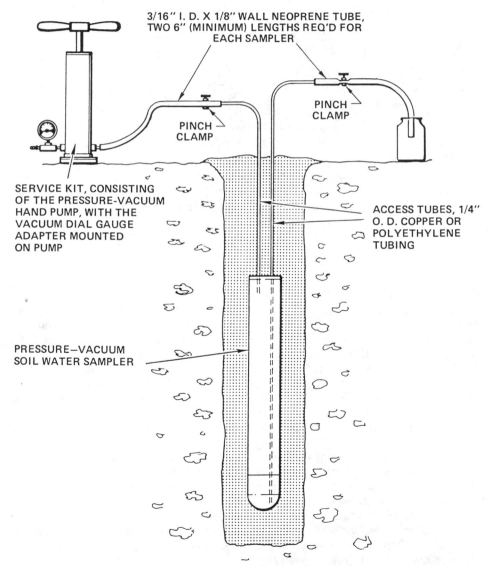

FIG. 5—*Vacuum-pressure suction cup sampler* [22].

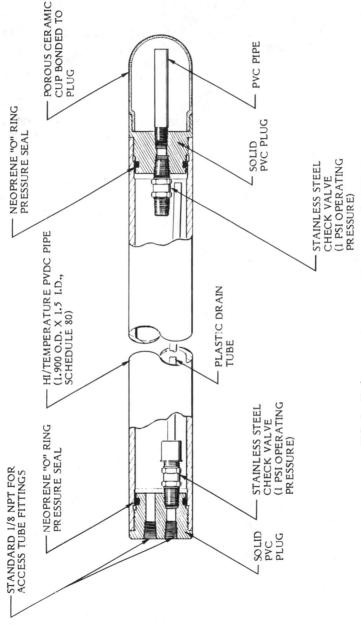

POROUS CERAMIC CUP BONDED TO PLUG

NEOPRENE "O" RING PRESSURE SEAL

PVC PIPE

SOLID PVC PLUG

STAINLESS STEEL CHECK VALVE (1 PSI OPERATING PRESSURE)

HI/TEMPERATURE PVDC PIPE (1.900 O.D. X 1.5 I.D., SCHEDULE 80)

PLASTIC DRAIN TUBE

NEOPRENE "O" RING PRESSURE SEAL

STAINLESS STEEL CHECK VALVE (1 PSI OPERATING PRESSURE)

STANDARD 1/8 NPT FOR ACCESS TUBE FITTINGS

SOLID PVC PLUG

FIG. 6—*High-pressure/vacuum suction cup sampler* [22].

sampler consists of a lower and an upper chamber. The upper chamber has a capacity of about 1 L. The lower chamber includes the suction cup. A sample tube, containing a stainless steel check valve, connects the lower and upper chambers. The sample delivery tube also contains a stainless steel check valve.

Applying a vacuum opens the one-way check valve to the lower chamber, drawing in pore liquid through the cup. The vacuum also shuts the valve in the discharge line and forces the collected fluid from the lower chamber into the upper chamber. Subsequently, the vacuum is released and a pressure of nitrogen gas is applied. This shuts the lower valve, opens the upper valve, and forces the sample through the delivery line to the surface. High pressures do not damage the cup or cause backflow of the sample from the cup. These units sample from depths as low as 91.5 m (300 ft) below the land surface [22].

Preparation of the ceramic cups involves leaching several pore volumes of acid (for example, 1 N HCl) through the cups, followed by flushing with deionized water. This process reduces the concentrations of major cations and trace elements released from the cup during sampling [19]. Care is required when installing a sample line into the base of a sampling cup to avoid having the tubing catch on the inside lip of the cup (see Fig. 7). Figure 7 shows that bevelling the end of the tube reduces this problem [23].

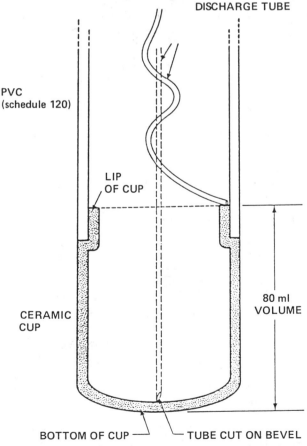

FIG. 7—*Creation of dead space in suction samplers* [10].

Installating suction samplers includes constructing a borehole to the depth of interest, pouring in a slurry of silica flour (454 g (1 lb) of 200-mesh silica flour per 150 mL of water [23]), using a tremie, to cover the sampler and 0.6 m (2 ft) or 0.915 m (3 ft) of the borehole, then backfilling with bentonite. Technicians adjust the pressure-vacuum in the system using a hand pump.

Round-bottom sampling cups are commercially available in a number of sizes and air-entry values [24]. Body lengths are available up to 25.4 cm (10 in.), and even longer upon request. Greater lengths may be desirable for units installed in fractured media to increase the opportunity for contacting fractures.

Operational Features of Porous Suction Samplers

Table 1 summarizes the operational features and constraints of suction cup samplers.

Factors affecting the operation of suction cup samplers include the air-entry, or bubbling, pressure of the cups; the effective operating range of the cups; the physical conditions of the soil, including hydraulic factors; cup-fluid interactions; and climatic factors [3,23]. Everett and McMillion [23] obtained laboratory data on air entry pressures ("bubbling pressures") for ceramic and PTFE cups (see Table 1). Air entry pressures are related to the pore size relationships of the cups. Table 1 summarizes pore size relationships for these cups.

The most critical operational feature of the porous cup is the range of negative pore pressures for effective sampling. Generally, "effective sampling" can be thought of in terms of sampling times and sample volume. Basically, as the suction increases in the medium being sampled, the unsaturated hydraulic conductivity of the medium becomes so low that the rates of delivery become vanishingly small. This occurs even though the air entry value of the sampler has not been exceeded. Table 1 summarizes the effective sampling ranges for ceramic and PTFE samplers.

Amter [25] evaluated the injection/recovery technique for extracting fluids from the vicinity of suction samplers beyond the normal operating range. His technique involves injecting deionized water through the cup into the surrounding medium and subsequently drawing the mixture of injected water and native fluid back into the cup. Because of dilution, the method is best suited when looking for target chemicals. The technique can be used with vacuum-type samplers and pressure-vacuum samplers.

The likelihood that suction cups will operate in the wet range is enhanced by terminating units near the interface between layers of differing texture. Information on layered conditions is obtained during solids sampling. For example, Fig. 8 shows a textural and water content profile from a sampling hole near a dry well in Tucson, Arizona. Three regions of high water content are present in fine-grained sediments. Suction samplers should be located above the contacts between the coarse and fine sediments.

Another important operational feature of suction cup samplers is loss of vacuum. Everett and McMillion [23] found that ceramic samplers were able to maintain a high vacuum (about 93 cbar) for 20 days of laboratory testing. The vacuum dropped off in all PTFE units, which must be embedded in silica flour to minimize vacuum loss. The joints in the units with screw fittings must be sealed.

Effect of the Soil Physical Properties on the Operation of Porous Suction Samplers

The use of suction samplers to obtain pore-liquid samples for determining a "representative" water chemistry in a relatively unstructured soil has been questioned [3,4,26–29].

TABLE 1—*Operational features and constraints with suction samplers.*

PORE SIZE	OPERATING RANGE	AIR ENTRY
1. Low-flow ceramic samplers: 1.04 microns	1. Ceramic samplers: 0 to 60 centibars	1. Low flow ceramic samplers: 233 centibars
2. High flow ceramic samplers: 2.8 microns	2. PTFE samplers packed in silica sand: 0 to 20 centibars	2. High flow ceramic samplers: 100 centibars
3. PTFE cups vary from 14 to 300 microns		3. PTFE cups: 20 centibars

SOIL PHYSICAL PROPERTIES	HYDRAULIC PROPERTIES	CUP-FLUID INTERACTIONS
1. Maintaining contact between cups and pores is difficult in coarse-textured or fractured materials. Solution: place cups in silica sand.	1. The hydraulic properties of the sphere sampled by a cup vary. Thus, point samplers provide only an indication of relative and not absolute changes in pore liquid composition. Solution: use samplers to detect arrival of pollutants but not absolute concentrations.	1. Clogging by particulates does not appear to be a problem. Precipitation of chemical compounds in pores is possible in reducing environments.
2. The composition of pore liquid from sequences of large pores and cracks differs from the composition of liquid from micropores.	2. A two-domain flow system exists in structured and fractured media; gravity flow occurs in cracks; unsaturated flow occurs inside blocks. Solution: use free-drainage samplers to sample from blocks. Alternative, long cups embedded in silica sand will enhance sampling from cracks.	2. Suction cups filter out bacteria but not viruses.
		3. Sorption or repulsion of molecules by ceramic cup samplers may change the concentrations of negatively charged contaminants present in low concentrations. The problem is accentuated by long sampling times.
		4. Ceramic, alundum, and PTFE cups release some trace elements and metals to solution flowing through cups. A problem only if concentrations in pore fluids are low.

CLIMATIC FACTORS
1. In frozen soils the tension of unfrozen water may prohibit flow into a cup.
2. Water films in extraction line may freeze. Solution: install a bleed valve in the extraction line.

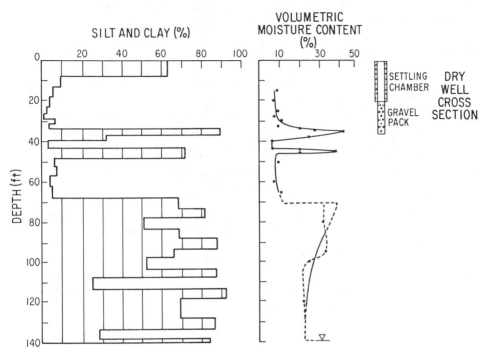

FIG. 8—*Geological and water-content profiles near a dry well in Tucson, Arizona.*

As pointed out by England [26], the composition and concentration of pore liquid is not homogeneous throughout a given mass. Accordingly, pore liquid extracted from sequences of large pores at low suction may have a composition that is quite different from that extracted from sequences of micropores. Moreover, "a point source of suction, such as the porous cup, samples roughly a sphere, draining different-sized pores as functions of distance from the point, the amount of applied suction, the hydraulic conductivity of the medium, and the soil-water content" [26]. The concentrations of various ions in solution do not as a general rule vary inversely with the soil solution [26].

Variability of hydraulic properties within the sampling sphere also affects sampling. Accordingly, "point samples" only provide an indication of relative changes in pore liquid composition but not absolute concentrations, unless the spatial variability of the vadose zone properties are quantified [4]. Thus, suction samplers are best suited for detecting the arrival of pollutants at the sampling depth and not the absolute amounts or concentrations.

Structured and fractured media accentuate the problems described in the last two paragraphs [28]. In such media, liquid movement occurs in a two-domain system [29]. One domain consists of the porous matrix. The conventional, Darcian-based unsaturated flow equation describes flow in this domain. The second domain includes a system of fractures or macropores through which water flows by gravity [29]. Free-drainage samplers are best suited for sampling saturated drainage in the larger cracks and macropores [27]. Suction samplers are best suited for sampling from the unsaturated flow regime within the blocks. The opportunity for suction samplers to extract liquid from fractures and macropores may be enhanced by embedding sampling units in sections of silica sand.

Cup-Fluid Interactions

There are two classes of interactions: (1) those affecting the operation of cups by plugging and (2) those affecting the composition of liquids flowing through the cups. Clogging by particulate matter does not appear to be a problem [*30*]. Formation of chemical precipitates within the cup pore space is always a possibility, for example, in the reducing environments of perched ground water.

Suction cups appear to be very effective in filtering out bacteria. During laboratory studies, Dazzo and Rothwell [*31*] observed a 100-fold to 10 000 000-fold reduction in fecal coliform in manure slurry during sampling, and 65% of the cups yielded coliform-free samples. They concluded that suction samplers do not yield valid water samples for fecal coliform analysis. In contrast, suction samplers are effective in sampling for viruses.

Anderson [*28*] found that low concentrations of negatively charged contaminants present in the soil solution may be changed during sampling because of sorption or repulsion of molecules in ceramic cup samplers. The long time delay required to obtain sufficient sample for laboratory analyses in slowly permeable sediments may also alter the chemical composition. Similarly, reducing conditions within a cup may alter the concentration of metals in a sample [*28*]. During studies at a tannery disposal site, Anderson [*28*] found that chromium precipitated in the samplers because of changes in redox conditions between the soil and sampler. She estimates that the quantity precipitated was equivalent to 4 to 40 times the measured dissolved concentrations.

Climatic Factors

Climatic conditions may affect the operation of suction samplers (see Table 1). For example, in frozen soils the tension of unfrozen water may be great enough to limit flow into the cups [*3*]. Water films in the extraction line may freeze [*17*]. This problem may be eliminated by installing a bleeder valve along the extraction line [*17*].

Alternative Suction Sampler Designs

According to Baier, Aljibury, and Meyer [*32*], tensiometer units, used to determine water potentials in the vadose zone, may be converted to vacuum-pressure samplers by draining the tensiometer fluid during a sampling cycle. This is possible because tensiometer cups are also made of ceramics. Baier, Aljibury, and Meyer [*32*] used this approach to sample the vadose zone beneath sewage sludge ponds at Sacramento, California. When interpreting the results, care is required to account for the presence of residual tensiometer fluid in the collected sample.

Nightingale, Harrison, and Salo [*33*] described a subsurface water extractor that can be used under both saturated and unsaturated conditions, for example, in the vicinity of fluctuating water tables. Figure 9 shows the basic design of their sampler. The interior of the body tube serves as a sample reservoir. Applying a vacuum to the vacuum-pressure line draws the sample up the standpipe into the reservoir. The sample remains in the reservoir when the vacuum is removed. Pressurizing the pressure-vacuum line with nitrogen gas forces the sample through the outlet line into the sample bottle. The backflow of sample through the ceramic cup is minimal. The maximum sampling depth is estimated to be 28 m (91.8 ft). The reservoir capacity is from 1.5 to 2.0 L.

Haldorsen, Petsonk, and Tortstensson [*34*] reported on the filter-tip sampler system for obtaining samples from both saturated and unsaturated regions of the vadose zone. The system includes a downhole filter tip connected to an access tube, and aboveground adap-

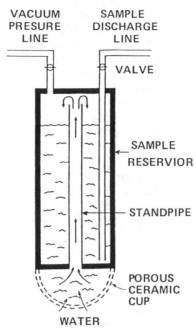

FIG. 9—*Alternative suction sampler design* [33].

tors. The hollow filter tip consists of a porous filter section above a pointed drive point, a threaded section to be attached to the access pipe, and an upper, necked-down section containing a rubber septum. The pore size of the filtered section is about the same as that for the suction samplers, approximately 3 μm. This ensures that the porous segment remains saturated even though the suction in the surrounding medium increases [*34*].

The first phase of sampling with the filter-tip sampler system involves lowering a sampling adaptor into the access hole to make contact with the filter tip. The adaptor incorporates an air-evacuated sampling vial with a rubber septum and a sliding section with a concentric, double-ended hypodermic needle. The second phase involves forcing the hypodermic needle through the rubber septums into the filter tip and sampling vial, causing fluid, drawn through the porous section by vacuum, to fill the sample vial.

Pore-Liquid Sampling from Saturated Regions of the Vadose Zone

Two classes of saturation develop in an otherwise unsaturated vadose zone: (1) free-drainage water, and (2) perched ground water. Free drainage occurs when fluid applied at the land surface flows downward in a profile under saturated or near-saturated conditions through macropores and through fractures and cracks. Perched ground water develops at the interface between vadose zone layers of differing texture, for example, in a gravel layer overlying a clay lens.

Free-Drainage Samplers

Free-drainage samplers intercept and collect water flowing in saturated pores or fractures for delivery to a sample container. Everett and Wilson [*10*], Hornby, Zabcik, and Crawley

[27], and Barbee [35] reviewed alternative designs for free-drainage samplers, including stainless steel troughs, sand-filled funnels, and hollow glass blocks. In each case, gravity drainage creates a slightly positive pressure at the soil-sampler interface causing fluid to drip into the sampler. Shaffer, Fritten, and Baker [36] designed a 20-cm (7.9-in.)-pan lysimeter with a tension plate capable of pulling a six-centibar suction.

Recently, Hornby, Zabcik, and Crawley [27] presented the design of a wicking-type sampler (see Fig. 10).

Typically, free-drainage samplers are installed in tunnels, extending from trenches or buried culverts [2,37], constructed to the total desired sampling depth. The total depth of sampling is limited only by the availability of construction equipment. Similarly, the horizontal extent of the samplers depends on the availability of tunneling equipment. Hornby, Zabcik, and Crawley [27] described a barrel lysimeter comprising an encased soil monolith of undistributed soil. Above the sealed base of the monolith is a system of porous ceramic cups. The monolith is placed in a tight-fitting hole beneath a monitoring site (for example, in the treatment zone beneath the zone of incorporation of a land treatment unit). The lysimeters collect fluid draining through the monolith.

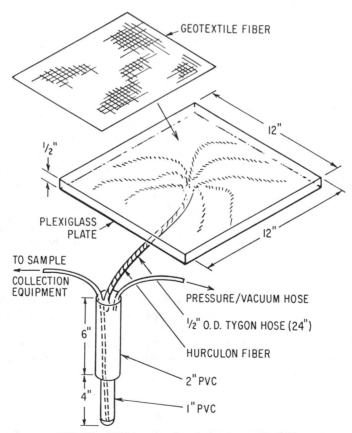

FIG. 10—*Wicking-type free-drainage sampler* [27].

Sampling from Perched Ground Water

The presence of perched ground-water regions in the vadose zone provides the opportunity to obtain larger volumes of fluid for analysis than is possible with core samples and suction samplers. There are three basic sampling alternatives: (1) sampling tile drainage, (2) sampling cascading water, and (3) sampling from wells.

In agricultural areas with high water tables, a common practice is to install buried tile lines to maintain water table levels below the rooting depths of crops. Samples of drain water are accessible at the point of discharge of a tile line into an open ditch. The samples may or may not be representative of the "average" conditions, depending on the distribution of soil mapping units in the drained area.

Piezometers or multilevel samplers collect samples from shallow perched ground-water regions. Pickens et al. [38] described a mulitlevel sampler suitable for sampling in cohesionless soils in which flow is predominantly in the horizontal direction.

Wilson and Schmidt [39] described techniques for sampling from cascading wells. Cascading water occurs through cracks in casing joints and through dewatered perforations. Cascading water samples are obtained from abandoned wells near the site being monitored or from operating wells after the removal of pumps for servicing. A bucket lowered into the cascading stream collects the water samples. Alternatively, water is collected at the wellheads during the start-ups of wells that have been shut down for a prolonged period.

Dedicated wells provide perched ground-water samples from deeper regions of the vadose zone. The construction techniques and well designs are identical to those used for ground-water monitoring wells [8]. A problem in some areas is that perched ground-water systems tend to be ephemeral [39]. Backup systems, for example, suction samplers and the extractor design by Nightingale, Harrison, and Salo [33], are recommended for these conditions.

References

[1] Everett, L. G., Wilson, L. G., and McMillion, L. G., *Ground Water,* Vol. 20, No. 3, May–June 1982, pp. 312–324.

[2] Morrison, R. D., *Ground Water Monitoring Technology,* Timco Mfg., Inc., Prairie DuSac, WI, 1983.

[3] Everett, L. G., Wilson, L. G., and Hoylman, E. W., "Vadose Zone Monitoring for Hazardous Waste Sites," Noyes Data Corp., Park Ridge, NJ, 1984.

[4] Rhoades, J. D. and Oster, J. D. in *Methods of Soil Analysis,* Part I, *Physical and Mineralogical Methods,* 2nd ed., A. Klute, Ed., American Society of Agronomy, Madison, WI, 1986, pp. 985–1006.

[5] Gardner, W. H., in *Methods of Soil Analysis,* Part I, *Physical and Mineralogical Methods,* 2nd ed., A. Klute, Ed., American Society of Agronomy, Madison, WI, 1986, pp. 493–544.

[6] Hvorslev, M. J., *Subsurface Exploration and Sampling of Soils for Civil Engineering Purposes,* U.S. Army Corps of Engineers Waterways Experimental Station, Vicksburg, MS, 1949.

[7] Driscoll, F. G., *Groundwater and Wells,* 2nd ed., Johnson Division, St. Paul, MN, 1986.

[8] "RCRA Ground-Water Monitoring Technical Enforcement Guidance Manual," U.S. Environmental Protection Agency, Washington, DC, 1986.

[9] Riggs, C. O. in *Proceedings,* Conference on Characterization and Monitoring of the Vadose (Unsaturated) Zone, Las Vegas, NV, 8–10 Dec. 1983, National Water Well Association, Worthington, OH, pp. 611–622.

[10] Everett, L. G. and Wilson, L. G., "Permit Guidance Manual on Unsaturated Zone Monitoring for Hazardous Waste Land Treatment Units," EPA/530-SW-86-040, U.S. Environmental Protection Agency, Environmental Monitoring Systems Laboratory, Las Vegas, NV, 1986.

[11] Hackett, G., *Ground Water Monitoring Review,* Vol. 7, No. 4, 1987, pp. 51–62.

[12] "Soil Sampling Tubes," Product Bulletin A23, Soilmoisture Equipment Corp., Santa Barbara, CA, 1981.

[13] Chong, S. K., Khan, M. A., and Green, R. E., *Journal of the Soil Science Society of America,* Vol. 46, 1982, pp. 433–434.

[14] Sharma, P. K. and De Datta, S. K., *Journal of the Soil Science Society of America,* Vol. 49, 1985, pp. 1069–1070.

[15] Fenn, D. E., Isbister, J., Briads, O., Yare, B., and Roux, P., "Procedures Manual for Ground Water Monitoring at Solid Waste Disposal Facilities," EPA/530/SW, United States Geological Survey, Denver, CO, 1977.

[16] Jackson, D. R., Brinkley, F. S., and Bondietti, E. A., *Journal of the Soil Science Society of America,* Vol. 40, 1976, pp. 327–329.

[17] Morrison, R. D., *Soil Science,* Vol. 134, No. 3, 1982, pp. 206–210.

[18] Stevenson, C. D., *Environmental Science and Technology,* Vol. 12, 1978, pp. 329–331.

[19] Creasey, C. L. and Dreiss, S. J. in *Proceedings,* Conference on Characterization and Monitoring of the Vadose (Unsaturated) Zone, Denver, CO, 19–21 Nov. 1985, National Well Water Association, Worthington, OH, 1985, pp. 173–181.

[20] Parizek, R. R. and Lane, B. E., *Journal of Hydrology,* Vol. 11, 1970, pp. 1–21.

[21] Wood, W. W., *Water Resources Research,* Vol. 9, No. 2, 1973, pp. 486–488.

[22] "About Our Soil Water Samplers," Soilmoisture Equipment Corp., Santa Barbara, CA, 1980.

[23] Everett, L. G. and McMillion, L. G., *Ground Water Monitoring Review Journal,* Vol. 5, No. 3, 1985.

[24] "Porous Ceramics by Soilmoisture," 600 Series Catalog, Soilmoisture Equipment Corp., Santa Barbara, CA, 1987.

[25] Amter, S., "Injection/Recovery Lysimeter Technique for Unsaturated Zone Soil-Water Extraction," M.S. thesis, University of Arizona, Tucson, AZ, 1987.

[26] England, C. B., *Water Resources Research,* Vol. 10, No. 6, 1974, p. 1049.

[27] Hornby, W. J., Zabcik, J. D., and Crawley, W., *Ground Water Monitoring Review,* Vol. 6, No. 2, 1987, pp. 61–66.

[28] Anderson, L. D., *Ground Water,* Vol. 24, No. 5, 1986, pp. 761–769.

[29] Nielsen, D. R., van Genuchten, M. T., and Biggar, J. W., *Water Resources Research,* Vol. 2, No. 9, 1986, pp. 89S–108S.

[30] Johnson, T. M., Cartwright, K., and Schuller, R. M., *Ground Water Monitoring Review,* Vol. 1, No. 3, 1981, pp. 55–63.

[31] Dazzo, F. B. and Rothwell, D. F., *Applied Microbiology,* Vol. 27, No. 6, 1974, pp. 1172–1174.

[32] Baier, D. C., Aljibury, F. K., Meyer, J. K., and Wolfenden, A. K. in *Proceedings,* Conference on Characterization and Monitoring of the Vadose (Unsaturated) Zone, Las Vegas, NV, 8–10 Dec. 1983, National Water Well Association, Worthington, OH, pp. 476–491.

[33] Nightingale, H. I., Harrison, D., and Salo, J. E., *Ground Water Monitoring Review,* Vol. 5, No. 4, 1985, pp. 43–50.

[34] Haldorsen, S., Petsonk, A. M., and Torstensson, B.-A. in *Proceedings,* Conference on Characterization and Monitoring of the Vadose (Unsaturated) Zone, Denver, CO, 19–21 Nov. 1985, National Water Well Association, Worthington, OH, pp. 159–172.

[35] Barbee, G. C., "A Comparison of Methods for Obtaining "Unsaturated Zone" Soil Solution Samples," M.S. thesis, Texas A & M University, College Station, TX, 1983.

[36] Shaffer, K. A., Fritton, D. D., and Baker, D. E., *Journal of Environmental Quality,* Vol. 8, 1979, pp. 241–246.

[37] Kmet, P. and Lindorff, D. E. in *Proceedings,* Characterization and Monitoring of the Vadose (Unsaturated) Zone, Las Vegas, NV, 8–10 Dec.1983, National Water Well Association, Worthington, OH, pp. 554–579.

[38] Pickens, J. F., Cherry, J. A., Coupland, R. M., Grisak, G. E., Merrit, W. F., and Risto, G. A., *Ground Water,* Vol. 6, No. 5, 1987, pp. 322–327.

[39] Wilson, L. G. and Schmidt, K. D., "Monitoring Perched Ground Water in the Vadose Zone," *Establishment of Water Quality Monitoring Programs,* American Water Resources Association, Minneapolis, MN, 1978, pp. 134–149.

Thomas P. Ballestero,[1] Susan A. McHugh,[1] and Nancy E. Kinner[1]

Monitoring of Immiscible Contaminants in the Vadose Zone

REFERENCE: Ballestero, T. P., McHugh, S. A., and Kinner, N. E., **"Monitoring of Immiscible Contaminants in the Vadose Zone,"** *Ground Water and Vadose Zone Monitoring, ASTM STP 1053,* D. M. Nielsen and A. I. Johnson, Eds., American Society for Testing and Materials, Philadelphia, 1990, pp. 25–33.

ABSTRACT: The monitoring of nonaqueous-phase fluids in the unsaturated zone can be accomplished with tensiometers or suction lysimeters if care is taken before the instrumentation is installed. In this instance, the porous material used to monitor fluid pressure, or to sample a fluid, should be prewetted with the fluid of interest. Experiments with three-phase fluid flow of immiscible fluids in a porous medium have shown that there is a small window of fluid contents at which all three fluids are mobile. With this in mind, monitoring/sampling of each phase was accomplished for various liquid contents for each of the liquids. More often than not, though, liquid sampling is preferential towards the wetting fluid. For petroleum hydrocarbons, water, and air in porous media and in porous ceramic cups, the dominant wetting fluid has been found to be water.

KEY WORDS: ground water, vadose zone, immiscible contaminants, gasoline, heating oil, tensiometers, lysimeters, monitoring, sampling

A common field endeavor in ground-water science is the evaluation of contamination at a particular site. Because of the very nature of human-caused aquifer contamination, most contaminants move from at or near the ground surface down to the ground-water table. It is in this very basic process that the vadose zone becomes affected as well as the saturated zone. This paper deals with vadose zone contamination. Of interest here is the evaluation of the presence of immiscible contaminants within the vadose zone itself. That is to say, in order to delineate the degree of contamination, the fluids present in the vadose zone must be monitored or sampled, or both. In lieu of destructive sampling (i.e., coring), an *in situ* probe would deal best with the changing conditions in the vadose zone, especially in light of treatment or cleanup methodologies. Thus, the objective of the present research is to identify whether prewetting of porous ceramic cups will make possible sampling or monitoring of non-aqueous-phase liquids in the unsaturated zone.

Characteristics of Multiphase Vadose Zone Flow

Fluids flowing in aquifers conveniently subscribe to Darcy's law

$$q = -K\frac{dh}{dx} \tag{1}$$

[1] Assistant professor, graduate research assistant, and associate professor, respectively, University of New Hampshire, Department of Civil Engineering, Durham, NH 03824.

where q is the Darcian velocity, which is the volume of flow per unit of time per bulk unit of area (L/T); h is the total energy of the fluid at the particular location in space (L); x is the one-dimensional position or space coordinate (L); and K is the hydraulic conductivity. When a porous medium is fully saturated with one fluid, K represents the saturated hydraulic conductivity. Given a porous medium and two different fluids, it is obvious that, to maintain a fixed, constant hydraulic gradient (dh/dx) in each case of saturated flow for each fluid, the discharge or Darcian velocity will change owing to changes in the fluid characteristics; specifically, how shear is transmitted through the fluid.

With dh/dx constant and q changing, K must change. This signifies that K is not only a function of the type of porous medium (formation property) but also a function of the fluid property of viscosity (kinematic viscosity, ν). If the fluid property is abstracted from hydraulic conductivity, permeability (k) results

$$k = \frac{K\nu}{g} \tag{2}$$

where k is the permeability (L^2), here at saturated conditions, and g is the acceleration due to gravity. Permeability may be used in Darcy's law as was hydraulic conductivity. This results in

$$q = -\frac{kg}{\nu}\frac{\partial h}{\partial x} \tag{3}$$

In the case of a porous medium with one fluid present, the value of permeability found from measurement of flows, fluid properties, and gradients is the permeability at saturated conditions. When two fluids are present at the same time in a porous medium, there is competition for the void spaces. As the fluids now have more tortuous paths to follow than under saturated conditions, the multiphase porous medium presents more resistance to flow. This directly results in a reduction of permeability. The reduction in permeability of the porous medium to fluid flow is proportional to the content of that fluid in the porous medium. When the content of the pore spaces is entirely filled with a fluid, the permeability is maximum and the magnitude is that of the saturated value. When the pore space volume is filled less and less with the fluid of interest (i.e., water) and more and more with another (i.e., air), then the permeability of the fluid of interest (in this case water) decreases, while at the same time the permeability of the other fluid, in this case air, increases because its content is increasing. These changes in permeability can occur over numerous orders of magnitude.

In cases of three immiscible fluids in a porous medium (gasoline-water-air), it has been cited by Corey [1], Leverett [2], and Schwille [3] and experimentally observed by Ballestero, Johnson, and Kinner [4] that there are threshold fluid contents which must be reached before a fluid can move. The fluid content of the liquids must be significant in order to obtain any three-phase flow. As air is a nonwetting fluid (preferentially, air would rather be next to the water or gasoline than to the solids composing the porous medium), it takes the most advantageous pore space, leaving very sinuous, tortuous paths for the water and gasoline. Thus, to get three-phase flow of gasoline, water, and air in a porous medium, the air content should be low, and both the gasoline and water content should be high.

On a percentage basis, the sum of all three fluid contents will equal 100% of the pore volume. Thus, the key to identifying when multiphase flow occurs is to identify the percentage of pore volume taken up by each fluid—in other words, the fluid content. There are numerous techniques for performing fluid content measurements directly and indirectly, as well as destructively and nondestructively. For field monitoring purposes, one technique is to measure the subsurface pressure in the fluid phases and then relate this to the fluid content. This presumes that there is a unique relationship between fluid pressure and fluid content for a particular porous medium and then another unique relationship between fluid content and permeability. The pressure measurements also aid in delineating subsurface energy gradients. Thus, by combining all of this information, the subsurface content and mobility of immiscible contaminants can be evaluated. Figure 1 delineates the typical relationship between fluid pressure and fluid content and that between fluid content and permeability.

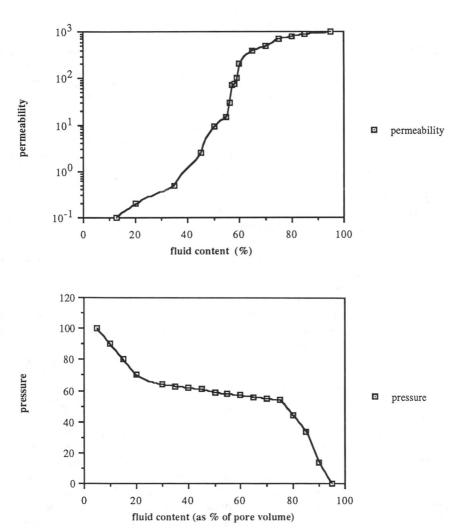

FIG. 1—*Interrelationships between fluid pressure and permeability and between fluid content and permeability in a porous medium.*

Capillary Pressure

When immiscible fluids come in contact with one another, their interface acts as a membrane. This membrane tends to form geometrically in a form other than a plane surface. Because fluid molecules would rather be next to like molecules than at the interface, there is a certain amount of energy required to keep molecules at the interface. This energy is described by the fluid property of surface tension. The end result of the energy necessary to maintain molecules at the interface is a force imbalance, at the interface, which is perpendicular to the interface. The action of this force over the area of the interface results in what is termed capillary pressure. The capillary pressure requires that there be a difference in pressure at the interface in going from one fluid to the other. The difference in these pressures is the capillary pressure (p_c). As defined by Corey [1]

$$p_c = p_{nw} - p_w \qquad (4)$$

where p_c is the capillary pressure, p_{nw} is the fluid pressure in the nonwetting phase at the interface, and p_w is the fluid pressure of the wetting phase at the interface. It is p_w that is measured in the vadose zone and that is related to the fluid content. The capillary pressure is also related to the total energy at any one point, and thus can be used to delineate hydraulic gradients.

In soils, air is the nonwetting phase and water is the wetting phase: p_{nw} is on the order of atmospheric pressure (or zero gage pressure), and the water pressures are negative and a function of the distance from the water table. Thus, p_c is computed as a positive value.

Tensiometers measure p_w. By assuming atmospheric conditions in the subsurface air, p_c is equal to the negative of the water pressure reading form the tensiometers. A perverse feature of laboratory experiments of three-phase flow in closed porous media systems is that, because of the volatility of gasoline, p_{nw} is greater than atmospheric pressure (Ballestero, Johnson, and Kinner [4]). This factor obviously must be taken into account in such experiments.

The surface tension property of fluids is a function of temperature, the purity of the fluid of interest, and the other fluid existing at the interface. In the case of a water-air interface, surface tension is on the order of 70 dyne/cm; for a gasoline-water interface, this may be reduced to less than 10 dyne/cm [2].

In the three-fluid problem of gasoline-water-air, water wets, gasoline preferentially wets over air, and air is the nonwetting fluid. Thus, water will most likely coat the soil particles, gasoline will fill small crevasses between the water-coated particles, and air will remain in the large void spaces.

Monitoring and Sampling in the Vadose Zone with Lysimeters and Tensiometers

As presented in the last section, multiphase fluid flow in a natural porous medium, in which air is one of the fluids, occurs under negative liquid (gasoline, water) pressures and results in positive capillary pressures. Sampling any of the fluids in the vadose zone can be accomplished by utilizing a device that operates under negative pressures which are lower than the fluid pressures. Tensiometers and lysimeters fit this description. In each case, the part of the instrument inserted into the vadose zone is a porous cup. For operational purposes, the porous cup is prewetted (presoaked), usually with water, allowing the prewetting liquid to pass through the porous cup in either direction once the cup is inserted in the porous medium. It is at this point that the present investigation began. As many immiscible fluids can exist in the vadose zone—fluids other than water—it is reasonable to ask whether fluids other than water can be sampled, preferentially, when the porous cups are prewetted with the fluid of interest.

Experimental Design

In order to answer this question, a laboratory experiment was designed. A flume (Fig. 2) of the dimensions 30 by 30 by 122 cm was constructed of wood and filled with a clean sand of known porosity (34%) and gradation [the percentages of sand particles passing through the following sieve sizes are as follows: $D_{10} = 0.17$ mm; $D_{30} = 0.25$ mm; $D_{50} = 0.32$ mm; $D_{60} = 0.35$ mm; and $D_{90} = 0.63$ mm (see Ref 7 for an explanation of the terms)]. During sand filling, porous ceramic cups, connected to gasoline-resistant plastic tubing, were installed at various depths. No slurry was used around the ceramic cups.

The porous cups were presoaked for at least 24 h. Depending on the trial, the presoaking occurred in either water, gasoline, or fuel oil. The cups were dedicated: thus, no one cup was used more than once. Presoaking occurred by immersing the porous cups in at least 20 cm of fluid and then pulling a small vacuum on the tube leading to the porous cups. The porous cups were rated by the manufacturer as 20.6×10^5 dyne/cm^2 but, when bubble tested, proved to be in the range of 27.5 to 33.0×10^5 dyne/cm^2.

Initially the flume was saturated with water and then allowed to drain until the water in the sand had a level of only 3 cm. Approximately 35% of the water added remained in the flume.

The plastic tubing to each porous cup was left filled with the soaking fluid and capped. At this point, the flumes were left for another 24 h in order to let the porous cups come to equilibrium.

At this time, various trials were run. In the first trial, a mixture of half gasoline, half water was added to the flume by pouring it over the top surface. The volume added was on the order of 70% of the pore volume. This added fluid mixture was immediately drained until the level of fluid at the bottom of the flume was 3 cm. Of the drained volume, approximately 85% of each fluid was recovered. The flume was left for another 24 h to equilibrate. No measurements of residual saturations were made, but previous work with this material [4] shows residual liquid contents on the order of 15% for water and 18% for gasoline, which are in the ranges found by Hoag and Marley [5]. Sampling for this run began on the next day by drawing a suction on the plastic tubes to the porous cups and recording the volume and type of fluid obtained from the cups.

Trial I—In this trial, the cups were presoaked with water. The suction pressure on each cup was started at 0 dyne/cm^2 and increased in steps. At each step, the pressure was held constant for 15 to 20 min and the fluid sample was recorded. The pressures (in dynes per square centimetre) tested included: 345 000; 690 000; 1.03×10^6; 1.38×10^6; 1.72×10^6; and 1.93×10^6.

Trial II—This trial mimicked Trial I, except that the porous cups were presoaked with gasoline.

Trial III—This trial was similar to Trial I only in that the fluid mixture added was water and home heating oil, and the porous cups were presoaked with water.

Trial IV—This trial was similar to Trial III, except that the porous cups were presoaked with the home heating oil.

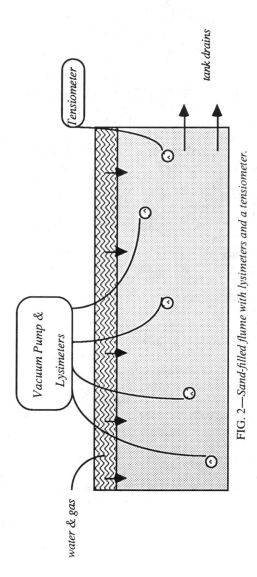

FIG. 2—Sand-filled flume with lysimeters and a tensiometer.

Trial V—In this trial, which started with the initially dry-packed flume, the flume was saturated with gasoline, drained, and then sampled with porous cups that had been presoaked with gasoline.

Trial VI—This trial mimicked Trial V, except that heating oil was used instead of gasoline. A summary of the experimental design can be found in Table 1.

After each trial, the flume was emptied of the sand and repacked with new sand. This limited any contamination from trial to trial.

Results and Conclusions

Tables 2 through 6 exhibit the qualitative summary of the results of these experiments. The results do appear to be mixed and in part reflect some difficulty with the equipment.

TABLE 1—*Summary of experiment design.*

Trial Run No.	Porous Cup Presoaking Fluid	Fluid Mixture Added and Drained	Porous Cup Location Below Top Surface of Sand, cm
1	water	water/gasoline	10, 12, 15, 22
2	gasoline	water/gasoline	10, 15, 20, 25
3	water	water/heating oil	10, 15, 20, 25
4	heating oil	water/heating oil	10, 15, 20, 25
5	gasoline	gasoline	15
6	heating oil	heating oil	15

TABLE 2—*Data from Trial I.*

Porous Cup Location Below Top Surface of Sand, cm	Vacuum Pressure Applied to Porous Cups, dyne/cm^2 $\times$ 10^5					
	3.45	6.90	10.3	13.8	17.2	19.3
10	water	air	air	water/air	water/air	water/air
12	water	water	water	water/air	water/air	water/air
15	water	water	water/air	water/air	water/air	water/air
22	air	water/air	water/air	water/air	water/air	water/air

TABLE 3—*Data from Trial II.*

Porous Cup Location Below Top Surface of Sand, cm	Vacuum Pressure Applied to Porous Cups, dyne/cm^2 $\times$ 10^5					
	3.45	6.90	10.3	13.8	17.2	19.3
10	no sample	no sample	gas/water/air	air	air/water	air/water
15	no sample	no sample	no sample	no sample	no sample	no sample
20	no sample	air	trace water	gas/air	gas/air	water/air
25	no sample	gas/water	gas/water	gas/water	air/water	air/water

TABLE 4—*Data from Trial III.*

Porous Cup Location Below Top Surface of Sand, cm	Vacuum Pressure Applied to Porous Cups, dyne/cm^2 × 10^5					
	3.45	6.90	10.3	13.8	17.2	19.3
10	trace water	water/air	water/air	air	air	air
15	trace water	water	water/air	water/air	water/air	water/air
20	trace water	water	water/air	water/air	water/air	water/air
25	water	water/air	water/air	water/air	water/air	water/air

TABLE 5—*Data from Trial IV.*

Porous Cup Location Below Top Surface of Sand, cm	Vacuum Pressure Applied to Porous Cups, dyne/cm^2 × 10^5					
	3.45	6.90	10.3	13.8	17.2	19.3
10	trace water	air	air/trace water	air/trace water	air/water/ oil	air/water/ oil
15	no sample	water/air	water/air	water/air	air/water	air/water
20	water/air/ oil	air/water	air/water	oil/water	air/water	air/water
25	no sample	water	water/air	water/air	water/air	air/water

TABLE 6—*Data from Trials V and VI.*[a]

	Vacuum Pressure Applied to Porous Cups, dyne/cm^2 × 10^5					
Trial	3.45	6.90	10.3	13.8	17.2	19.3
5	gas	gas	gas	air/gas	air/gas	air/gas
6	oil	oil	oil	oil/air	oil/air	oil/air

[a] The porous cups were located approximately 15 cm below the top surface of the sand.

The most basic conclusion that can be drawn (from Trials I and III) is that when the porous cups were presoaked with water, no hydrocarbon fluid phase was sampled by the ceramic cups. In addition, in Trials II and IV, when the porous cups were presoaked with the hydrocarbon, both liquid phases were sampled, but there was a trend in which the heating oil sample diminished, ultimately yielding only water. During the experiments, the yield of the hydrocarbon diminished more rapidly at higher suction values than at lower values.

In Trials II, IV, V, and VI, it was not entirely apparent whether the air which was entering the suction lysimeters resulted from equipment failure at the connections or from volatilization of the hydrocarbon within the porous cup.

The significance of these results lies in their effect on field monitoring protocols. At a gasoline spill site, lysimeters presoaked with gasoline can pick up some of the pure gasoline when operating at low vacuum. If cups presoaked with water were to be used, only those constituents which readily dissolve in water would be detected. As gasoline is a mixture of as many as 200 constituents [6], the absence of a vadose zone gasoline sample from a spill location hinders the evaluation of the spill magnitude and the effectiveness of remediation strategies.

Some recommendations for field utilization of this methodology include low-vacuum sampling and multiple porous cup arrays. In the experiments, since water was ultimately obtained in the porous cups initially presoaked with the hydrocarbon, water was most likely preferentially wetting the porous cups. Thus, the porous cups that had been initially soaked with gasoline or heating oil were, in time, incapable of providing gasoline or heating oil samples. If sample pressures are kept low, or possibly the cups are kept under positive hydrocarbon liquid pressures until in use, water will not be drawn into the porous cups as fast as under higher vacuum pressures. Multiple porous cup arrays will allow for cups to be fouled with time before others are used, thus affording consistent spatial data. Tensiometers will have to be filled with the prewetting fluid and may necessitate redesign of the instrument altogether.

Acknowledgments

This project was funded in part by the State of New Hampshire Department of Environmental Services and in part by the New Hampshire Water Resources Research Center. The help of Fritz Fiedler in the laboratory experiments aided the project completion. Also, the authors would like to thank Kyle McElroy Dallaire for manuscript review.

References

[1] Corey, A. T., *Mechanics of Heterogeneous Fluids in Porous Media,* Water Resources Publications, Fort Collins, CO, 1977.
[2] Leverett, M. C., "Capillary Behaviour in Porous Solids," Petroleum Technology Conference, Tulsa, OK, 1940, pp. 152–168.
[3] Schwille, F., "Groundwater Pollution in Porous Media by Fluids Immiscible with Water," *Science of the Total Environment,* Vol. 21, 1981, pp. 173–185.
[4] Ballestero, T. P., Johnson, M. J., and Kinner, N. E., "Gasoline-Water-Air Relative Permeabilities," University of New Hampshire, Durham, NH, 1989.
[5] Hoag, G. E. and Marley, M. C., "Gasoline Residual Saturation in Unsaturated Uniform Aquifer Materials," *Journal of Environmental Engineering,* Vol. 112, No. 3, 1986, pp. 586–604.
[6] Shepherd, W. D., "Petroleum Hydrocarbons and Organic Chemicals in Ground Water—Prevention, Detection, and Restoration," National Water Well Association Conference, Houston, TX, 1984.
[7] De Alba, P., Baldwin, K., Janoo, V., Roe, G., and Celikkol, B., "Elastic-Wave Velocities and Liquefaction Potential," *Geotechnical Testing Journal,* Vol. 7, No. 2, June 1984, pp. 77–87.

David I. Stannard[1]

Tensiometers—Theory, Construction, and Use

REFERENCE: Stannard, D. I., **"Tensiometers—Theory, Construction, and Use,"** *Ground Water and Vadose Zone Monitoring, ASTM STP 1053,* D. M. Nielsen and A. I. Johnson, Eds., American Society for Testing and Materials, Philadelphia, 1990, pp. 34–51.

ABSTRACT: Standard tensiometers are used to measure matric potential as low as −870 cm of water in the unsaturated zone by creating a saturated hydraulic link between the soil water and a pressure sensor. The direction and, in some cases, quantity of water flux can be determined using multiple installations.

A variety of commercial and fabricated tensiometers are commonly used. Saturated porous ceramic materials, which form an interface between the soil water and the bulk water inside the instrument, are available in many shapes, sizes, and pore diameters. A gage, manometer, or electronic pressure transducer is connected to the porous material with small- or large-diameter tubing. Selection of these components allows the user to optimize one or more characteristics, such as accuracy, versatility, response time, durability, maintenance, extent of data collection, and cost.

Special designs have extended the normal capabilities of tensiometers, allowing measurement in cold or remote areas, measurement of matric potential as low as −153 m of water (−15 bars), measurement at depths as deep as 6 m (recorded at land surface), and automatic measurement using as many as 22 tensiometers connected to a single pressure transducer.

Continuous hydraulic connection between the porous material and soil, and minimal disturbance of the natural infiltration pattern are necessary for successful installation. Avoidance of errors caused by air invasion, nonequilibrium of the instrument, or pressure-sensor inaccuracy will produce reliable values of matric potential, a first step in characterizing unsaturated flow.

KEY WORDS: ground water, infiltration, instrumentation, moisture content, moisture tension, Richards apparatus, tensiometers, underground waste disposal, unsaturated flow

Movement of water in the unsaturated zone is of considerable interest in studies of hazardous-waste sites [*1–3*], recharge studies [*4,5*], irrigation management [*6–8*], and civil-engineering projects [*9,10*].

Unsaturated flow obeys the same laws that govern saturated flow: Darcy's law and the equation of continuity, which have been combined in the Richards equation [*11*]. Baver et al. [*12*] present Darcy's law for unsaturated flow as

$$q = -K\nabla(\psi + Z) \tag{1}$$

where

q = the flux density, in metres per second;
K = the unsaturated hydraulic conductivity, in metres per second;

---

[1] Hydrologist, U.S. Geological Survey, Denver, CO 80225.

ψ = the matric potential of the soil water at a point, in metres;
Z = the elevation at the same point, relative to some datum, in metres; and
∇ = the gradient operator, in inverse metres.

The sum of $\psi + Z$ commonly is referred to as the hydraulic head.

Unsaturated hydraulic conductivity, K, can be expressed as a function of either matric potential, ψ, or water content, θ (cubic metres of water per cubic metre of soil), although both functions are affected by hysteresis [13]. If the wetting and drying limbs of the $K(\psi)$ function are known for a soil, time series of on-site matric-potential profiles can be used to determine the following: (1) which limb is more appropriate to describe the on-site $K(\psi)$, (2) the corresponding values of the hydraulic-head gradient, and (3) an estimate of flux using Darcy's law. If, instead, K is known as a function of θ, on-site moisture-content profiles (obtained, for example, from neutron-scattering methods) can be used to estimate K and can be combined with matric-potential data to estimate flux. In either case, the accuracy of the flux estimate needs to be assessed carefully. For many porous media, $dK/d\psi$ and $dK/d\theta$ are large, within certain ranges of ψ or θ, making estimates of K particularly sensitive to on-site measurement errors of ψ or θ. (On-site measurement errors of ψ also have a direct effect on $\nabla(\psi + Z)$ in Darcy's law.) Other sources of error in flux estimates can result from the following: (1) inaccurate data used to establish the $K(\psi)$ or $K(\theta)$ functions (accurate measurement of very small permeability values is particularly difficult) [14]; (2) use of an analytical expression for $K(\psi)$ or $K(\theta)$ that facilitates computer simulation, but only approximates the measured data; (3) an insufficient density of on-site measurements to define adequately the θ or ψ profile, which can be markedly nonlinear; (4) on-site soil parameters that are different from those used to establish $K(\psi)$ or $K(\theta)$; and (5) invalid assumptions about the state of on-site hysteresis. Despite the possibility of large errors, certain flow situations occur in which these errors are minimized and fairly accurate estimates of flux can be obtained [5,15]. The method has a sound theoretical basis, and refinement of the theory to match measured data would markedly improve the reliability of the estimates.

Matric-potential and elevation data can be used to determine direction of flow [9], a valuable piece of information. If the moisture-characteristic curve is known for a soil, matric-potential data can be used to determine the approximate water content of the soil [10]. The tensiometer is used to measure matric potential between the values of 0 and approximately -870 cm of water; this range includes most values of saturation for many soils [16]. In theory, these techniques can be applied to almost any unsaturated-flow situation, whether it is recharge, discharge, lateral flow, or combinations of these situations.

In this report, the theoretical and practical considerations pertaining to successful on-site use of commercial and fabricated tensiometers are described. Measurement theory and on-site objectives are used to develop guidelines for tensiometer selection, installation, and operation.

The tensiometer, formally named by Richards and Gardner [17], has undergone many modifications for use in specific problems [19,18–29]. However, the basic components have remained unchanged. A tensiometer is comprised of a porous surface (usually a ceramic cup) connected to a pressure sensor by a water-filled conduit. The porous cup, buried in a soil, transmits the soil-water pressure to a manometer, a vacuum gage, or an electronic-pressure transducer (referred to in this report as a pressure transducer). During normal operation, the saturated pores of the cup prevent bulk movement of soil gas into the cup.

Measurement Theory

The concept of fluid tension refers to the difference between standard atmospheric pressure and the absolute fluid pressure. The values of tension and pressure are related as

$$T_F = P_{atm} - P_F \tag{2}$$

where

T_F = the tension of an elemental volume of the fluid,
P_{atm} = the absolute pressure of the standard atmosphere, and
P_F = the absolute pressure of the same elemental volume of fluid.

(All the pressures and tensions for Eq 2 are expressed in the same units.)

Soil-water tension (or soil-moisture tension), similarly, is equal to the difference between soil-gas pressure and soil-water pressure. Thus

$$T_W + P_G = P_W \tag{2a}$$

where

T_W = the tension of an elemental volume of soil water,
P_G = the absolute pressure of the surrounding soil gas, and
P_W = the absolute pressure of the same elemental volume of soil water.

(All these pressures and tensions are expressed in the same units.)

In this report, for simplicity, soil-gas pressure is assumed to be equal to 1 atm, except as noted. Various units are used to express tension or pressure of soil water and are related to each other by the equation

1.000 bar = 100.0 kPa = 0.9869 atm = 1020 cm of water at 4°C
$$= 1020 \text{ g/cm}^2 \text{ in a standard gravitational field} \tag{3}$$

A standard gravitational field is assumed in this report; thus, centimetres of water at 4°C are used interchangeably with grams per square centimetre.

The negative of soil-water tension is known formally as matric potential [14]. The matric potential of water in an unsaturated soil arises from the attraction of the soil-particle surfaces for water molecules (adhesion), the attraction of water molecules for each other (cohesion), and the unbalanced forces across the air-water interface. The unbalanced forces result in the concave water films typically found in the interstices between soil particles. Baver et al. [12] present a thorough discussion of matric potential and the forces involved.

An expanded cross-sectional view of the interface between a porous cup and soil is shown in Fig. 1. Water held by the soil particles is under tension; that is, the absolute pressure of the soil water, P_W, is less than atmospheric. This pressure is transmitted through the saturated pores of the cup to the water inside the cup. Conventional fluid statics relates the pressure in the cup to the reading obtained at the manometer, vacuum gage, or pressure transducer.

In the case of a mercury manometer (Fig. 2a)

$$T_W = P_A - P_W = (\rho_{Hg} - \rho_{H2O}) r - \rho_{H2O} (h + d) \tag{4}$$

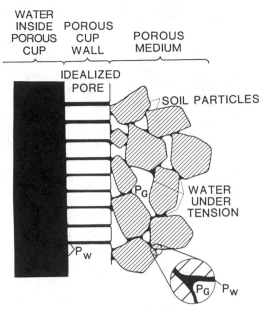

Pw –Absolute pressure of soil water
PG –Absolute pressure of soil gas

FIG. 1—*Enlarged cross section of the porous cup/porous medium interface.*

where

T_W = the soil-water tension relative to atmospheric pressure, in centimetres of water at 4°C;

P_A = the atmospheric pressure, in centimetres of water at 4°C;

P_W = the average pressure in the porous cup and soil, in centimetres of water at 4°C;

ρ_{Hg} = the average density of the mercury column, in grams per cubic centimetre;

ρ_{H2O} = the average density of the water column, in grams per cubic centimetre;

r = the reading, or height of the mercury column above the mercury-reservoir surface, in centimetres;

h = the height of the mercury-reservoir surface above the land surface, in centimetres; and

d = the depth of the center of the cup below the land surface, in centimetres.

Although the density of mercury and water both vary about 1% between 0 and 45°C, Eq 4 commonly is used, with ρ_{Hg} and ρ_{H2O} constant. Using ρ_{Hg} = 13.54 and ρ_{H2O} = 0.995 (the median values for this temperature range) yields about a 0.25% error (1.5 cm H$_2$O) at 45°C, for $T_W \approx 520$ cm H$_2$O. This small, but needless, error can be removed by using the following density functions

$$\rho_{Hg} = 13.595 - 2.458 \times 10^{-3}\,(T) \tag{5}$$

$$\rho_{H2O} = 0.9997 + 4.879 \times 10^{-5}\,(T) - 5.909 \times 10^{-6}\,(T)^2 \tag{6}$$

where ρ_{Hg} and ρ_{H2O} are as defined above, and (T) is the average temperature of the column, in degrees Celsius.

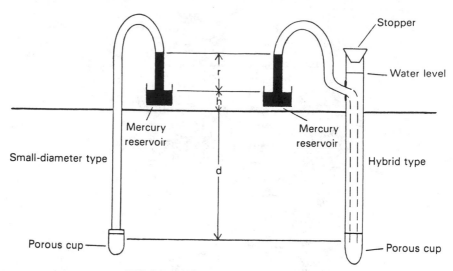

FIG. 2a—*Manometer type of tensiometer.*

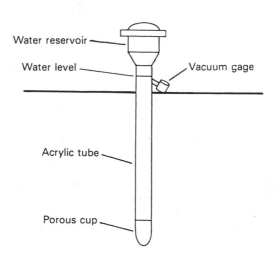

FIG. 2b—*Gage type of tensiometer.*

The average temperature of the buried segment of the water column can be estimated with a thermocouple or thermistor in contact with the tubing, buried at about 45% of the depth of the porous cup. Air temperature is an adequate estimate for exposed segments.

Most vacuum gages used with tensiometers are calibrated in bars (and centibars) and have an adjustable zero reading. The zero adjustment is used to offset the effects of altitude, the height of the gage above the porous cup (Fig. 2b), and changes in the internal characteristics of the gage with time. The adjustment is set by filling the tensiometer with water and then setting the gage to zero while immersing the porous cup to its midpoint in a container of water. This setting is done at the altitude at which the tensiometer will be used and it needs to be repeated periodically after installation, either by removing the tensiometer from the soil or by unscrewing the gage and measuring a tension equal to that used

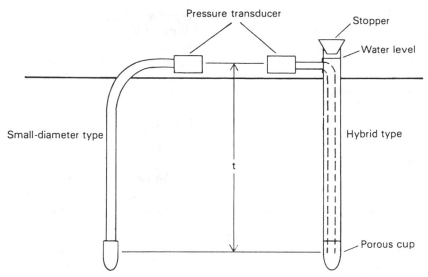

FIG. 2c—*Transducer type of tensiometer.*

in the original calibration. The gage then directly reads the tension in the porous cup. Use of a vacuum gage without an adjustable zero reading could result in inaccurate measurements because the zero reading could become negative and, therefore, would be indeterminate.

Pressure transducers convert pressure, or pressure difference, into a voltage (or current) signal. The pressure transducer can be connected remotely to the porous cup with tubing [22], attached directly to the cup [21,30], or transported between sites [28]. An absolute-pressure transducer measures the absolute pressure (P_P) in its port. A gage-pressure transducer measures the difference between ambient-atmospheric pressure (P_A) and the pressure in its port (P_P), known as gage pressure. When $P_P < P_A$, the gage pressure is identical to the tension. A differential-pressure transducer measures the difference between two pressures, one in each of its two ports. When used with tensiometers, the second port usually is connected to the atmosphere; the unit is used as a gage-pressure transducer and it measures tension.

A calibration equation supplied by the manufacturer is used to convert the measured signal into pressure or tension at the pressure-transducer port. The tension in the porous cup and soil is then (Fig. 2c)

$$T_W = T_P - t \cdot \rho_{H2O} \tag{7}$$

where

T_W = the average tension in the porous cup and soil, in centimetres of water at 4°C;
T_P = the tension in the pressure-transducer port, in centimetres of water at 4°C;
t = the difference in elevation between the pressure-transducer port and the center of the porous cup, in centimetres; and
ρ_{H2O} = the average density of the water column connecting the porous cup and transducer, in grams per cubic centimetre.

At 15°C, pure liquid water begins to cavitate (vaporize) if its tension exceeds approximately 970 cm H_2O. If cavitation happens in a tensiometer, liquid continuity is interrupted and the tension readings are invalid. Water used in tensiometers is deaerated as completely as practicable, but some impurities and dissolved gases remain that decrease the tension sustainable by liquid water to about 870 cm H_2O [31]. Thus, the operating range of tensiometers is described by the equation

$$T_C + \Delta h < 870 \text{ cm} \tag{8}$$

where

T_C = the tension in the porous cup, in centimetres of water at 4°C; and
Δh = the elevation of the highest point in the hydraulic connection between the porous cup and the pressure sensor, minus the elevation of the porous cup, in centimetres.

Equation 8 indicates that a trade-off occurs between the depth of installation of the porous cup and the maximum tension measurable. Equation 8 is approximate; if the water is insufficiently deaerated, the value 870 would be replaced with a smaller value.

The only tensiometer described thus far that measures absolute soil-water pressure (P_W) directly is the absolute-pressure-transducer type. The others—differential tensiometers—measure the quantity $P_A - P_W$, where P_A is ambient atmospheric pressure. The driving forces for liquid water in the unsaturated zone are the absolute pressure gradient in the liquid-water phase and gravity (Eq 1). If air moves readily down through the unsaturated zone, then differential tensiometers can be used directly to determine pressure gradients. However, if a barometric-pressure change is transmitted readily to one differential tensiometer porous cup and not to another (because of an intermediate confining layer), the calculated gradient between the two porous cups would be in error. If a porous cup is isolated from the atmosphere by a confining layer, then a time series of soil-water pressures at the porous cup, calculated with P_A constant, will indicate fluctuations that correlate well with barometric fluctuations. In this case, a recording barometer will provide a record of ambient atmospheric pressure from which absolute soil-water pressure and pressure gradients can be determined. The resulting time series of absolute soil-water pressures at the isolated porous cup will be a smoother curve that will indicate real pressure changes in the water phase.

Richards [16] defined the time constant of a tensiometer as

$$\tau = \frac{1}{K_c S} \tag{9}$$

where

τ = the time constant, or time required for 63.2% of a step change in pressure to be recorded by a tensiometer when the cup is surrounded by water, in seconds;

K_c = the conductance of the saturated porous cup, or the volume of water passing through the cup wall per unit of time per unit of hydraulic-head difference, in square centimetre per second; and

S = the tensiometer sensitivity, or change in pressure reading per unit volume of water passing through the porous-cup wall, per square centimetre.

Also, the porous-cup conductance may be expressed as

$$K_c = \frac{kA}{W} \tag{10}$$

where

K_c = the cup conductance, in square centimetres per second;
k = the permeability of the cup material to water at the prevailing temperature, in centimetres per second;
A = the average surface area of porous-cup material, estimated as the mean of the inside area and the outside area, in square centimetres; and
W = the average wall thickness of the porous cup, in centimetres.

Richards' [16] definition does not apply to a tensiometer buried in soil, because the soil conductance, K_s, is in series with K_c, and usually $K_s \ll K_c$. In fact, an on-site time constant cannot be defined [21] because the response is not logarithmic because of varying K_s during equilibration. However, the phrase "response time" is used to describe the rate of on-site response to pressure changes [31]. The term is not to be confused with the time constant because two tensiometers with equal time constants placed in the same soil can have different response times. For example, if $K_{c1} = 10\ K_{c2}$ and $S_2 = 10\ S_1$, then $\tau_1 = \tau_2$, but if $K_s \approx K_{c2}$, then response time$_1 >$ response time$_2$. Nonetheless, τ as defined here can be used comparatively to help evaluate tensiometer design. Greater sensitivity, large porous-cup surface area and permeability, and thin porous-cup walls are characteristics of a tensiometer with a short response time. Use of a sensitive pressure transducer is the most effective way to decrease response time in a soil of low hydraulic conductivity.

A bubble that interrupts hydraulic continuity between the porous cup and the pressure sensor will cause a change in the calculated value of P_W

$$\Delta = (E_P - E_C)\rho_{H2O} \tag{11}$$

where

Δ = the change in the calculated value of P_W, in centimetres of water at 4°C;
E_P = the elevation of the end of the bubble nearest the pressure sensor, in centimetres;
E_C = the elevation of the end of the bubble nearest the cup, in centimetres; and
ρ_{H2O} = the density of water adjacent to the air bubble, in grams per cubic centimetre.

If bubbles are detected and measured, these corrections can be made to P_W, as calculated in Eq 4 or 7. Small bubbles that cling to the wall of the tubing and do not block the entire cross section do not affect the calculated value of P_W.

Construction and Applications

The following definitions are used to describe the quality of a measurement and are used in Table 1 to compare types of tensiometers. The accuracy of a measurement is the difference between the value of the measurement and the true value. Precision (repeatability) refers to the variability among numerous measurements of the same quantity. Resolution refers to the smallest division of the scale used for a measurement and it is a factor in

TABLE 1—Tensiometer characteristics.

| | Commercial Tensiometers | | Constructed Tensiometers | | | |
| | | | Manometer | | Pressure Transducer | |
Characteristic	Vacuum Gage	Manometer (Hybrid)	Small Diameter	Hybrid	Small Diameter	Hybrid
Accuracy	poor	excellent	excellent	excellent	good to excellent	good to excellent
Precision[a]	poor	good	good	good	excellent	excellent
Hysteresis	poor	excellent	excellent	excellent	fair to excellent	fair to excellent
Response time	poor to excellent	fair	fair	fair	excellent	excellent
Versatility of application	fair	fair	excellent	fair	excellent	fair
Durability	good	good	good to excellent	good	good	good
Purging	seldom	occasionally	often	occasionally	often	occasionally
Recalibration	occasionally	never	never	never	often	often
Data collection method	manual	manual	manual	manual	manual or automatic	manual or automatic
Cost for five[b]	$260	$200	$120[c]	$150	$410[d]	$440[d]

[a] Precision (repeatability) is rated for either a wetting or a drying cycle to distinguish from hysteresis effects.
[b] Estimated for five 0.9-m (3-ft)-deep tensiometers.
[c] Does not include the cost of deaerating the water.
[d] Does not include the cost of deaerating the water or the recording equipment.

determining precision and accuracy. Hysteresis is that part of inaccuracy attributable to the tendency of a measurement device to lag in its response to environmental changes. Parameters affecting pressure-sensor hysteresis are temperatures and measured pressure.

The operating characteristics of commonly available tensiometers vary (Table 1), and they need to be matched to the specific installation, cost constraints, and the desired quality of data collection. Complete tensiometers may be purchased from soils and agricultural research companies, made entirely from parts, or made from parts of commercial units modified to suit the user's needs. The advantages and disadvantages of the different types are discussed in the following paragraphs and in Table 1.

Commercially available vacuum-gage-type units (Fig. 2b) usually have a large-diameter porous cup cemented to a rigid acrylic tube of equal diameter (19 or 22 mm). A vacuum gage that indicates from 0 to 100 cbars of tension is screwed into the side of the tube, several centimetres below the top. The space between the vacuum gage and the top of the tube is a reservoir for air (the water is not deaerated beforehand) to collect. When the water level inside the tube approaches the vacuum-gage inlet, the tube cap is unscrewed and the air space is refilled with water. Some vacuum-gage tensiometers have a large water reservoir connected to the top of the tube with a spring-loaded valve to simplify refilling.

The major advantage of a vacuum-gage tensiometer is the maintenance of a hydraulic connection between the porous cup and the gage, even with large quantities of air present. However, this advantage typically is offset by the use of a vacuum gage with a resolution of 0.5 cbar (5 cm H_2O) and an overall accuracy of 3 cbars (31 cm H_2O). The response time is excellent immediately after all air has been removed, but it slows rapidly as the air reservoir fills up. The construction is fairly durable, but its rigidity can transfer shock and actually damage the porous cup, cup-tube bond, or hydraulic connection with the soil if the top is impacted after installation. Although the tube usually is installed vertically, it can be inclined to a nearly horizontal orientation as long as the zero adjustment of the vacuum gage is made at the same inclination.

A vacuum-gage tensiometer is used predominantly for irrigation scheduling where extreme accuracy is not necessary. It is not recommended for measurement of unsaturated hydraulic gradients [31]. However, replacement of a standard vacuum gage with a more accurate, higher-resolution gage, or with an accurate pressure transducer, would improve the usability of the tensiometer.

In this paper, a tensiometer with a large-diameter cup-tube assembly connected to the pressure sensor with small-diameter (3.2 mm, for example) tubing is referred to as a hybrid tensiometer (Fig. 2a). Hybrid tensiometers, like vacuum-gage tensiometers, have a space at the top of the large tube to collect air. Hydraulic continuity is not broken, unless air bubbles block an entire cross section of the small-diameter tubing.

Commercial manometer-type tensiometers commonly are hybrid types. Almost all of the air that enters the tensiometer through the porous cup collects harmlessly at the top. However, air also tends to be liberated from solution near the top of the manometer, where the maximum tension occurs; use of deaerated water minimizes air production.

A mercury manometer probably is the most accurate pressure-sensor commonly used in tensiometers and it never needs calibrating (a water manometer, usable only in special cases, is more accurate because of a better resolution). Hysteresis in a manometer tensiometer (from surface tension at the interface) is much less than that in a vacuum-gage or pressure-transducer tensiometer. Thus, the hybrid-manometer tensiometer combines fairly maintenance-free operation with excellent accuracy.

The major advantages of constructing a manometer tensiometer with small-diameter tubing are its versatility of on-site application, accuracy, and low cost. Flexible nylon tubing can be routed around obstructions, connecting a porous cup to a gage hundreds of feet

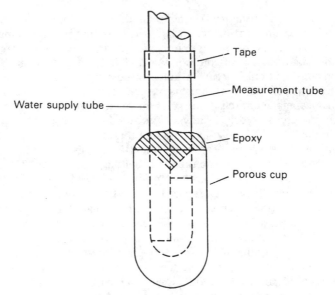

FIG. 3a—*Small porous cup design.*

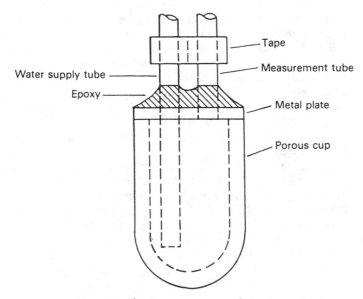

FIG. 3b—*Large porous cup design.*

away. The installation orientation is limited only by the backfilling capabilities. A typical design [29] employs two 3.2-mm-diameter (nominal ⅛-in.-diameter) nylon tubes directly cemented with epoxy to a 9.5-mm-diameter (nominal ⅜-in.-diameter) porous cup (Fig. 3a). The water-supply tube is connected via a shutoff valve to a deaerated water supply and the measurement tube is routed directly to the mercury reservoir. Manometer sensitivity (S, Eq 9) is the reciprocal of the cross-sectional area of the manometer tubing; response time

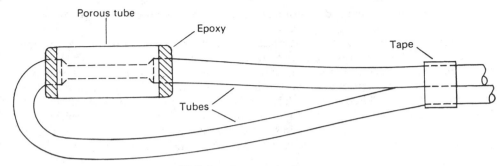

FIG. 3c—*Porous tube design.*

is minimized by using small-diameter tubing. The design is simple, but the epoxy-nylon-tube bond is somewhat susceptible to rupture from differential movement of the nylon tubes. A more robust design (Fig. 3b) uses a larger porous cup and a metal plate to separate the nylon tubes, allowing the epoxy to form a stronger bead around the tubes. A third design (Fig. 3c) uses a porous tube, made by cutting the rounded end off a 6.4-mm-diameter (nominal ¼-in.-diameter) porous cup. One of the nylon tubes is molded into a "U" shape with heat, to produce the design in Fig. 3c. The third design is extremely durable and versatile. The porous tube can be purged of air in any orientation, which is not entirely true of the first two designs (Figs. 3a and 3b).

Tensiometers tend to collect air bubbles on site that originate from the following: (1) insufficiently deaerated water; (2) air that diffuses through the water-filled pores of the porous cup; (3) dissolved gases in the soil moisture that flow into the porous cup during a wetting cycle; and (4) air that diffuses through the tubing material.

The major disadvantage of the small-diameter designs is that air bubbles can easily block the entire cross section of the tubing, interrupting hydraulic continuity. Thus, small-diameter designs require frequent purging of air—especially at large tensions. Use of thick-walled tubing decreases diffusion through the tube wall. Some plastics (such as polyethylene) are relatively permeable to air and are unsatisfactory tubing material. Use of metal tubing nearly eliminates air diffusion through the tube wall.

A hybrid-manometer tensiometer can be constructed at a material cost of two thirds to three fourths of the price of a commercial unit. The cost in Table 1 was determined for five "cup-tube kits" purchased at a supply company and outfitted with a manometer made from parts. A cup-tube kit consists of a porous cup cemented to a stoppered acrylic tube with a fitting that accepts 1.6-mm-diameter (nominal ¹⁄₁₆-in.-diameter) manometer tubing. The manometer tubing extends to the bottom of the acrylic tube. The cost could be decreased further by assembling the entire unit from basic parts.

Tensiometers equipped with pressure transducers are well suited to collect large quantities of data. Measurements can be made often and recorded automatically by a data logger or a strip-chart recorder. The extreme sensitivity of the pressure transducer results in the shortest time constant obtainable, making this type of tensiometer ideal for tracking wetted fronts. However, the extreme sensitivity also causes pressure-transducer tensiometers to be particularly susceptible to transient temperature effects [32] caused by thermal expansion and contraction of water in the tensiometer. Water that freezes inside a pressure-transducer cavity will affect the calibration, and it can rupture the unit. Pressure transducers (and above-ground connecting tubing) can be enclosed in an insulated (and, if need be, heated) shelter or surface pit to minimize temperature fluctuations. The shelter or pit needs to be located so that it does not disturb the natural flow field at the porous cup or cups.

The sensing elements of a typical pressure transducer are semiconductive resistors, embedded in a diaphragm that moves from applied pressure. As the resistors shorten or lengthen, their resistances change. Inherently, the resistance is a nonlinear function of pressure (or pressure difference) and temperature. The resistors are included in a modified Wheatstone bridge, excited by a regulated voltage (or current) source. The output of the bridge is a nearly linear function of pressure that is nearly independent of temperature; however, all pressure transducers retain some nonlinearity and a slight temperature dependence [33]. In addition, the zero offset and, to a lesser extent, the sensitivity may change with time (known as drift), possibly requiring recalibration at regular intervals [33]. Other sources of error are repeatability and pressure and temperature hysteresis effects [33]. On-site application determines the required pressure-transducer specifications; for example, gradient measurement normally warrants a more accurate transducer than simple pressure measurement does.

A large degree of accuracy may be achieved in a variety of ways. A sophisticated pressure transducer costing $700 to $800 typically has an overall accuracy of 1 cm H_2O at 23°C, and a temperature coefficient of 0.07 cm H_2O per degree Celsius. Without temperature correction, a worst-case error (at 0°C) of 2.6 cm H_2O could result. At the other extreme, a simple pressure transducer can be purchased for about $50, and its output can be corrected for nonlinearity and temperature dependence by measuring the temperature and applying a second-order polynomial fit to the measured data [34]. The decreased linearity and temperature errors, combined with a typical repeatability and hysteresis error of 1.5 cm H_2O, produce a root-sum-square error of 1.8 cm H_2O. Such a transducer is listed in Table 1. Of course, the same curve-fitting procedure can be applied to a pressure transducer with better repeatability and hysteresis to achieve greater accuracy.

The accuracies determined in the two examples just discussed are degraded further by drift or lack of long-term stability. If a long-term-stability specification cannot be supplied by the pressure-transducer manufacturer, much of the data collected may be inaccurate. A long-term-stability specification is used to determine how often a transducer needs to be recalibrated to maintain desired accuracy. A hanging column of water or mercury (or a calibrated vacuum source) is used for periodic recalibration.

Air that collects in an automatic-tensiometer system can be purged manually or automatically at regular intervals using solenoid valves triggered by a data logger. Air collects more rapidly when tension is greater; on-site experience will determine the necessary time interval to maintain desired accuracy. The time interval can be maximized by using horizontal sections of tubing at high points in the system. Bubbles that collect in these horizontal sections do not cause errors but they do increase the response time.

Porous cups used in tensiometers remain saturated during normal operation. If the difference between air pressure outside the porous cup and water pressure inside the cup exceeds the bubbling pressure of the cup, air will displace the water in the largest pores and eventually will enter the interior of the cup. The most common bubbling pressure of tensiometer cups is 1 bar (1020 cm H_2O). Porous cups with bubbling pressures of 2, 3, and 5 bars are available, but they only have applications in the laboratory. If on-site tensions are known not to exceed 0.5 bar, a porous cup with a 0.5-bar bubbling pressure can be used to decrease response time. However, a better way to decrease response time is by the use of commonly available, "high-flow" porous cups, made with a pore-size distribution that emphasizes the larger pores. (The cost of a porous cup is between $2 and $15; larger porous cups are more expensive than smaller ones, and high-flow cups are more expensive than standard-flow cups.)

Specialized tensiometers have been developed to address specific on-site problems. Peck and Rabbidge [24] extended the upper limit of measurable tension by using a reference solution with a low osmotic potential and a porous cup coated with a semipermeable membrane. The large solute molecule, polyethylene glycol, cannot pass through the membrane, but smaller molecules and ions typically found in soil water can pass through. Using a pressure transducer, soil-water tensions as large as 153 m H_2O (15 bars) can be measured. The solution in the tensiometer is usually under positive pressure, thus decreasing the problem of air invasion. However, the pressure can be quite large and it could cause permanent "creep" of the transducer diaphragm. Depolymerization of the solute and subsequent loss through the membrane also could cause creep. Ambient temperature affects the osmotic potential directly (as qualitatively predicted by van't Hoff's law) and indirectly by causing flow of water through the membrane. Thus, an osmotic tensiometer is valuable for measuring extremely low matric potentials, but some sources of measurement error are unique to it.

The U.S. Forest Service developed an inexpensive recording tensiometer for use in remote areas where electric power is unavailable [25]. This instrument can record as much as one month of continuous data on a battery-driven rain-gage chart. Oaksford [27] designed a unit based on a coaxial water manometer that provides maximum sensitivity at depths as deep as about 6 m. This unit uses a calibrated wire and ohmmeter to sense a below-ground, free-water surface inside the unit.

Fluid-scanning switches have been used successfully [23,32,35,36] to connect as many as 22 tensiometers sequentially to a single pressure transducer. The approach minimizes cost and removes bias between tensiometers. Equally relevant, this network has the capability of measuring a zero and a full-scale calibration tension before each scan. This capability removes measurement errors from hysteresis, temperature dependence, and long-term drift. However, if the pressure transducer fails, data from all its tensiometers are lost. Also, the scanning switch is made with precise tolerances and it may develop leaks over time.

Another approach to efficient data collection with large numbers of tensiometers [28,37] is to connect a portable pressure transducer to each tensiometer using a hypodermic needle and septum. The needle tip is inserted in an air space above the tensiometer fluid. A small change in tensiometer pressure is caused by the connection, probably affecting measurements by a few centimetres of water. Also, changes in the fluid-surface elevation, although small, affect measurements directly unless the changes are accounted for.

A small proportion of water in a frozen soil may remain liquid and, therefore, mobile. Measurement of tension in frozen soils can be accomplished using a pressure transducer and ethylene glycol [26] as a tensiometer fluid. The pressure transducer minimizes exchange of fluid between the soil and tensiometer, but slight bulk flow and diffusive flux do occur, which decrease the freezing point of the soil fluid. The osmotic tensiometer probably will work well in frozen soils.

Porous cups have been connected directly to pressure transducers; the internal cavity has been filled with deaerated fluid and the entire assembly has been buried in a soil, with no provisions for purging of the fluid [21,30]. Although this approach provides a stable temperature environment, the pressure transducer cannot be recalibrated readily for drift. Also, soil gas diffusing through the porous cup creates an air pocket in the cavity. Because this pocket is at a pressure less than that of the soil gas, air will continue to diffuse through the porous-cup wall, eventually emptying the cup of water. A purging system is needed for extended undisturbed operation.

The specialized tensiometers developed thus far have resulted from insight and persistence of researchers faced with particular problems. Most of these solutions require extra effort, care, and expense; review of the cited report or reports is needed before implementation.

Installation

Continuous hydraulic connection between the porous cup and the soil [17] and minimal disturbance of the natural flow field are essential to collection of accurate tensiometric data. When a hole is made to accept the tensiometer, the cuttings need to be preserved in the order in which they were removed if the hole is to be backfilled.

Hydraulic connection can be established in several ways. Commercial tensiometers fit snugly into holes made with a coring tool (available as a tensiometer accessory) or made with standard iron pipe. The porous cup is forced against the hole bottom and no backfill is used. When the soil is rocky, or when a small-diameter tensiometer is installed, a hole larger in diameter than the porous cup is excavated. If the soil at the bottom of the hole is soft, the porous cup may be forced into the soil. A hard soil sometimes can be softened with water. If the porous cup will not penetrate the moistened soil, the last cuttings from the hole need to be used to backfill around the cup, either by making them into a slurry or by careful tamping. A tremie pipe ensures clean delivery of cuttings or slurry to the bottom of the hole. Gaps between the porous cup and soil increase the tensiometer response time by reducing the effective area of the porous cup. In the worst case, no hydraulic connection occurs and the tensiometer will not indicate the soil-water tension.

If water is used to establish hydraulic connection, the tension adjacent to the porous cup will be reduced, and it will recover asymptotically as the added water is dispersed in the soil. The rate of dispersal will depend on the K (Eq 1) of the soil, and a time series of tension will indicate when natural conditions are restored sufficiently.

If vertical profiles or gradients of pressure are to be measured, multiple small-diameter tensiometers can be installed in a single hole. Ideally, the original lithology is duplicated (except that gravel larger than 6 mm in diameter and cobbles and boulders can be removed) by using the cuttings in reverse order for backfilling. A backfill that is more compacted than the undisturbed soil will tend to shed infiltrating water, but, after a short time, the tension in the undisturbed soil and in the backfill will be in equilibrium. A backfill that is less compacted than the surrounding soil, or one with excessive gaps, will be a conduit for infiltrating water, resulting in abnormally low tensions in the backfill and in the undisturbed soil. Therefore, compaction of the backfill to a slightly greater bulk density than that of the undisturbed soil is desirable. Less permeable layers (such as clay lenses) need to be reproduced or even exaggerated by importing a fine-grained material.

Installation of tensiometers at depths where Eq 8 is not satisfied can be accomplished by drilling horizontal holes radially from a central caisson hole. This method also preserves undisturbed conditions above and below each porous cup. Backfilling horizontal holes with a tamping rod is painstaking. An alternative is to backfill in the vicinity of the porous cup and to fill the remainder of the hole with an expanding insulation foam. Pressure transducers work particularly well in caisson holes. If the entire length of tubing connecting the porous cup to the pressure transducer is horizontal, air bubbles do not cause errors. Also, the problem of transducer sensitivity to temperature change is minimized because the temperature in the caisson hole remains relatively constant. A caisson hole allows use of water manometers for improved precision because the entire manometer can be placed below the level of the porous cup.

Operation

Testing the porous cup, the porous cup-tubing interface, and all fittings before installation is desirable. After the porous cup is saturated, air pressure is applied to the interior of the tensiometer while the parts to be tested are submersed. If bubbles appear at a gage pressure substantially less than the bubbling pressure of the porous cup, the unit is faulty and the appropriate parts need to be repaired or replaced.

Water used in tensiometers is deaerated in a carboy by applying a vacuum or heat or both. Excellent results have been obtained using a pump that generates a 970-cm H_2O vacuum with a heated magnetic stirrer for 48 h. Insufficiently deaerated water requires frequent on-site purging or, in the worst case, allows bubbles to form before a tensiometer has reached equilibrium, thus preventing accurate data collection. The deaerated water is siphoned (to minimize reaeration) from the carboy to a collapsible plastic container (available at a sporting goods store) for on-site use. Although the air above the water is forced out the spout immediately, the water reaerates slowly by diffusion of air through the container wall. Proper fittings connect the container spout to the tensiometer supply tube.

A small-diameter tensiometer is purged by connecting the water supply container to the water supply tube and raising it above the top of the tensiometer while opening the supply valve. Several tensiometers can be connected in a "T" network to simplify multiple purging. Purging instructions for vacuum-gage and hybrid tensiometers are supplied by the manufacturer. These instructions can be modified if use of deaerated water is desired. Purging time needs to be short to minimize wetting of the soil immediately surrounding the porous cup. When purging is complete, the system is closed and the soil draws water through the porous cup until equilibrium is established. The pressure inside the porous cup approaches the soil-water pressure asymptotically at a rate determined by the time constant and the unsaturated hydraulic conductivity of the soil. When equilibrium is reached, the measurement is made. A single value of pressure is recorded when using a vacuum-gage or pressure transducer; a manometer requires measurement of the mercury column and reservoir elevations.

The most reliable data are obtained by purging a tensiometer and allowing it to equilibrate before recording the measurement. However, wet soils, lack of wetting fronts, low-permeability tubing, or thoroughly deaerated water tend to prevent air accumulation for long periods; these conditions, either singly or in combination, permit reliable data collection without purging.

Dry soils or inadequate manometer design, or both, occasionally result in mercury being pulled over the top of the manometer and into the porous cup. The porous cups shown in Figs. 3a, 3b, and 3c may be purged of mercury by applying pressure to the measurement tube, thus forcing mercury out the supply tube. Pressure applied to the top of a hybrid tensiometer will force the mercury out the measurement tube.

Porous cups that are removed from a soil in order to be reused need to be washed with warm water to prevent plugging of the pores. A porous cup with plugged pores possibly can be restored by sanding or rinsing in a weak HCl solution [30].

Summary

Measurement of soil-moisture tension has extensive applications in quantification of flow in the unsaturated zone. Tensiometers directly and effectively measure soil-water tension, but they require care and attention to detail. In particular, installation needs to establish hydraulic connection with minimum disturbance, and air invasion needs to be nullified continually.

A variety of tensiometer designs are available [38]; some can be purchased commercially and some can be constructed from parts. The three basic components of a tensiometer are a porous cup, a connecting tube, and a pressure sensor. Component selection is determined by on-site constraints, the required accuracy, and the cost.

Pressure transducers allow the automated collection of large quantities of data. However, the user needs to be aware of the pressure-transducer specifications, particularly temperature sensitivity and long-term drift. On-site measurement of known zero and "full-scale" readings is probably the best calibration procedure; however, on-site temperature measurement or periodic recalibration in the laboratory may be sufficient. The normal range of application of tensiometers can be extended to include measurements of extremely large tensions, in subfreezing temperatures, and at depth, but not without a substantial investment of effort and money.

References

[1] Healy, R. W., Peters, C. A., DeVries, M. P., Mills, P. C., and Moffett, D. L., "Study of the Unsaturated Zone at a Low-Level Radioactive-Waste Disposal Site near Scheffield, Ill.," *Proceedings, National Water Well Association Conference on Characterization and Monitoring of the Vadose Zone,* Las Vegas, NV, December 1983, pp. 820–831.

[2] McMahon, P. B. and Dennehy, K. F., "Water Movement in the Vadose Zone at Two Experimental Waste-Burial Trenches in South Carolina," *Proceedings,* National Water Well Association Conference on Characterization and Monitoring of the Unsaturated (Vadose) Zone," Denver, CO, November 1985, pp. 34–54.

[3] Ripp, J. A. and Villaume, J. F., "A Vadose Zone Monitoring System for a Fly Ash Landfill," *Proceedings,* National Water Well Association Conference on Characterization and Monitoring of the Unsaturated (Vadose) Zone," Denver, CO, November 1985, pp. 73–96.

[4] Lichtler, W. F., Stannard, D. I., and Kouma, E., "Investigation of Artificial Recharge of Aquifers in Nebraska," U.S. Geological Survey Water-Resources Investigations Report 80-93, U.S. Geological Survey, Denver, CO, 1980.

[5] Sophocleous, M. and Perry, C. A., "Experimental Studies in Natural Groundwater Recharge Dynamics: Analysis of Observed Recharge Events," *Journal of Hydrology,* Vol. 81, 1985, pp. 297–332.

[6] Richards, L. A. and Neal, O. R., "Some Field Observations with Tensiometers," *Proceedings, Soil Science Society of America* (1936), Vol. 1, 1937, pp. 71–91.

[7] Anderson, E. M., "Tipburn of Lettuce: Effect of Maturity, Air and Soil Temperature, and Soil Moisture Tension," Cornell Agricultural Experimental Station Bulletin 829, Cornell University, New York, 1946.

[8] Richards, S. J., Willardson, L. S., Davis, S., and Spencer, J. R., "Tensiometer Use in Shallow Ground-Water Studies," *Proceedings of the American Society of Civil Engineers,* Vol. 99, No. IR4, 1973, pp. 457–464.

[9] Richards, L. A., Russell, M. B., and Neal, O. R., "Further Developments on Apparatus for Field Moisture Studies," *Proceedings, Soil Science Society of America,* Vol. 2, 1938, pp. 55–64.

[10] McKim, H. L., Walsh, J. E., and Arion, D. N., "Review of Techniques for Measuring Soil Moisture In Situ," Special Report 80-31, U.S. Army Corps of Engineers Cold Regions Research and Engineering Laboratory, Hanover, NH, 1980.

[11] Richards, L. A., "Capillary Conduction of Liquids Through Porous Mediums," *Physics,* Vol. 1, 1931, pp. 318–333.

[12] Baver, L. D., Gardner, W. H., and Gardner, W. R., *Soil Physics,* Wiley, New York, 1972, pp. 291–296.

[13] Brooks, R. H. and Corey, A. T., "Hydraulic Properties of Porous Media," Hydrology Papers No. 3, Colorado State University, Fort Collins, CO, 1964.

[14] Marshall, T. J., "Relations Between Water and Soil," Technical Communication No. 50, Commonwealth Bureau of Soil Science, London, England, 1960.

[15] Haverkamp, R., Vauclin, M., Touma, J., Wierenga, P. J., and Vachaud, G., "A Comparison of Numerical Simulation Models for One-dimensional Infiltration," *Soil Science Society of America Journal,* Vol. 41, 1977, pp. 285–294.

[16] Richards, L. A., "Methods of Measuring Soil Moisture Tension," *Soil Science,* Vol. 68, 1949, pp. 95–112.

[17] Richards, L. A. and Gardner, W., "Tensiometers for Measuring the Capillary Tension of Soil Water," *Journal of the American Society of Agronomy,* Vol. 28, 1936, pp. 352–358.

[18] Richards, L. A., "Soil Moisture Tensiometer Materials and Construction," *Soil Science,* Vol. 53, 1942, pp. 241–248.

[19] Colman, E. A., Hanawalt, W. B., and Burck, C. R., "Some Improvements in Tensiometer Design," *Journal of the American Society of Agronomy,* Vol. 38, 1946, pp. 455–458.

[20] Hunter, A. S. and Kelley, O. J., "Changes in the Construction of Soil Moisture Tensiometers for Field Use," *Soil Science,* Vol. 61, 1946, pp. 215–218.
[21] Bianchi, W. C., "Measuring Soil Moisture Tension Changes," *Agricultural Engineering,* Vol. 43, 1962, pp. 398–404.
[22] Klute, A. and Peters, D. B., "A Recording Tensiometer with a Short Response Time," *Proceedings, Soil Science Society of America,* Vol. 26, 1962, pp. 87–88.
[23] Bianchi, W. C. and Tovey, R., "Continuous Monitoring of Soil Moisture Tension Profiles," *Transactions of the American Society of Agricultural Engineers,* Vol. 11, No. 3, 1968, pp. 441–447.
[24] Peck, A. J. and Rabbidge, R. M., "Design and Performance of an Osmotic Tensiometer for Measuring Capillary Potential," *Proceedings, Soil Science Society of America,* Vol. 33, 1969, pp. 196–202.
[25] Walkotten, W. J., "A Recording Soil Moisture Tensiometer," U.S. Forest Service Research Note PNW-180, U.S. Department of Agriculture, Washington, DC, 1972.
[26] McKim, H. L., Berg, R. L., McGraw, R. W., Atkins, R. T., and Ingersoll, J., "Development of a Remote-Reading Tensiometer Transducer System for Use in Subfreezing Temperatures," *Proceedings,* American Geophysical Union Second Conference on Soil-Water Problems in Cold Regions, Edmonton, Alberta, Canada, 1976, pp. 31–45.
[27] Oaksford, E. T., "Water-Manometer Tensiometers Installed and Read from the Land Surface," *Geotechnical Testing Journal,* Vol. 1, No. 4, 1978, pp. 199–202.
[28] Marthaler, H. P., Vogelsanger, W., Richard, F., and Wierenga, P. J., "A Pressure Transducer for Field Tensiometers," *Soil Science Society of America Journal,* Vol. 47, 1983, pp. 624–627.
[29] Stannard, D. I., "Theory, Construction, and Operation of Simple Tensiometers," *Ground Water Monitoring Review,* Vol. 6, No. 3, 1986, pp. 70–78.
[30] Watson, K. K., "A Recording Field Tensiometer with Rapid Response Characteristics," *Journal of Hydrology,* Vol. 5, 1967, pp. 33–39.
[31] Cassel, D. K. and Klute, A., "Water Potential: Tensiometry," Agronomy Monograph No. 9 (2nd ed.), "Methods of Soil Analysis: Part I—Physical and Mineralogical Methods," American Society of Agronomy-Soil Science Society of America, Madison, WI, 1986, pp. 563–596.
[32] Rice, Robert, "A Fast-Response, Field Tensiometer System," *Transactions of the American Society of Agricultural Engineers,* Vol. 12, 1969, pp. 48–50.
[33] "Omega 1987 Complete Pressure, Strain and Force Measurement Handbook and Encyclopedia," Omega Engineering, Inc., Stamford, CT, December 1986.
[34] "Pressure Sensors, Catalog 15," Micro Switch, Freeport, IL, 1985.
[35] Fitzsimmons, D. W. and Young, N. C., "Tensiometer-Pressure Transducer System for Studying Unsteady Flow Through Soils," *Transactions of the American Society of Agricultural Engineers,* Vol. 15, No. 2, 1972, pp. 272–275.
[36] Williams, T. H. L., "An Automatic Scanning and Recording Tensiometer System," *Journal of Hydrology,* Vol. 39, 1978, pp. 175–183.
[37] Haldorsen, S., Petsonk, A. M., and Torstensson, B. A., "An Instrument for In Situ Monitoring of Water Quality and Movement in the Vadose Zone," *Proceedings,* National Water Well Association Conference on Characterization and Monitoring of the Unsaturated (Vadose) Zone, Denver, CO, November 1985, pp. 158–172.
[38] Johnson, A. I., "Methods of Measuring Soil Moisture in the Field," U.S. Geological Survey Water Supply Paper 1619-U, U.S. Geological Survey, Denver, CO, 1962.

Drilling, Design, Development, and Rehabilitation of Monitoring Wells

Stephen A. Smith[1]

Monitor Well Drilling and Testing in Urban Environments

REFERENCE: Smith, S. A., **"Monitor Well Drilling and Testing in Urban Environments,"** *Ground Water and Vadose Zone Monitoring, ASTM STP 1053,* D. M. Nielsen and A. I. Johnson, Eds., American Society for Testing and Materials, Philadelphia, 1990, pp. 55–63.

ABSTRACT: Experiences at Superfund sites in the metropolitan area of Phoenix, Arizona, have shown that installing and testing monitor wells in urban environments present special problems for the hydrogeologist and the well driller. Access is difficult to obtain, and noise, dust, water, and mud must be carefully controlled. Underground utilities must be accurately located prior to drilling, and overhead power lines are safety hazards to drilling rigs and pump rigs. Disposal of drilling fluids and cuttings, site cleanup, and well termination require special attention. Special permits or agreements may be required for drilling in urban areas; these include a permit to work in the public right-of-way, agreements with private land owners, a permit to discharge water from aquifer testing into the sewer system or surface waters, and a permit to obtain drilling water from fire hydrants. Traffic control barricades, police, and private security guards may be necessary to protect public safety. Some drilling methods are not practical in urban areas, and others may have to be modified. In comparison with rural areas, drilling and testing programs in urban areas require significantly more advance planning and are more costly.

KEY WORDS: ground water, monitor wells, drilling, testing, urban areas, permits, cable tool, auger, mud rotary, air rotary, dual-wall reverse circulation

Many sites of contaminated ground water are in urban areas or cities. Industries and businesses use large quantities of chemicals, and spills or leaks are inevitable. Contaminant plumes extend downgradient beneath buildings, streets, parking lots, and residential areas.

Characterizing the magnitude and extent of ground-water contamination in urban environments presents a new set of challenges to drillers and geologists, who are more accustomed to working in undeveloped or rural areas. Technical concerns may become secondary to such issues as the following:

- Site access
- Right-of-way permits
- Control of noise, dust, water, and mud
- Public safety

Careful advance planning is required to address these issues satisfactorily. The objective of this paper is to discuss some of the more important considerations. This discussion is largely based on experiences of the author's company, Dames & Moore, in the Phoenix metropolitan area. Completely satisfactory solutions have not yet been found to some of

---

[1] Associate hydrogeologist, Dames & Moore, Phoenix AZ 85020.

the problems, but a description of the methods we have used to try to resolve these issues may help others who are undertaking their own programs.

Special Considerations in Drilling

Site Access

Access and room to work are significant problems when drilling in urban areas. Sites must be carefully selected in advance, and questionable locations should be inspected by the driller. Available clearances should be measured and compared with the clearance requirements for the drill rig and support vehicles. Selecting a location for a monitor well in an urban area is more complicated than simply drawing a circle on a map.

Monitor wells cannot legally be installed on private land without a license, lease, contract, or some form of agreement with the land owner. In residential areas, numerous agreements may have to be negotiated with individual homeowners. In commercial or industrial areas, the occupant of the property may not be the owner. Dealing with absentee landlords can be particularly time consuming; negotiations can take months.

The agreement with the landowner can be a simple letter which states the terms of the agreement and is signed by each party (or both parties). It should clearly state the anticipated purpose of the monitor well, the drilling schedule, and the proposed monitoring frequency. Terms of indemnification and limits of liability should be included. The biggest concern of most private landowners is the impact on resale value, so the proposed surface completion should be stipulated. Disposition of cuttings and water should also be addressed.

Guidelines for payment are difficult to define. Some homeowners view a monitor well as a status symbol, but others worry about the negative impact on resale value and demand a substantial fee. One approach to negotiations is not even to discuss payment; if the owner brings it up, then a low, but reasonable offer can be made. The author's company has found that a "sign-up bonus" and an annual payment is an effective negotiating strategy with some landowners. The bonus is paid as soon as the agreement is signed; the first annual payment is made when the well is completed, and annual payments are made for as long as the agreement is in effect.

The length of time during which the agreement is in effect should be stated in the agreement. Two to five years with annual renewals has worked well for Dames & Moore. Our agreements also state the cancellation provisions. Either party can usually cancel the agreement with 60 days notice to the other party, and we agree to decommission the well if the agreement is canceled and the property owner wants the well abandoned.

Public lands are sometimes more favorable as drilling locations than private lands, particularly in areas where contaminated ground water is a divisive issue. Access to private lands may not be available at any reasonable price. If state or federal regulatory agencies are involved in the investigation, drilling sites on private land can be obtained by condemnation, but public lands may be a more attractive alternative.

Public lands include streets, sidewalks, alleys, parks, government-owned buildings and adjacent parking lots, drainage facilities, and canals and canal banks. However, to obtain access to these lands, approval from a government agency is necessary and significant delays may occur.

Most municipalities have a special department to deal with requests to work in public rights-of-way such as streets or alleys. However, procedures to deal with requests to work on other public-owned lands are not well established. No bureaucrat may have the authority to grant permission. For example, Dames & Moore made a request to drill a monitor well in a city-owned equipment storage yard. However, the request was not acted on for several months. The delay took place even though the storage yard was used by the City Water Department, which had a representative on the Ground-Water Study Review Committee. No one in the Water Department was willing to accept the responsibility to grant permission. As the drilling rig was mobilizing, the company finally had to contact the mayor of Phoenix. Permission was obtained.

Right-of-Way Permits

Permits are required to drill on public rights-of-way, and municipalities in the Phoenix area have special permitting departments to deal with requests. Requests to drill monitor wells are handled in the same manner as requests to repair a sewer or perform other work. A permit application is filed along with a plan showing where and how the work will be conducted. A processing fee and a bond are also required. In Phoenix, the bond is in the form of a cash payment equal to one-tenth the estimated value of the work. The purpose of the bond is to ensure that the work is completed in accordance with the plans. If not, the bond is forfeited, and the city will correct the work.

Public rights-of-way include streets, sidewalks, and alleys. Most streets do not occupy the full public right-of-way, so a strip of unpaved public land is usually available next to the street. A sidewalk may be present, and the remaining land is usually used by the private landowner. In residential areas, homeowners generally regard this strip of land as their own.

The unpaved portion of the public right-of-way can provide favorable drilling locations. Monitor wells in streets or alleys must be installed below grade in traffic-rated vaults. Sometimes, a sewer manhole with a cast-iron lid is required by the city traffic department. The expense of this type of installation can be avoided by drilling on public land adjacent to the street. Furthermore, installation and subsequent monitoring can be conducted without the traffic hazard associated with a location in a street.

Drilling in the strip of public land adjacent to the street has some disadvantages. Underground and overhead utilities are frequently present. Underground utilities can be carefully located and avoided, but the presence of overhead power lines may disqualify a site. The width of the publicly-owned strip is usually only 3 to 6 m, which is too narrow to provide adequate clearance between the mast of a drill rig and overhead power lines. Homeowners may be opposed to monitor well installations in what they regard as their front yards. A good public relations program with newsletters, fact sheets, and public meetings can help reduce opposition to these installations.

Utilities

The presence of utilities is another source of delays and problems when drilling in urban areas. Overhead utilities can be readily identified and avoided. If necessary, they can be rerouted, although the expense of rerouting even a short section of an overhead power line can greatly exceed the cost of drilling and installing a monitor well. Underground utilities are less obvious. However, utility companies will identify the locations of underground utilities when requested. A 24 to 48-h advance notification may be required.

In urban areas, most utility companies participate in a notification center. In Phoenix, this center is called the Blue Stake Center, and it relays underground location requests to the participants who have utilities in the area. Not all utilities participate, however. In Phoenix, the company has to notify the city sewer and water departments individually. Underground traffic signal facilities also may not be included in the Blue Stake service.

Interstate utilities such as pipeline companies and AT & T may not participate. During one drilling project in Phoenix, phone service to several million users was interrupted when a transcontinental phone cable was damaged.

Utility companies will only identify the locations of underground utilities up to the point where their service ends. On industrial or commercial sites, private underground piping is present which must be located without the assistance of the utility companies. Plant records and drawings may be out-of-date or inaccurate. Under some conditions, the best way to avoid these pipes is to dig the upper part of the hole carefully by hand. However, in cold climates, where pipelines are buried 1.5 m or more to avoid freezing, this approach may not be practical.

Noise, Dust, Water, and Mud

Noise, dust, water, mud, and other emissions are associated with drilling, testing, and sampling operations at monitor well locations. These emissions may be a nuisance or they may pose a health risk. If ground water is contaminated, then cuttings, drilling fluids, and water produced during sampling and testing may be a hazardous waste.

Noise is a special problem is residential neighborhoods. Impact noise and engine noise are both associated with drilling. Of the two, impact noise associated with operations such as driving casing and casing hammers is the most difficult to control. Mufflers, shrouds, and sound-absorbing curtains or walls can be used to control noise, but they can decrease the efficiency of drilling operations and increase the cost. A shrouded engine can overheat; some enclosed air compressors cannot be operated in hot weather with the shrouds in place.

In some municipalities, noise ordinances are used to control noise. Construction work which generates noise may be restricted to certain daylight hours. In other municipalities, noise is covered by nuisance ordinances. These are less specific than noise ordinances, which establish standards for noise levels.

The author has found that the best approach to the noise issue in urban areas is (1) to select the most quiet drilling method possible, particularly in residential neighborhoods; (2) not to operate at night, when the level of background noise is low; and (3) to select drilling locations as far from residences as possible.

Control of dust, mud, and water is relatively easy and is largely a matter of good housekeeping at the drill site. Water injection will control dust when drilling with air. Portable pits are usually used when drilling with mud or water in urban areas, and spills can be cleaned up with sand or other adsorbent material.

Disposal of the cuttings, drilling mud, and water produced during drilling and testing of monitor wells can be costly and time consuming. If contaminant levels are high, these materials may have to be disposed of as hazardous waste. At sites where ground water is contaminated with volatile organics, the author's company has used all of the following procedures at one time or another to dispose of drilling fluids and cuttings:

- Spreading contaminated cuttings and drilling mud in a thin layer on bare soil; the contaminants volatilize.
- Decanting the drilling mud and discharging the water to the municipal sewer system with or without prior treatment by forced aeration and disposing of the remaining sludge as hazardous waste.
- Disposing of all of the drilling mud and cuttings as hazardous waste using gondolas or drums.

These procedures are listed in order of increasing cost; however, use of the first two methods may violate hazardous waste disposal regulations.

Public Safety

Drilling operations on public rights-of-way usually interfere with pedestrian or vehicular traffic. City traffic departments have regulations regarding the placement of temporary traffic control devices, such as barricades, flashers, and warning signs. These regulations are designed to protect workers as well as the general public.

Traffic department regulations may prohibit or severely restrict working in certain areas. These areas include signalized intersections and multiple-lane streets. Completely blocking local access should also be avoided, particularly if access to a fire station, police station, school, or hospital is affected. In some situations, the regulations may require a police officer or flagman. If a police officer is needed, an off-duty officer will have to be hired; the city will generally not provide one.

Special precautions are normally required to protect the public if a drill rig is left overnight at a well location. An unattended drill rig is an attractive nuisance, especially to children. Vandalism and theft are also likely to occur in some urban areas. Therefore, some form of security is required. Temporary construction fencing is one way to provide drill site security, but few drill sites in urban areas are large enough to allow use of a fence, and the equipment is usually not on one location very long. Setting up and taking down fences could become time consuming. The company has had reasonably good success with security guards provided by commercial security services. Guards can usually be provided on a few hours notice, and normal rates in the Phoenix area are about $10 per hour. The security benefit that they provide is sometimes questionable, however. The best that can be expected is that their presence at the well location will deter casual visitors and that they will report any unusual occurrences.

Drilling Water

Some drilling operations require several thousand gallons of drilling water daily. A water truck is usually used to haul water to the rig, and if one wishes to avoid a lengthy wait while filling the truck, a source which can supply water at a rate of about 0.5 L/s or more is necessary.

In urban areas, an adequate supply of drilling water is relatively easy to obtain with proper advance planning. City water departments allow the use of municipal fire hydrants to obtain water for construction purposes. When using a fire hydrant, a 10 000-L water truck can be filled in about 15 min. To use a fire hydrant, a permit application must be submitted to the city's water department. The application will have to be accompanied by a processing fee, and most cities also require a water meter deposit for a temporary meter. City crews install the meter on the fire hydrant. The permit allows the driller to obtain water only from the designated hydrant; use of any others is illegal. The water department will usually charge a nominal fee for water usage.

At industrial facilities, a private fire system may be present. These systems may be connected to the plant fire alarm, which will activate if the pressure drops. Therefore, for fire safety purposes, use of these systems to obtain drilling water may not be possible.

Well Termination

In urban areas, wells are normally completed so that casing does not extend above the existing grade. Many wells are drilled in parking lots, streets, or alleys; therefore, well casing which extends above the existing ground surface is a traffic hazard. An at-grade or below-grade completion also reduces the problem of vandalism.

Many techniques are available for completing wells below the grade. In a street or alley, the city traffic department may require that the well be completed in a manhole. However, manholes are expensive, and access to such wells is difficult. Manhole covers are heavy and awkward to handle.

In lightly traveled areas, the author's company has successfully used precast concrete meter boxes to enclose the wells. These are available in a wide variety of sizes. The lids are usually constructed of steel which is 3 to 6 mm thick; precast concrete lids are available for some boxes, but they do not stand up well under heavy use. The main disadvantages of the boxes are that they require a relatively large hole for installation and they tend to collect water and trash. However, they are relatively inexpensive.

Very neat and attractive at-grade completion is possible by using the combination of a guard pipe with a tamper-proof cap, which is available from some equipment suppliers. These caps are watertight and secured to the guard pipe with various types of tamper-resistant bolts. They are easy to install and relatively inexpensive. However, they cannot be locked, and for this reason, they are not used in public areas.

Selection of Drilling Methods

Drilling equipment for monitor wells is selected on the basis of its ability (1) to obtain accurate representative samples of soils and rock and (2) to construct a well that allows collection of representative ground-water samples. In urban areas, drilling equipment must also be compatible with the environment. Therefore, in some instances, equipment and methods may be selected that are less than optimum in terms of speed or cost. The flexibility and adaptability of the equipment and the crew are also critical factors to consider. To a driller who is used to working in rural areas, the frustrations of drilling in the city may be insurmountable. In this section of the paper, the relative advantages and disadvantages of different types of drilling equipment are discussed. These are summarized in Table 1.

TABLE 1—*Comparison of drilling methods, showing their advantages and disadvantages.*

Factor	Auger	Mud Rotary	Air Rotary	Cable Tool	Dual-Wall Reverse Circulation
Maneuverability and size	good	poor	fair	good	fair
Noise	good	fair	poor	good	very poor
Dust, water, mud	good	poor	poor	fair	poor
Use of water[a]	good	poor	fair	fair	good
Disposal of cuttings and fluids	good	poor	fair	good	fair
Public safety	good	poor	poor	fair	poor
Drilling in bedrock	poor	good	good	good	poor
Drilling in boulders	poor	fair	poor	good	good
Drilling below the water table	good	good	fair	good	fair
Deep drilling (>50 m)	poor	good	good	good	fair

[a] Refers to the need for large quantities of water for drilling. This is a disadvantage in urban areas.

Cable Tool

Cable tool drilling is one of the oldest drilling methods which is still in widespread use. The method is slow, but it has several advantages for monitor well drilling in urban areas. The engine noise is low; no compressors or mud pumps are required. Driving casing does generate noise, but the levels are moderate in comparison with pneumatic or diesel casing hammers. Furthermore, cable tool rigs are small, and their clearance requirements are not excessive. Many are mounted on short, maneuverable, semitrailers. Cable tool drilling also does not generate large amounts of cuttings or drilling fluid, and the method is adaptable to most subsurface conditions, from granite to caving boulders.

The disadvantages of cable tool rigs are well known. They are slow; a drilling rate of 5 to 8 m a day is typical in alluvium in the Phoenix area. Undisturbed soil samples are impossible to obtain because the bit crushes soils several feet below the bottom of the hole. Some cable tool rigs require guy lines for support of the mast, and space may not be available in urban areas.

Auger Drilling

Auger drilling is the most practical method for soil sampling and monitor well installation in urban areas where drilling conditions are suitable. The method is relatively fast and allows accurate sampling; with hollow-stem augers, small-diameter monitor wells can be readily constructed. The noise levels are low, the equipment is small and maneuverable, and the drilling method does not generate excessive amounts of cuttings or fluids. The biggest disadvantage to auger drilling is that it cannot be used to drill rock, well-developed caliche, or boulders. In the alluvial basins in the Southwest, its inability to drill through caliche and boulders is a significant shortcoming.

Mud Rotary

Mud rotary is the cheapest method of drilling a hole in the ground in many circumstances. It is fast and can be adapted to most subsurface conditions. However, it is generally not the most suitable method for monitor well drilling because of several factors: (1) undisturbed soil samples cannot be easily obtained; (2) the use of drilling fluid presents an opportunity for extensive cross-contamination in the drill hole; and (3) the drilling mud invades water-bearing formations and affects the quality of subsequent water samples. Additional disadvantages in urban areas are the noise level and the use of relatively large quantities of drilling fluids. Furthermore, the method is inherently messy, and the space requirements are greater than those of other methods. If a casing hammer is used to control caving, it also adds to the noise level. However, under some drilling conditions, mud rotary may be the only practical drilling method to use.

Air Rotary

Where it is practical, air rotary drilling is generally preferred to mud rotary for monitor well construction because the problems associated with mud invasion and cross contamination are eliminated or minimized. However, the method is not necessarily well suited for drilling in urban areas. Air compressors are noisy, and they require additional space at the drill location if they are not mounted on the rig. Air rotary drilling can be very messy when drilling below the water table, and large volumes of contaminated water may be produced.

Dual-Wall Reverse Circulation

In the Phoenix area, the percussion method of dual-wall reverse circulation drilling has become a popular method for monitor well installation in the past three or four years. The drilling equipment, which is also known as "center stem recovery," "center sample recovery," or "Becker rig," is used extensively for sampling and monitor well construction in alluvial soils. It can rapidly penetrate a wide variety of unconsolidated materials using a dual-wall drill pipe and a diesel hammer. The pipe is equipped with an open-faced bit, and air is forced down the annulus between the drill pipes. Cuttings are blown up the center, and boulders nearly as large as the inside diameter of the drill pipe can be lifted to the surface.

Although the rig is fast and efficient, it is not particularly well suited to drilling in residential areas. The diesel hammer is smoky, and a fine layer of soot is deposited on surfaces near the rig. The casing hammer and the air compressor are also extremely noisy. However, the method is very fast, and under some drilling conditions, such as the presence of boulders, it may be the method of choice. For example, one day of the noise associated with the diesel hammer may be preferable to four or five days of cable tool drilling.

The method cannot be used to drill rock, and it is practical only at depths of about 60 m or less.

Aquifer Testing

Aquifer testing in urban areas can be even more of a logistical challenge than well drilling, especially when tests are conducted by pumping. In Phoenix, the alluvial aquifer is extremely productive, and many monitor wells are tested by pumping at rates of 1 to 10 L/s for periods of several hours to several days. Large quantities of water are produced, and the water is usually contaminated to some degree.

At higher flow rates, extensive treatment of test water prior to discharge is impractical. Limited aeration can be induced at high flow rates by constructing baffles and drop structures in the discharge channel. This has been used successfully in Phoenix to remove low concentrations of volatiles from water that was discharged from a dewatering system at a rate of several thousand gallons per minute. For water containing concentrations of volatile contaminants in the range of a few hundred to several thousand parts per billion, aeration towers are necessary to provide sufficient stripping. These can be rented, but the cost may be difficult to justify for aquifer testing.

In urban areas, the municipal sewer is a possible option for disposal of contaminated water. Large municipal sewage treatment plants receive flows of 600 L/s or more, and a temporary discharge of contaminated water from an aquifer test can usually be treated. However, permission must be obtained from the sewer department, and they will probably want an analysis of the water prior to granting permission to allow a discharge. The treatability of the water and the discharge rates are potential concerns. Many municipal sewers flow at or above design capacity; additional flow will cause the system to back up.

Discharge to the municipal sewer system is not always practical. It requires access to a manhole, and these are frequently in the middle of a street. The cost of installing a temporary sewer tap is difficult to justify, and problems with permits could be time consuming. Therefore, the author's company has also discharged water from pump tests into irrigation canals after pretreatment. Water analyses are necessary to show that the treated water will not be harmful to crops. Discharge into surface waters via storm sewers or a natural drainage is also possible, but a permit from the National Pollutant Discharge Elimination System (NPDES) is required, and pretreatment requirements are stringent.

TABLE 2—*Permits and agreements required.*

Permit or Agreement	Reason for Requirement
Well permit and aquifer testing permit	Required for drilling and testing a well or boring anywhere, not just in urban areas.
Right-of-way permit	Required for drilling on public rights-of-way. Obtained from the city, county, flood control district, irrigation district, etc. May need more than one.
Discharge permit	Needed to discharge water or mud to sewer system or surface waters.
Water permit	Used to obtain water from a fire hydrant.
Property owner's agreement	Needed to drill on private property.

In addition to aeration, the company has also used granular activated carbon (GAC) to treat highly contaminated water from aquifer tests. Carbon removes volatile organic compounds and also reduces the concentration of hexavalent chromium. Portable GAC units can be rented, but they are rather costly. We have purchased small units to treat flows of about 2 L/s. Units with a 0.6-L/s capacity were connected in parallel. Each 0.6-L/s unit costs about $4,000.

Removal efficiencies of 99% or more are possible with GAC. When the carbon is spent, the units are disposed of as hazardous waste.

Summary

In urban areas, monitor well construction is considerably more complicated than in rural environments. Advance planning is required to deal with issues such as site access, permits, and public safety. Seemingly insignificant items, such as disposal of water, can cause very significant delays in the drilling schedule. Table 2 summarizes some of the permit requirements. The net effect is that costs for drilling will tend to be higher in urban areas than in nondeveloped areas.

Drillers and geologists who are not used to working in the city will tend to underestimate the time and costs required for construction and testing of monitor wells. The discussion in this paper should help identify some of the more important factors to consider when preparing estimates.

Monitor well construction and testing in urban areas does have some advantages. The job site can be close to home, repairs and replacement parts for drilling equipment are readily obtainable, and communication between the client, owner, consultant, and driller is enhanced. As more ground-water monitoring programs are initiated in cities, drilling in urban areas will probably become a speciality field for drillers and geologists.

Ronald Schalla[1] and Wallace H. Walters[1]

Rationale for the Design of Monitoring Well Screens and Filter Packs

REFERENCE: Schalla, R. and Walters, W. H., **"Rationale for the Design of Monitoring Well Screens and Filter Packs,"** *Ground Water and Vadose Zone Monitoring, ASTM STP 1053,* D. M. Nielsen and A. I. Johnson, Eds., American Society for Testing and Materials, Philadelphia, 1990, pp. 64–75.

ABSTRACT: Well screens and filter packs are used extensively in the water well industry. Water supply wells are designed with large diameters to accommodate high-capacity pumps for municipal, industrial, and irrigation uses. Monitoring wells serve a different purpose, and therefore have some different design requirements. Monitoring wells are used to collect ground-water samples for chemical evaluation and are typically smaller in diameter and have shorter screened intervals. Monitoring well design, particularly for well screens and filter packs, must meet specific requirements. Unlike water wells, monitoring wells usually have an artificial filter pack between the formation and the well screen and often a secondary filter above the filter pack. The designs of well screens and filter packs are more critical for monitoring wells than for water wells, because monitoring wells serve as sampling ports in an aquifer and must minimize disturbance of the water chemistry and hydrology. At present, screen and filter pack requirements for monitoring wells have been only partially addressed by the technical community. Specific technical requirements should include filter pack parameters (i.e., uniformity coefficient, effective size, kurtosis, skewness, roundness, sphericity, and mineralogy). The method of filter pack placement, which involves particle settling through borehole fluids, is also important, particularly in relation to the nature of the geologic materials, the slot type and size of the screen, and the water level in the well.

KEY WORDS: ground water, monitoring, aquifer, formation, well, screen, filter pack, secondary filter, roundness, effective size, uniformity coefficient

Well screens and filter packs have been used extensively by the water well industry to construct efficient, large-diameter water wells for providing water for irrigation, municipal, and industrial use. Monitoring wells, on the other hand, serve a different purpose and thus have different design requirements for screens and filter packs. Monitoring wells are sampling ports in an aquifer and therefore are typically smaller in diameter and screened in only a portion of the aquifer. Unlike water wells, monitoring wells usually have an artificial filter pack and often a secondary filter. The filter pack is a permeable envelope that surrounds the well screen to filter out the fine particles from the adjacent formation and stabilize it. The secondary filter, which is placed above the filter pack, serves as a barrier to prevent grout or sealant slurries from migrating into the filter pack. Therefore, the secondary filter is a much finer grained sand than the filter pack sand around the well screen. Some design properties for monitoring-well filter packs are different from those for water-well filter packs because sampling of monitoring wells must minimize disturbance of the water chemistry and hydrology.

---

[1] Senior research engineer and senior research scientist, respectively, Pacific Northwest Laboratory, Richland, WA 99352.

Well Screen and Filter Pack Design

Currently, water well practices are used to determine the design of well screens and filter packs for monitoring wells, because all of the major available design texts pertain almost exclusively to water well technology [1–4]. Information on monitoring well design appears only in specific papers [5] and in the U.S. Environmental Protection Agency's technical enforcement guidance document [6], and none of these publications fully or adequately addresses the subject of well screen and filter pack requirements for monitoring wells.

Unless a formation is fairly coarse grained, developing a natural sand pack for a monitoring well is difficult because (1) the small diameter limits the size and capacity of development pumps and equipment, (2) the short screen length limits the rate of withdrawal, and (3) the removal of formation sand from the well is difficult and creates stabilization problems. Therefore, the remainder of this discussion on filter packs is limited to the design and placement of artificial filter packs in relation to monitoring well screens. The following are the filter pack design principles that monitoring and water wells have in common:

- Filter packs are installed to create a permeable envelope around the well screen.
- The grading of the filter pack should be based on the grain size of the finest layer to be screened.

In water wells, the filter pack, or permeable envelope, separates the screen from the formation material, thus reducing drawdown, encrustation, and transport of formation sediment into the well. In monitoring wells, the filter pack minimizes the amount of formation sediment transported into the well (where it could interfere with water chemistry analyses of metals), reduces clogging of the well screen by highly contaminated aqueous and non-aqueous solutions, and provides sufficient permeability to direct the flow of contaminated ground water through the well rather than around it.

Well Screen Requirements

The correct slot size for a well screen is determined by the distribution of grain sizes in the filter pack. The grain-size distribution of the filter pack is determined by the particle-size distribution of the aquifer to be screened. Primary considerations in the selection of a well screen are the chemical resistance and strength of the material type, the casing diameter, slot type and design, slot size, maximum open area during development, and length.

Material Type

In general, the well screen should be Schedule 40 or 80, meet applicable ASTM standards, and be constructed of polyvinyl chloride (PVC), stainless steel, epoxy with fiberglass reinforcement, fluorocarbon resins (i.e., Teflon[2]-type polymers), or whatever other material [e.g., acrylonitrile-butadiene-styrene (ABS) or glass] is most suited to the monitoring environment. The selection of well screen and casing materials should focus on the material's structural strength, ease of handling, chemical durability in long-term exposures to potentially hostile subsurface conditions (particularly below the water table), and potential impact on the chemical integrity or "representativeness" of ground-water samples.

---

[2] Teflon is a trademark of E. I. du Pont de Nemours and Co., Wilmington, DE.

Well Diameter

The diameters of monitoring wells are typically smaller than those of water wells or wells used in baseline or water resource studies. The advantages and disadvantages of small-diameter versus large-diameter casings have been debated in the literature [7–11]. Important reasons for using large-diameter wells include determining large-scale aquifer characteristics (i.e., transmissivity, ability to store) and boundary conditions of high-yield aquifers. However, when such high-yield conditions do not exist, the reasoning has been that small is better [10]. Small-diameter wells are also less expensive, because smaller quantities of materials are installed; drilling costs per foot are lower, because borehole diameters are smaller and less costly drilling methods can be used; and the quantities of contaminated purge water and drill cuttings for disposal at an approved hazardous waste disposal site are much lower. However, small-diameter well screens can be more difficult to develop, because of the limitations of the few methods available for effectively developing small-diameter wells.

Slot Type and Size

Although there are several well screen configurations, basically two types of well screen are used in monitoring wells. One type is a pipe with horizontal slots cut into it at uniform vertical spacings (typically 0.64 or 0.32 cm [0.25 or 0.125 in.]). The other type is a continuous-slot screen that is formed by wire wound around and bonded (typically welded) onto vertical rods. The primary advantage of the wire-wound design over the slotted pipe is the larger open area, which allows more rapid and effective development (Fig. 1). This larger open area is especially important for small-diameter wells that require fine slot openings for fine grained formations. The wire-wound screen provides open areas >10% in small-slot well screens and percentages up to 52% in larger slot sizes. The percentage of open area is several times higher than that provided by standard slotted well screens in the smaller slot sizes (0.025 and 0.05 cm [0.010 and 0.020 in.]), even if the slots are placed 0.32 cm (0.125 in.) apart rather than the standard 0.64 cm (0.25 in.). It has been stated [12] that the hydraulic performance of well screens is independent of the screen design, provided that the open area of the screen exceeds a threshold of about 10% open area. However, other studies indicate that the screen design is important and that the open area threshold

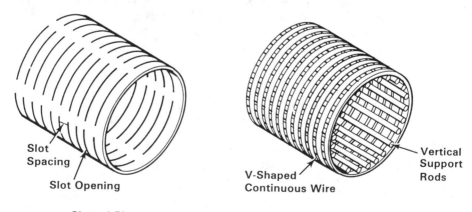

Slot Spacing

Slot Opening

V-Shaped Continuous Wire

Vertical Support Rods

Slotted Pipe   **Continuous-Slot Screen**

FIG. 1—*Two basic types of monitoring-well screens.*

can be as little as 8% [13,14]. Other views are offered [15], but they agree that an unusually high percentage of open areas (i.e., 20 to 50%) has been overemphasized in reducing head loss and entrance velocities, which is important to efficiency in high-production water-supply wells.

In choosing between types of well screen, another factor to consider is the speed and effectiveness of well development. Personal experience with both large- and small-diameter wells indicates that screens with a high percentage of open area reduce the time and effort required for well development [16]. Similar findings on the importance of the percentage of open area have been reported by others [12,17]. A high percentage of open area is particularly important where smaller slot sizes and fine-grained filter packs must be used to retain the bulk of the formation sediments.

Filter Pack Requirements

The choice of filter pack is based on the grain-size distribution of the finest layer of formation sediments to be screened. A few desirable filter pack characteristics for monitoring wells are presented in the water well reference book [4]; however, additional specifications are needed for monitoring wells. Specific technical requirements should be expanded to include filter pack parameters (i.e., uniformity coefficient, effective size, kurtosis, skewness, roundness, sphericity, and mineralogy) or at least the designer of the monitoring well should be aware of these parameters and their influence on the effectiveness of the filter pack. Criteria for these parameters are influenced by how the filter materials are transported through the water column during placement. The criteria for the grain size distribution properties are expressed in relationships of particle sizes expressed in terms of the percentage passing a given sieve size. For example, the D_{10} is the size at which 10% by weight of the total particles are smaller, as determined by mechanical sieve analysis.

Uniformity Coefficient, Effective Size, Kurtosis, Skewness

Because the well screen slots have uniform openings (actually they vary by about 0.05 mm [0.002 in.]), the filter pack should be composed of particles that are as uniform in size as is practical. Ideally, the uniformity coefficient (the quotient of the 60% passing, D_{60} size, divided by the 10% passing, D_{10} size [effective size] of the filter pack should be as close to 1.0 (i.e., the D_{60} and the D_{10} sizes should be identical) as possible. In theory, this statement would mean that a totally uniform size should result from all grains passing one size sieve and retained on the next smallest size. Although this uniformity is easily achieved on a laboratory scale, actual commercial sieve operations do not routinely achieve this ideal sorting of particles. The uniformity coefficient is determined by Eq 1

$$\text{Uniformity coefficient} = \frac{D_{60}}{D_{10}} \qquad (1)$$

A low uniformity coefficient is very desirable, particularly if the tails of the particle-size distribution curve are also uniform (i.e., mesokurtic or platykurtic). Kurtosis, which is defined by Eq 2, is the property that describes the relative uniformity of the distribution tails.

$$\text{Kurtosis} = \frac{D_{95} - D_5}{2.44(D_{75} - D_{25})} \qquad (2)$$

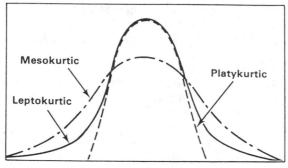

FIG. 2—*Types of kurtosis.*

In a normal (Gaussian) probability curve, the diameter interval between the 5 and 95% sizes should be exactly 2.44 times the diameter interval between the 25 and 75% sizes. Kurtosis is the quantitative measure used to describe this departure from normality, which is called mesokurtic. Such ideal filter pack materials are not readily available; however, they can be produced at a substantial increase in cost over materials normally available. Commercially available, high-quality filter pack materials, used primarily by oil and gas companies for hydraulic fracturing in boreholes and for the construction of monitoring well filter packs, typically have uniformity coefficients ranging from 1.1 to 1.9 and are usually mesokurtic or slightly platykurtic or leptokurtic, with a graphic kurtosis value of <1.7. These characteristics are a result of the sieving process. A schematic diagram illustrating these terms in exaggerated form is shown in Fig. 2.

Skewness is the measure of the degree of asymmetry from a normal distribution, as shown in Eq 3. Because of the nature of the washing and sieving process, the grain-size distribution of commercially available materials is typically symmetrical or negatively (i.e., coarse) skewed (Fig. 3). This coarse skewness is slight yet beneficial, because the percentage of the coarse material is higher than the percentage of the fine material. Fine-skewed filter pack material would contain a higher percentage of the finer sand particles, which would be lost through the well screen during well development.

$$\text{Skewness} = \frac{D_{16} + D_{84} - 2(D_{50})}{(D_{84} - D_{16})} \tag{3}$$

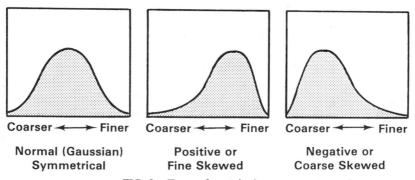

FIG. 3—*Types of particle skewness.*

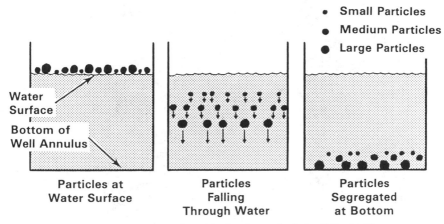

FIG. 4—*Segregation of particles by size and relative fall velocities.*

These characteristics, which describe the distribution of particle sizes, are important because particles falling at terminal velocity in the quiescent fluid surrounding the well screen will be influenced by a number of variables, including the fluid and particle density; the viscosity of the fluid; and the diameter, shape, and surface roughness of the particle. Assuming that all of the variables except particle size are constant, the fall velocity for sand-sized particles composed of quartz will be approximately proportional to the square root of the particle diameter [*18*]. For example, a coarse sand grain 4 mm (0.156 in.) in diameter falls twice as fast as a medium sand grain 1 mm (0.039 in.) in diameter. If the uniformity coefficient of a hypothetical coarse filter pack is 2.5, then the coarsest particles are probably about four times the size of the finest particles. In the example case, the 4-mm (0.156-in.) particles would fall at approximately 0.2 m/s (0.66 ft/s) in clean, distilled water, and the 1-mm (0.039-in.) particles would fall at approximately 0.1 m/s (0.33 ft/s). Therefore, particles will segregate rapidly according to size, as shown in Fig. 4.

If this filter pack material were placed in a well annulus through a few metres of water, the particles would segregate according to size, with the finest particles at the top of the screen and the coarsest particles at the bottom (Fig. 5). Fine formation particles would pass through the filter pack and well screen at the bottom, and become lodged in the slots, as detailed in the bottom window in Fig. 5. At the top of the screen, most, or possibly all, of the filter pack would be lost if the 10%-passing rule (or effective size) had been used to design the filter pack, because the 10%-passing size would be concentrated at the top 10% of the screened interval. Thus, each time water was pumped from the well, filter pack material would flow unabated through the top 10% of the well screen (see the upper window in Fig. 5). If sand was added incrementally, a series of fine to coarse layers would form at intervals along the length of the screen.

The problem with the fines would be eliminated if a more uniform filter pack were used and if the design of the filter pack allowed <1% of the particles to pass through the screen slots, rather than the 10% (i.e., 90% retained) in the criterion for water wells [*1,4*]. With a more uniform sand pack, the segregation problems with the coarse fraction would be diminished, because the tail of the coarse portion of the distribution curve would consist of a higher percentage of smaller, coarse-grained particles.

Through the use of elaborate and time-consuming circulation processes, the particles in a nonuniform filter pack can be distributed somewhat more evenly [*3*]; however, simply

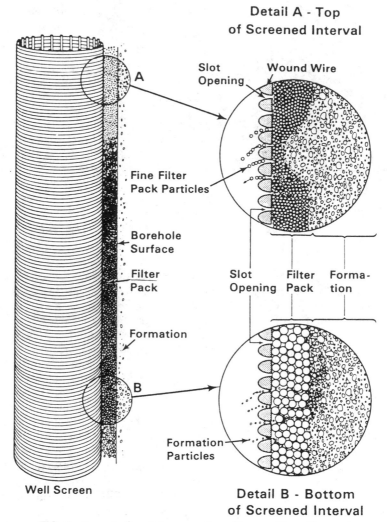

FIG. 5—*Impact of particle segregation of a nonuniform filter pack.*

obtaining more uniform filter pack materials is more cost effective and does not require the introduction of fluids into the well bore.

Roundness and Sphericity

Roundness and sphericity are also important parameters for filter pack design, because particles that are less round and less spherical tumble and oscillate as they fall through water. This tumbling and oscillation slows the rate at which the particles fall. Sand grains generally become less rounded and spherical as the particle size decreases. The greater angularity of the smaller grains tends to slow them, thus increasing the difference between the velocities of the large and small particles and assuring segregation.

Particle roundness (i.e., curvature of the edges of a particle) is important because minor changes in roundness can increase the potential for sand bridging (the adverse development of a friction bond between the temporary and permanent casings). The sand bridge may force removal of the casing and well screen from the borehole and, if caving occurs, may result in loss of the well. Bridging is more likely where the annular clearance is less than 5 cm (2 in.) or if the sand is added too rapidly, resulting in bridging above the water level in the annulus.

A common method used by geologists for defining particle roundness is the Powers scale (based on visual comparison of particles with photographic charts), which ranges from 1 (very angular) to 6 (well rounded) [19]. The potential for bridging is particularly critical between subangular (Powers scale 2 to 3) and subrounded (Powers scale 3 to 4) particles. The likelihood of bridging by particles that have a roundness of 2 to 3 is greater than that by particles with a roundness of 3 to 5. Bridging is a less serious problem in water wells than in monitoring wells, because water wells are typically constructed of Schedule 40 steel or stainless steel, whereas monitoring wells are usually constructed of much weaker PVC, Schedule 5 stainless steel, or Teflon. Therefore, stress-inducing techniques (e.g., holding the permanent casing down with the drill while pulling up on the temporary casing or auger) that allow a sand bridge to break in steel water wells either require greater care in the less durable monitoring well pipe or simply cannot be used in such pipe without damaging it.

Unlike roundness, sphericity can define quantitatively how nearly equal the three dimensions of a particle are. Sphericity, like roundness, can also reduce the potential for sand bridging in monitoring wells. Yet sphericity is rarely included in well specifications that define the shape of sand particles, and numerical values are rarely mentioned. By defining the desired sphericity, we can eliminate undesirable particle shapes such as platy particles (e.g., micaceous particles), bladed particles (e.g., shell fragments, volcanic glass fragments), and elongated particles that are contained in certain types of sand. Spherical particles are desirable, because they reduce the probability of bridging during pullback and because they settle faster than rounded, nonspherical particles (Fig. 6), thus minimizing particle segregation.

Maximum projection sphericity [20] is based on a comparison of the maximum projection area of a sphere, which has a sphericity of 1.0, with that of a given particle of the same volume. Thus, if a sand particle has a sphericity of 0.6, it means that a sphere of the same volume would have a maximum projection area only 0.6 that of the sand particle. Con-

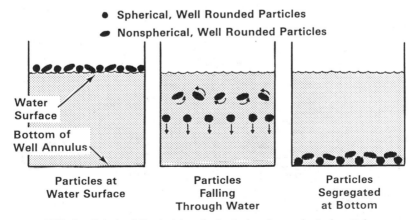

FIG. 6—*Relative fall velocities of spherical and nonspherical particles.*

sequently, the particle would settle about 0.6 times as fast as the sphere because of the increased surface area resisting the downward motion [21]. Most subrounded to rounded sand and gravel particles have sphericities ranging from 0.6 to 0.8.

Mineralogy

To minimize loss of the filter pack by dissolution and adsorption, the filter pack should be primarily quartz (i.e., 95% quartz), with less than 5% other siliceous sediments such as feldspars or oxides [22]. Sulfate and calcareous filter pack sediments should be less than 0.5%, particularly in environments where dissolution of or precipitation on the filter pack might occur. Acid solubility in weak solutions should be less than 0.2% and loss on ignition less than 0.05%.

Secondary Filter Requirements

The main purpose of the secondary filter, shown in Fig. 7, is to prevent slurries of bentonite or other annular sealants from migrating into the primary filter pack, thus sealing the well partially or totally from the formation to be monitored and contaminating the screened interval and the well. Three important surfaces must be considered in the design of the cylindrical secondary filter that surrounds the well casing: the bottom surface (or the interface with the filter pack), the top surface (or the interface with the grout), and the outer surface.

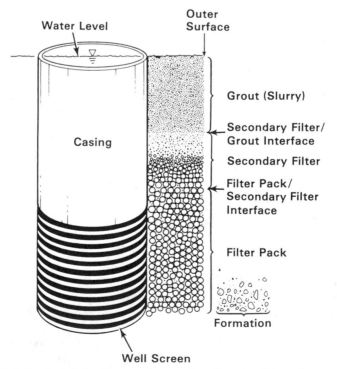

FIG. 7—*Particle distribution of a secondary filter placed through water.*

To prevent fine materials of the secondary filter from invading the underlying filter pack, the coarsest fraction (at least 10%) of the secondary filter must be larger than the average diameter of the voids (interstices) in the filter pack. Therefore, the finest fraction of the filter pack should be approximately equal in size to the D_{90} size (coarsest fraction) of the secondary filter. Quantifying the proper size of the secondary filter is readily accomplished, especially if the uniformity coefficient of the filter pack is 2.5 or less and the graphic kurtosis is <1.5: in this case, one should simply use the effective size, D_{10}, of the filter pack, because it represents 10% of the finest material in the filter pack. This design guide will maximize the effectiveness of the secondary filter, thus preventing grout from invading the primary filter without permitting loss of the secondary filter particles into the filter pack.

At the secondary filter/grout interface, the filter material should be as fine as is practical, so that the bentonite slurry or cement grout will not significantly invade the secondary filter and will not invade the filter pack at all. The particles need to be very fine grained, but not be so fine grained that the time required for them to settle is signficantly influenced by the fluid viscosity or minor turbulence caused by placement of the secondary filter. The smallest particle size should be about U.S. Standard Mesh No. 230 [ASTM Specification for Wire Cloth Sieves for Testing Purposes (E 71–87), which has a designation of 63 μm (0.0025 in.)] to prevent invasion of the grout. If the viscosity or weight of the fluid through the secondary filter is significantly greater than clear water, then the finest particles should be Mesh No. 140 (0.0041 in.) to accelerate settling of the secondary filter and reduce invasion of the grout. This recommendation is based not on the size of the cement or bentonite particles but on the viscosity of the grout slurry (which should have a Marsh funnel viscosity of at least 80 s) and the height (the hydraulic head less the density effect) of the grout column above the sand pack. Using a guideline of <2% by weight passing the Mesh No. 200 (74 μm [0.0029]) should achieve that goal. Several commercially available sand packs meet these criteria.

The grain size distribution may be symmetric or have a positive or negative skewness. However, a positive skewness would indicate a larger percentage of the finer sand particles and would be preferable for preventing grout invasion. In contrast to the filter pack, which should be very uniform, the secondary filter should have a uniformity coefficient of 2.5 to as much as 10. This lack of uniformity will increase particle segregation if the material is placed through standing water and, therefore, will increase the effectiveness of the material as a barrier to grout (Fig. 7). If the secondary filter is placed in an annulus where the water level is inside the filter pack, the mixture of various particle sizes will form a layer of relatively low permeability because the smaller particles will fill the voids between the larger particles (Fig. 8). However, if a graded secondary filter is desired to avoid losses of the finer secondary filter materials into the filter pack, a layered system can be created by placing layers of progressively finer materials above the filter pack. The resulting layering will be similar to the distribution of particles placed through water.

The secondary filter should extend at least 0.305 m (1 ft) above the top of the filter pack. The upper half of the secondary filter should be in contact with a lithologic layer of equal or lower permeability and thickness to prevent migration of the slurry seal around the secondary filter and into the coarser filter pack.

Conclusions

The design of a proper filter pack for monitoring wells is different from that for water wells, because monitoring wells serve a different purpose, are typically smaller in diameter, and are often composed of inert materials that are less strong than steel. Monitoring wells are usually constructed in chemically hostile environments where the removal of large

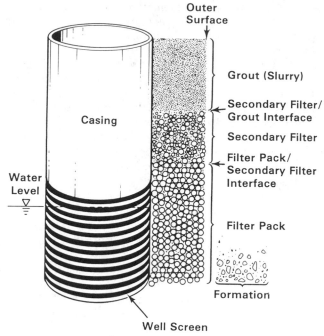

FIG. 8—*Particle distribution of a secondary filter placed above water level.*

quantities of sediment-laden liquids during well development is potentially damaging to the well, difficult to achieve, and costly.

Requiring the removal of up to 10% of the sand pack during development needlessly increases the time and difficulty of completing a monitoring well that will never be used for producing large quantities of drinking or irrigation water. We recommend that the amount of filter pack passing through the well screen be 1% or less; this reduction is easily achieved using commercially available filter pack materials that are uniform and adequately rounded.

The specific technical recommendations for monitoring well filter packs include (1) uniformity coefficients as close to 1.0 as possible; (2) platykurtic distribution tails for a uniform filter pack, to conform to the uniform well screen slot size; (3) particles in the filter pack that have a roundness ranging from 3 to 6 on the Powers scale; and (4) particle sphericity of 0.6 to 1.0 to eliminate rod-, disc-, or blade-shaped particles (because of their slow fall velocity and potential for bridging between the temporary and permanent casings during construction or passing through the well screen).

The secondary filter may be necessary, depending on the particle size distribution of the formation. It must be composed of well-graded (not uniform), preferably positively skewed sand, with the coarsest fraction equal to the 90%-retained size of the filter pack, and less than 2% by weight passing the Mesh No. 200 sieve. Although the particles in the secondary filter should be much less uniform, they should have mineralogical characteristics similar to or at least compatible with those of the filter pack. Typically these fine-grained sediments in the secondary filter will be much less round and spherical than those in the filter pack.

Acknowledgment

This work was supported by the U.S. Department of Energy under Contract DE-ACO6-76RLO 1830.

References

[1] Johnson, E. E., *Ground Water and Wells,* 6th ed., Johnson Division, UOP, Inc., St. Paul, MN, 1966.

[2] Anderson, K. E., *Water Well Handbook,* Missouri Water Well and Pump Contractors Association, Inc., Rolla, MO, 1971.

[3] Campbell, M. D. and Lehr, J. H., *Water Well Technology,* 4th ed., National Water Well Association, Dublin, OH, 1973.

[4] Driscoll, F. G., *Groundwater and Wells,* 2nd ed., Johnson Division, UOP, Inc., St. Paul, MN, 1986.

[5] Barcelona, M. J., Gibb, J. P., and Miller, R. A., *A Guide to the Selection of Materials for Monitoring Well Construction and Ground Water Sampling,* ISWS Report Manuscript, Contract Report 327, Illinois State Water Survey, Champaign, IL, 1983.

[6] U.S. Environmental Protection Agency, *Resource Conservation and Recovery Act Ground-Water Monitoring Technical Enforcement Guidance Document,* U.S. Government Printing Office, Washington, DC, 1986.

[7] Rinaldo-Lee, M. B., "Small- vs. Large-Diameter Monitoring Wells," *Ground Water Monitoring Review,* Vol. 3, No. 1, 1983, pp. 72–75.

[8] Schalla, R. and Oberlander, P. L., "Variation in the Diameter of Monitoring Wells," *Water Well Journal,* Vol. 37, No. 5, 1983, pp. 56–57.

[9] Schmidt, K. D., "How Representative are Water Samples Collected from Wells?" *Proceedings of the Second National Symposium on Aquifer Restoration and Ground Water Monitoring, Columbus, Ohio,* National Water Well Association, Dublin, OH, 1983, pp. 117–128.

[10] Voytek, J. E., Jr., "Considerations in the Design and Installation of Monitoring Wells," *Ground Water Monitoring Review,* Vol. 3, No. 1, 1983, pp. 70–71.

[11] Schmidt, K. D., "The Case for Large-Diameter Monitor Wells," *Water Well Journal,* Vol. 36, No. 12, 1982, pp. 28–29.

[12] Clark, L. and Turner, P. A., "Experiments to Assess the Hydraulic Efficiency of Well Screens," *Ground Water,* Vol. 21, No. 3, 1983, pp. 270–281.

[13] Jackson, P. A., *A Laboratory and Field Study of Well Screen Performance and Design,* M.S. thesis, Department of Geological Sciences, Ohio University, Athens, OH, *Ground Water,* Vol. 21, No. 6, 1983, pp. 771–772.

[14] Bikis, E. A., "A Laboratory and Field Study of Fiberglass and Continuous-Slot Screens, Abstract," *Ground Water,* Vol. 17, No. 1, 1979, p. 111.

[15] Ahmad, M. U., Williams, E. B., and Hamdan, L., "Commentaries on Experiments to Assess the Hydraulic Efficiency of Well Screens," *Ground Water,* Vol. 21, No. 3, 1983, pp. 282–286.

[16] Schalla, R., "A Comparison of the Effects of Rotary Wash and Air Rotary Drilling Techniques on Pumping Test Results," *Proceedings of the Sixth National Symposium and Exposition on Aquifer Restoration and Groundwater Monitoring, Columbus, Ohio,* National Water Well Association, Dublin, OH, 1986, pp. 7–26.

[17] Ericson, W. A., Brinkman, J. E., and Dar, P. S., "Types and Usages of Drilling Fluids Utilized to Install Monitoring Wells Associated with Metals and Radionuclide Ground Water Studies," *Ground Water Monitoring Review,* Vol. 5, No. 1, 1985, pp. 30–33.

[18] Simons, D. B. and Senturk, F., *Sediment Transport Technology,* Water Resources Publications, Fort Collins, CO, 1977.

[19] Powers, M. C., "A New Roundness Scale for Sedimentary Particles," *Journal of Sedimentary Petrology,* Vol. 23, 1953, pp. 117–119.

[20] Sneed, E. D. and Folk, R. L., "Pebbles in the Lower Colorado River, Texas: A Study in Particle Morphogenesis," *Journal of Geology,* Vol. 66, 1958, pp. 139–183.

[21] Folk, R. L., *Petrology of Sedimentary Rocks,* Hemphill Publishing, Austin, TX, 1968.

[22] Palmer, C. D., Keely, J. F., and Fish, W., "Potential for Solute Retardation on Monitoring Well Sand Packs and Its Effect on Purging Requirements for Ground Water Sampling," *Ground Water Monitoring Review,* Vol. 7, No. 2, 1987, pp. 40–47.

Charles A. Rich[1] and Bruce M. Beck[1]

Experimental Screen Design for More Sediment-Free Sampling

REFERENCE: Rich, C. A. and Beck, B. M., **"Experimental Screen Design for More Sediment-Free Sampling,"** *Ground Water and Vadose Zone Monitoring, ASTM STP 1053,* D. M. Nielsen and A. I. Johnson, Eds., American Society for Testing and Materials, Philadelphia, 1990, pp. 76–81.

ABSTRACT: An experimental double-walled screen design has been found to increase the reliability of a properly installed, monitoring well sand pack, intended to produce long-term sediment-free sampling. The screen design is most useful for shallow, small-diameter monitoring wells installed by the hollow-stem auger drilling method and can be produced in a wide range of lengths and diameters. The new screen design eliminates the need for state agencies to require "tremied" gravel packs for such wells when the tremie-pipe installation technique is difficult to complete successfully without compromising the intent behind the specification. The double-walled screen design contains an economically fabricated, preassembled, evenly graded sand pack that avoids the need for resin and that ensures emplacement of an *in situ* homogeneous sand pack in the field.

When installed directly against the formation wall, in conjunction with a conventional sand pack or the best efforts at a tremied pack, the preassembled double-walled screen and sand pack may help reduce the amount of development time necessary to obtain turbidity-free water (that is, water containing less than 50 nephelometric turbidity units). The presence of the preassembled homogeneous sand pack downhole improves long-term sample integrity in silty deposits and minimizes interference attributable to improperly placed or poorly settled packs. The sand pack example described is simple in design and economical, accommodates small-diameter submersible pumps and bailers, and is easily fabricated in the field.

KEY WORDS: ground water, screen design, sand pack, preassembled double-walled screen, filter pack, monitoring well construction

Monitoring well design is basically a derivation of water well design. However, the purposes and objectives of monitoring wells differ greatly from those of water wells. Water wells are primarily designed to serve as a reliable and sustained high-capacity source of drinking water. In contrast, monitoring wells are designed to provide samples representative of the ground-water quality at strategic aquifer horizons, in target locations, during a given period, and for a given reason. Thus, the "as-built" monitoring well is not utilized for production purposes and, therefore, does not need to reflect the design characteristics of a miniature production well.

One common specification inherent in present-day monitoring well design is the downhole emplacement of a graded filter pack (gravel or sand pack) envelope surrounding the screened zone. In the water well industry, the gravel pack is an integral part of a properly designed screen, prepared by tapping fairly thick sequences of fully saturated, unconsoli-

[1] President and environmental engineer, respectively, CA Rich Consultants, Inc., Sea Cliff, New York 11579.

dated permeable formation materials in relatively large-diameter rotary-drilled boreholes. Its purpose is to maximize the efficiency of a production well's pumped withdrawals by improving upon the formation's permeability around the screen. Therefore, the thickness of the pack and its resultant as-built permeability are the most important factors in its operational effectiveness.

In contrast, monitoring well screened zones are usually only 1.5 to 3 m (5 to 10 ft) in length and are not necessarily designed to intercept fully saturated, highly permeable aquifers. In addition, they are used to remove only very small volumes of ground water, and only periodically. In practice, monitoring wells are often screened in silty zones of low permeability or stratified zones of highly variable permeability. Because the initial sampling objectives are commonly aimed at the "uppermost water-bearing zone," the monitoring well can be of small diameter, usually 10.2 cm (4 in.) or smaller, and set at relatively shallow depths.

Nationwide, monitoring wells are commonly installed by the drive-and-wash or hollow-stem auger method to avoid the use of drilling muds, which can cause interference with subsequent ground water sample analyses. The resulting borehole is relatively clean, small, and presents several advantages for representative completions at target locations. Two of the advantages include the cost savings realized by drilling a smaller diameter borehole, and the lower volume of drill cuttings and formation fluids that may have to be handled as potentially hazardous materials.

Need for Concept Design Improvement

The smaller borehole diameter not only minimizes the volume, and therefore the time and handling, of potentially hazardous drill cuttings, but it also allows a reduction of the volume of sand pack material required to backfill the annular space around the screen. In addition, the smaller borehole increases the chances of advancing a plumb hole at difficult drilling sites (i.e., glacial till or landfills). To achieve proper emplacement of the sand pack around the screen in small boreholes, it has been a common, optional, or requisite practice to "tremie" the pack to its desired depth. At depths greater than 20 m (65 ft), the field procedure for conventional sand packing of a monitoring well 10.2 cm (4 in.-) in inside diameter (ID) virtually mandates the tremie-pipe method, which requires a larger borehole, perhaps multiple drilling methods, and much higher associated footage cost ($15 to $30 higher per foot).

The assumption that larger boreholes with thicker sand packs filter out fine materials better than smaller boreholes with thinner packs (i.e., 1.3 cm or thicker) is a misconception because sand pack thickness has nothing to do with the pack's capability to produce sediment-free samples. Rather, the more important variable in sand pack design is the ratio of the packing material grain size, and its uniformity coefficient, to that of the screened formation material.

In addition, a small [10.2-cm (4-in.)]-diameter monitoring well usually necessitates an auger head cutting diameter of 33 cm (13 in.), which results in a geologic formation disruption, due to auger rotation, that affects an area at least 41 cm (16 in.) in diameter. As a consequence, the cross section of the annulus requiring a sand pack—the area between the installed 10.2-cm (4-in.) ID, 11.68-cm (4.6-in.) outside diameter (OD) well screen and the adjacent undisrupted formation material—may be 13 cm (5 in.) or more. This disruption zone must be filled with natural materials (often heaved sands), mixed formation materials, sand pack materials, or a combination of these. The determination that a competent, uniformly graded sand pack has been properly installed around the screen cannot be made with complete confidence. Emplaced sand pack materials are subject to inhomogeneities caused by settling, as a result of high-stress well development, the entry of mixed nonrepresentative contaminated fluids, formation wall caving and heaving, and differential compaction from overlying materials.

In most situations, feeding the sand pack downhole through the annular space so that it arrives, settled and compacted, around the entire length of the screened zone without uneven grading, gaps, or mixed-in drill cuttings is difficult to accomplish in the field. In practice, the principal problems include a high potential for sand bridging and heaving, formation wall collapse, uncontrollable segregation of sand pack grain size through the water/fluid column, and settling, all of which can result in turbid ground-water samples. Sample filtration during collection may then be necessary, further compromising the sample integrity. These problems are minimized by careful drilling and by well installation contractors using proper well installation practices.

Today, the tremie-pipe installation of monitoring well sand pack materials, which consists of pumping clean sand through a funnel and small-diameter pipe temporarily installed in the annular space between the borehole and the casing/screen assembly, is usually preferred to the older construction practice of pouring and tamping sand downhole. However, installing a tremied sand pack properly may require a specialized or difficult operation in an otherwise unnecessarily oversized borehole. Consequently, even good techniques followed by experienced operators do not necessarily guarantee the integrity of artificially emplaced pack materials.

An alternative experimental solution to the problem is the placement of loose, evenly graded, sand pack material around the screen as an integral part of screen construction— *before the screen and casing are installed in the borehole.* This design avoids the irregularities induced from the sand pack "falling" through the fluid column in the well's annular space or its being pumped or injected as an aerated slurry through a small-diameter pipe to its ultimate destination: a relatively short screened zone. The prebuilt pack is particularly effective with deeper monitoring well installations because it negates the need for the larger diameter borehole necessary for a successful pack under conventional installation. To accomplish the installation of an already built screen and sand pack assembly downhole in the zone of interest without the use of bonding agents/resins, foreign material, or synthetic packing in the form of glued-on sleeves (which would adversely affect the quality of ground-water samples) requires a natural sand pack contained in a simple double-walled screen. The concept of a screen and sand pack installation that responds to these limitations and that can be applied widely, either as fabricated in the field or as specifically manufactured by a screen vendor, is outlined below.

Screen Design

A 1 to 1.5-m (approximately 3 to 5-ft)-long, double-walled, Schedule 40 polyvinyl chloride (PVC) screen [10.2 cm (4 in.) in ID and 11.7 cm or (4.6 in.) in OD] serves as a suitable experimental model for descriptive purposes. The interior screen wall has a smaller diameter screen [for example, 5 cm (2 in.) in ID and 6 cm (2.375 in.) in OD], with either a uniform slot size identical to that of the outer screen or a smaller slot size, concentrically fitted inside the outer 10.2-cm (4-in.)-ID screen with two or more sets of PVC centralizers. The annular space between the two screens is manually filled with a sieved, clean, uniformly graded, well-rounded quartz (at least 95% SiO_2) sand. The grain size of the pack should be keyed to the finest grain size present in the horizon screened.

In the example just described, for a 1.5-m (5-ft) screen, the resulting sand pack between the inner and outer screen walls is approximately 2 cm (0.8 in.) thick, filling a void space of approximately 0.002 m^3 (0.075 ft^3). The authors suggest that the top of the inner-sleeved flush-joint screen be tapered outward with a coupling that mounts flush to the flush-joint head coupling of the outer 10.2-cm (4-in.) screen. The completed screen and sand pack assembly is then threaded onto standard 10.2-cm (4-in.)-ID riser pipe and lowered into the hole as a standard casing/screen assembly. The approximate weight of a standard 1.5-m (5-ft) length of 10.2-cm (4-in.)-ID PVC screen is 3.86 kg (8.5 lb), and the approximate weight of the double-walled screen with the sand pack is 15.88 kg (35 lb).

The design of the model outlined above should remain flexible and dependent on the user's site-specific sampling requirements. For example, a split double-walled screen installation in the same well may be useful for sampling composite ground-water quality over a fairly thick aquifer sequence. Alternatively, the installation of relatively short double-walled screens can be used to monitor effectively a series of strata having varying horizontal permeabilities at one specific monitoring location. However, the most widespread application of this new screen technology will probably be for hollow-stem auger drilled monitoring wells that must be advanced into caving or heaving sand formations.

A similar preassembled screen and sand pack can be machined with Type 304 stainless steel screen, resulting in only very minor variation in the sand pack thickness. For example, a 1.5-m (5-ft)-long, 9.84-cm (3.875-in.)-ID, stainless steel screen with an inner 6-cm (2.375-in.)-OD concentric screen sleeve allows a 1.5-cm (0.60-in.)-thick sand pack between the screens. The weight of the 1.5-m stainless steel screen is 13.6 kg (30 lb) and the weight of the double-walled assembly, with sand pack, is 34 kg (75 lb).

Optimal Open Area

The permeability of the assembled double-walled screen and sand pack is dependent upon the open area available and the condition of the interface between the screen and the geologic formation. The permeability of both the interior and the exterior screen must be greater than the permeability of the contained sand pack so that formation fluids entering through the outer screen do not have to overcome additional resistance. Therefore, the "ideal" filter pack would be comprised of uniformly graded spherical sand grains, packed together either as loosely or as tightly as the individual sand grain contact points permit. That is, a loosely packed filter can be expected to provide a free fluid flow area equal to approximately 20% of the entire cross section of the sand pack, whereas a consistently tightly packed filter provides about half as much, or up to only 10% free flow area across the same cross section [1]. In actuality, a combination of loosely packed and tightly packed sand pack material is achieved.

The gravity-filled containment of a uniform sand pack between two rigidly separated screen surfaces (centralizers) presents a compromise between the naturally tightly packed system that occurs against the screen surface in a naturally packed well following well development and the ideal loosely packed system outlined above.

Of additional significance in this new design application is the solution to the problem of formation fluid (often contaminated) becoming retained between the grains by capillary forces at the contact points of the individual sand grains. The presence of residual formation fluid in the sand pack can affect initial and subsequent sample integrity and can hydraulically impede the flow of formation fluid through the pack. The preassembled uniform pack does not contain the fines which are normally found migrating or running into a natural uneven or bridged pack; it is the presence of these fines that radically reduces flow (the increased surface area and contact points cause greater capillary retention). Therefore, in practice, the optimal open area of the screen (a design criterion) should be keyed to both the contained sand pack and the natural formation materials expected to be screened. By meeting this criterion, the double-walled screen will meet all of the necessary requirements common to the conventionally designed screen and granular sand pack. Ideally, the optimal open area of the screen can be decided in the field if a variety of screen slot sizes and diameters is available, along with the results of split-spoon samples or formation grain size curves. For example, a finer sand pack requires smaller screen slot widths and less open area, and so forth. Thus, the design of the double-walled screen must remain flexible enough to allow customized fabrications.

Some effort must also be directed toward minimizing the sand grain clogging anticipated with the inner sand pack resting loosely against the generally wedge-shaped slots inside the outer screen. Wedge-shaped or V-shaped screen slots are a means of maximizing the effectiveness of the open space area available on the screen surface [2]. Such a slot design is an improvement over straight slots because it reduces the clogging or lodging of grains in the slots that occurs during well development. Aggressive well development and agitation of the formation materials around the screened zone is necessary to maximize efficient production in water wells. Such procedures are not necessarily required for small-diameter monitoring wells, in which obtaining representative ground-water samples, rather than achieving maximum production, is the key concern. Further studies related to differences between water well screen design and monitoring well screen design technology are available and are recommended [3–8].

Conclusions

Four advantages of the double-walled, sand-packed screen, when used in conjunction with natural sand packing, are given:

1. A preassembled sand pack ensures appropriate placement of a uniformly graded pack around the screen.
2. The sand pack supports monitoring well completions in smaller boreholes because it precludes the need for reliance upon the potentially costly downhole packing by the tremie-pipe method.
3. An evenly graded, preassembled pack of loose clean sand surrounding the screen helps to avoid the cross-contamination interference introduced by residual formation fluids, sediment mixing and bridging, and caving or heaving. This facilitates a more effective installation, particularly in deeper auger-drilled wells.
4. The preassembled sand pack should help facilitate low-stress well development and turbidity-free water samples because of the uniform filtration of suspended solids through both an outer and inner screen and a sand pack.

Future adaptations and improvements upon this conceptual screen and filter design will improve the efficiency and effectiveness of monitoring representative ground-water quality.

References

[1] "Casing and Screens for Water Wells," Technical Data Sheets Nos. 1–9, Preussag AG, West Germany, 1987.
[2] Johnson, E. E., "Ground Water and Wells," 6th ed., Johnson Division, United Oil Products, Inc., St. Paul, MN, 1966.
[3] Barcelona, M. J., Gibb, J. P., and Miller, R. A., "A Guide to the Selection of Materials for Monitoring Well Construction and Ground Water Sampling," *Illinois State Water Survey,* No. 327, 1983.
[4] *Procedures Manual for Ground Water Monitoring at Solid Waste Disposal Facilities,* EPA 530/ SW-611.616, U.S. Environmental Protection Agency, Washington, DC, 1977.

C9 1.17   611

[5] "The Principles and Practical Methods of Developing Water Wells," Bulletin No. 1033, Johnson Division, United Oil Products, Inc., St. Paul, MN, 1975.
[6] "Methods of Setting and Pulling Johnson Well Screens," Bulletin No. 933, Johnson Division, United Oil Products, Inc., St. Paul, MN, 1957.
[7] National Water Well Association and the Plastic Pipe Institute, *Manual on the Selection and Installation of Thermoplastic Water Well Casing,* National Water Well Association, 1981.
[8] Palmer, C. D., Keely, J. F., and Fish, W., "Potential for Solute Retardation on Monitoring Well Sand Packs and Its Effect on Purging Requirements for Ground Water Sampling," *Ground Water Monitoring Review,* Vol. 7, No. 2, 1987.

David L. Kill[1]

Monitoring Well Development—Why and How

REFERENCE: Kill, D. L., "**Monitoring Well Development—Why and How,**" *Ground Water and Vadose Zone Monitoring, ASTM STP 1053,* D. M. Nielsen and A. I. Johnson, Eds., American Society for Testing and Materials, Philadelphia, 1990, pp. 82–90.

ABSTRACT: The objectives of monitoring well development are these:

(*a*) to permit taking of sediment-free samples,
(*b*) to allow water to flow freely into the well so representative samples can be taken rapidly, and
(*c*) to remove all traces of the drilling fluid used in order to minimize any interference with the water quality.

Several development methods have been well documented as to their practicality and effectiveness in developing water supply wells. These methods are mechanical surging, overpumping, air lift, high-velocity water jetting, and several combinations of these methods. All of these methods can be used in monitoring well development but have limitations. Their limitations are usually related to the well diameter, restrictions on placing foreign materials in the well, low-permeability formations, and poor well intake design.

The methods and limitations are reviewed in this paper.

KEY WORDS: ground water, well development, surging, overpumping, air lift, water jetting

Objectives

The objectives of monitoring well development are the following:

(*a*) to permit taking of sediment-free samples,

(*b*) to allow water to flow freely into the well so representative samples can be taken rapidly, and

(*c*) to remove all traces of the drilling fluid used in order to minimize any interference with the water quality.

Accurate laboratory analysis requires sediment-free fluid. Taking sediment-free samples from a monitoring well is, therefore, most important because filtering the water sample may change its chemistry. If soil particles are filtered from the sample, they may take with them absorbed contaminants, thus distorting the sample. Also, additional handling of a sample increases its chances of being aerated, thus causing even further change in chemistry.

Taking representative samples rapidly is best accomplished by improving the permeability of the zone around the well intake. The improved permeability makes it easier for fluid to get to the monitoring well. Improving permeability usually requires removing some of the natural formation during development. Also, it is necessary to repair drilling damage at the borehole face to restore even the original *in situ* permeability.

---

[1] Sales manager, Recovery Equipment Supply Inc., Maple Grove, MN 55369.

Removal of drilling fluids that have been deposited in the monitored zone is very important in order to minimize their effect on the water sample's chemistry. The drilling fluids may originate from any outside water supply or may include water plus additives, such as those used in fluid rotary drilling. Even the use of hollow-stem auger drilling does not entirely eliminate the use of drilling fluid, as occasionally the auger is filled with a fluid to prevent the water-bearing formation from entering the auger. Removal of the drilling fluid is best accomplished during well development.

Factors

The factors of monitoring well development that determine how effective the development process will be are these:

(a) damage to the formation,
(b) the method of development, and
(c) the intake open area.

Damage to the formation is that caused during the drilling process. This can be due to compaction and invasion of fine-grained particles into the formation near the well borehole. Invasion of the drilling fluid solids, such as bentonite particles, will also cause formation damage. The extent of this formation damage greatly affects how extensive the well development process must be.

The effectiveness of different development methods is well documented. Effectiveness is usually in direct relationship to the amount and concentration of energy that can be directed into the formation. Each of the methods used will be discussed.

All development methods work best in wells equipped with a high amount of inlet area in the intake. The high inlet area structure permits hydraulic forces exerted inside the well intake to be directed efficiently into the surrounding formation (Fig. 1). More fine material can be removed quickly if all the available energy can be directed at most or all of the surrounding formation.

Methods

Different monitoring well development procedures have evolved in different regions, primarily because of the type of drilling rig used to drill the well. Unfortunately, some development techniques are still used where other, more recently developed procedures would produce better results. Newer techniques, especially those using compressed air, should be considered by contractors when they buy and equip a new drill rig.

Overpumping

The simplest method of well development is by overpumping, that is, pumping at a higher rate than will be used on the well later when it is purged and sampled. The theory is that any monitoring well that can be pumped free of sediment at a high pumping rate can then be pumped free of sediment at a lower rate. Overpumping rates may be 19 to 38 L (5 to 10 gal)/min or, in a low-permeability formation, as low as 0.95 L (0.25)/min. These pumping rates can be achieved by an electric submersible pump, such as is commonly put in 10-cm (4-in.)-diameter wells, or any of the portable monitoring well sampling pumps that will fit into a 5-cm (2-in.)-diameter well and provide the very low pumping rates.

The main limitation of the overpumping method is that it seldom produces full devel-

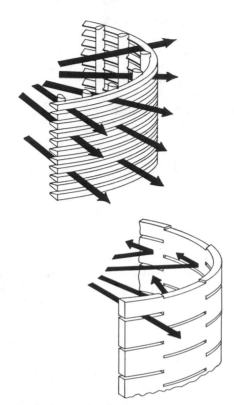

FIG. 1—*Screens with a high amount of open area* (above) *permit freer access to the formation than other well intake designs* (below) *which have less open area. The result is more effective well development.*

opment because most of the development action takes place in the most permeable zone, closest to the top of the screen. The longer the well intake, the less development will take place in the lower part of the intake. After development of the most permeable zone, water entering the intake moves preferentially through this developed zone, leaving the rest of the well poorly developed and not able to contribute water to the monitoring well.

Another objection to overpumping is that the water flows in only one direction toward the well intake (Fig. 2), and some particles may be left in a bridged condition. If the formation is agitated later during the purging and sampling procedure, these bridges may collapse and sediment may enter the well during each sampling. Pumping this sediment may also subject the pump to excessive wear and reduce its operating efficiency, which in turn increases the sampling and purging time. This sediment will also damage the flexible bladders in pneumatic sampling pumps.

Backwashing

Backwashing is a development method that can be used in conjunction with overpumping. The backwashing development method will cause reversal of flow (Fig. 2) through the well intake and agitate the surrounding sediment. This agitation and flow reversal breaks down the sediment bridges. Overpumping then moves the fine particles toward the well and through the intake.

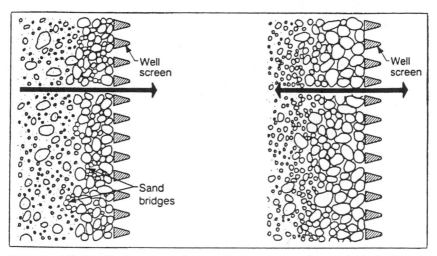

FIG. 2—*Effective development action requires movement of water in both directions through screen openings. Reversing flow helps break down bridging of particles. Movement in only one direction, as when pumping from the well, does not produce the proper development effect.*

Backwashing in a monitoring well can best be accomplished by adding water to the well. The problem with the added water is that it may not be totally removed during the overpumping. The concern then is how it may affect water samples taken from the monitoring well.

Full development with the backwashing method is very unlikely. As in the case with overpumping, the backwashing may produce effects only in the most permeable zone and near the top of the intake. Other parts of a long intake may thus remain relatively undeveloped.

Mechanical Surging

Mechanical surging as a method of development forces water into and out of the well intake by operating a plunger up and down in the casing, similar to a piston in a cylinder (Fig. 3). The plunger used is called a surge block (Fig. 4). A sampling bailer or monitoring well pump may be used, but these will not be as effective as a close-fitting surge block.

The proper procedure for mechanical surging is to bail the well first to make sure that water will flow into it. Then the surge block is lowered until it is below the static water level and a relatively gentle surging action is started. As water begins to move easily in and out of the well intake, the surge block is lowered farther into the well, thus increasing the force of the surging movement. (The force exerted on the formation depends on the length of the stroke and the vertical velocity of the surge block.)

The limitation of mechanical surging is much the same as that for overpumping and backwashing; i.e., it may affect only the most permeable zone. Surging will be more effective if the surge block is operated in the intake. Operating in the well intake will concentrate its action at various levels.

How well the surge block fits in the well casing also has a bearing on the effectiveness of mechanical surging. The fit of a surge block in stainless steel casing will be poor if it is sized to fit through the thread joints (Schedule 40), which will have a smaller inside diameter

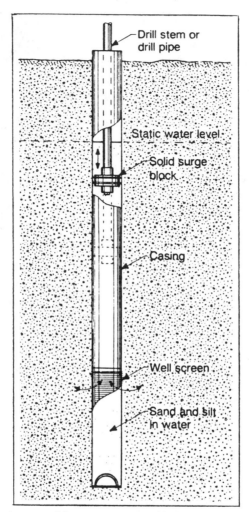

FIG. 3—*For certain types of formations, a surge block is an effective tool for well development. On the downstroke, water is forced outward into the formation; water, silt, and fine sand are then pulled into the well screen during the upstroke.*

than the casing (Schedule 5) barrel. A way to minimize this problem is to have a short length [1.5 m (5 ft)] of Schedule 40 casing directly above the well intake.

Compressed Air Surging and Air Lift

The use of compressed air in monitoring well development has grown with the increase in the number of drill rigs equipped with air compressors and the availability of portable air compressors. Compressed air can be used to alternately surge and air-lift pump the well to remove sediment from the well. In air surging, air is injected into the well to lift the water column. As the water reaches the top of the casing, the air supply is shut off. The water column then faills, causing a surging action in the well intake.

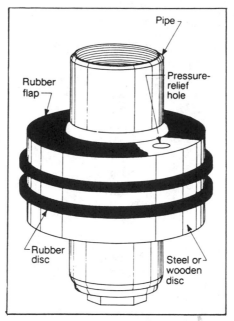

FIG. 4—*A typical surge block consisting of two leather or rubber disks sandwiched between three steel or wooden disks. The blocks are constructed so that the outside diameter of the rubber lips is equal to the inside diameter of the screen.*

In air-lift pumping it is often necessary to install an air line inside an eductor pipe in the well (Fig. 5). Eductor systems are generally useful in monitoring well development when limited volumes of air are available or when the static water level is low in relation to the well depth. Recommended eductor and air line sizes for different monitoring well diameters are shown in Table 1. The eductor pipe system will also minimize the chance of getting a large burst of air injected into the well intake area and into the formation.

Air surging and air-lift pumping are dependent on both air pressure and air volume. The air pressure must be enough to overcome the initial head created by the submergence of the air line. Once the pressure initiates flow, the air volume becomes the most important factor in successful air-lift pumping. The initial pressure or starting head is calculated here.[2]

$$\text{Minimum psi required} = \frac{\text{length of air line} - \text{static water level}}{2.31}$$

The volume of air required to operate an air lift efficiently depends on the total pumping lift, the air line pumping submergence, and the area of the annulus between the eductor pipe and air line. In air-lift pumping of monitoring wells the volume of air required will often be less than 1.42 m^3/min (50 ft^3/min) if the sizes of the eductor pipe and air line pipe used are as is shown in Table 1.

The recommended procedure for using compressed air in a monitoring well is first to set the eductor pipe and air line just below the static water level. After pumping at a moderate

---

[2] Note that 1 psi = 6.8948 kPa.

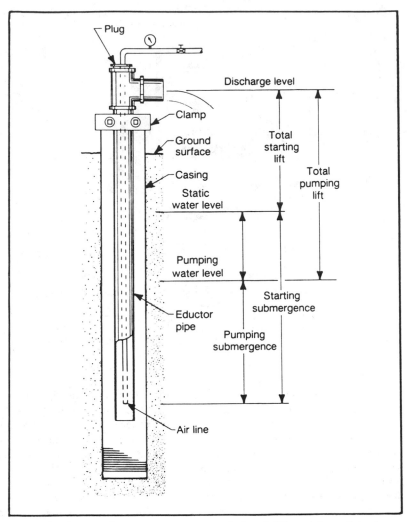

FIG. 5—*This diagram shows common terms applied to air-lift pumping and development.*

TABLE 1—*Recommended pipe sizes for air-lift pumping.*[a]

Size of Monitor Well Casing, in.	Size of Eductor Pipe, in.	Size of Air Line, in.
2	¾	¼
4	2	½
5	3	1
6	4	1¼

[a] One inch = 25.4 mm.

rate, the eductor pipe and air line can be lowered in steps, which will gradually increase the pumping rate. Once the eductor pipe is in the well intake, the air line should be placed so that its lower end is always up inside the eductor pipe. This will eliminate the chance of getting air injected into the formation. This also minimizes the chance of an uncontrolled discharge of water between the eductor pipe and the well casing.

The limitation of using compressed air in development of monitoring wells is generally related to pumping submergence. Quite often in a monitoring well this will cause intermittent flow rather than steady pumping. This is normal in the development method and does not indicate problems, nor does it interfere with eventual sample integrity.

High-Velocity Water Jetting

The jetting procedure consists of operating a horizontal water jet inside the well intake so that high-velocity streams of water shoot out through the intake openings. The equipment required for jet development includes a jetting tool (Fig. 6), a high-pressure pump, a string of pipe, and a water supply. The jetting tool is normally designed with two or four nozzles equally spaced around the circumference. The jetting tool should be constructed so that the nozzle outlets are as close to the inside diameter of the well intake as is practical.

For effective jet development the lowest nozzle velocity should be about 30 m (100 ft)/s. Velocities higher than this can be used in metallic well intakes. Care must be exercised when using jetting screens constructed of polyvinyl chloride (PVC) or other nonmetallic materials. Care must also be taken to jet only with clean water in order to minimize abrasion, especially when jetting in PVC intakes.

In jet development, the jetting tool is placed near the bottom of the monitoring well intake and slowly rotated while being pulled upward. With this procedure the entire surface

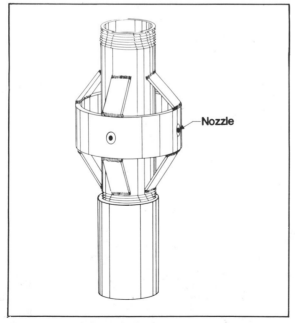

FIG. 6—*A four-nozzle jetting tool designed for jet development of well screens.*

of the intake is exposed to the vigorous action of the jets. Material loosened and brought into the well accumulates at the bottom of the intake and is later removed by bailing or air-lift pumping.

High-velocity water jetting is the most effective development method because the water force can be directed through the intake openings and into the formation. The disadvantage of jet development is that water is added to the well. If the water used is not from the specific well being developed, there will be a concern over whether the added water can be totally removed prior to sampling. In general, it is essential to remove this jetting water from the well as quickly as possible. This necessarily limits the procedure to wells of larger diameter, i.e., 10 cm (4 in.), in which higher capacity pumps can be used.

Summary

Patience, intelligent observation, and the right equipment are required to develop a monitoring well properly. Well development is not expensive, considering the importance of having sediment-free samples, reduced sampling time, and minimum water quality interference. The methods outlined are all effective to varying degrees. Each has limitations but all are easily adapted to monitoring well development.

Acknowledgment

All of the illustrations in this paper are taken from *Ground Water and Wells* published in 1986 by the Johnson Well Screen Co. and are used with permission.

John E. Sevee[1] and Peter M. Maher[1]

Monitoring Well Rehabilitation Using the Surge Block Technique

REFERENCE: Sevee, J. E. and Maher, P. M., **"Monitoring Well Rehabilitation Using the Surge Block Technique,"** *Ground Water and Vadose Zone Monitoring, ASTM STP 1053,* D. M. Nielsen and A. I. Johnson, Eds., American Society for Testing and Materials, Philadelphia, 1990, pp. 91–97.

ABSTRACT: Monitoring wells at a site where the ground water is contaminated with organic solvents were rehabilitated using the surge block method. Several different methods were reviewed for their ability to remove sediments from the well as well as from the soils adjacent to the outside of the well screen. The surge block technique was selected as a method that would accomplish both objectives. The surge block technique is described and is compared with other well rehabilitaton methods. The results of presurging and postsurging ground-water quality from selected monitoring wells are presented. The surge block method was shown to increase the hydraulic capacity of the wells and removed sediment from the bottoms of the wells. Continued monitoring of the wells will indicate whether these benefits are short term or are related to accumulations of sediments at the bottoms of the wells.

KEY WORDS: ground water, surge block, monitoring well rehabilitation

Because of erratic and questionable water quality data from wells at an organic chemical spill site, a decision was made to rehabilitate the wells. Various rehabilitation methods, including scrubbing, surge blocking, and jetting, were evaluated. The surge block method was selected because of its known success in rehabilitating water supply wells [1]. This method not only cleans the interior of the well but also effectively removes fine soil particles from the natural soils or fractures adjacent to the well screen. This method was easy to implement in the field and provided an understanding of the hydraulic behavior of the different wells throughout the site.

The site under investigation is located in southern Maine. The geologic setting consisted of 3 to 10 m of clay over a silty glacial till. The monitoring wells were screened in the till and underlying fractured bedrock.

Ground-water contamination at this site was caused by a variety of volatile organic chemicals, of which trichloroethylene (TCE) was the principal constituent. The chemical spills occurred during the period of 1970 through 1975. A total of 50 wells had been installed as part of hydrogeologic investigations at the site. Twenty-one wells were being routinely monitored for volatile organic chemicals.

Surge Block Method

The surge block method involves lowering into the wellhole a leather-collared steel block attached to the end of a metal rod. A schematic diagram of the surge block used at this site

---

[1] Sevee and Maher Engineers, Westbrook, ME 04092.

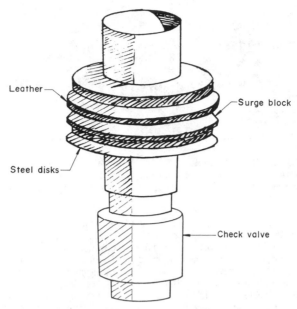

FIG. 1—*Valve-type surge plunger.*

is illustrated in Figs. 1 and 2. The surge block itself consists of a series of metal rings sandwiched with leather collars (Fig. 1). The diameters of the leather collars are cut to the same size as the interior diameter as the well. The purpose of the metal blocks is to provide rigid support to the leather collars except at their very edges. A check valve is located at the base of the surge block to allow water in the well to alternately pass through the block or go into tension as the block is lowered and raised.

During the cleaning operation, the rods that are attached to the block are raised and lowered using a cathead. The weight of the rods and block is generally sufficient, except at very shallow depths, to cause the block to drop after it has been raised. The block is raised and lowered approximately 1 m with each stroke. The rate of raising and lowering depends on the relative position of the surge block within the well and the amount of standing water within the well. Generally, as the level of water in the well decreases, the rate of surging increases.

As the block is lowered (Fig. 2), water within the well casing moves up through the check valve through the rods and discharges at the ground surface. When the block is raised, the check valve closes and creates a suction on the water within the well casing and the surrounding aquifer. The suction within the surrounding aquifer tends to cause small soil particles to migrate toward the well screen. Once within the well the particles move up and out of the well. This has the desirable effect of increasing the hydraulic efficiency of the geologic formation surrounding the well screen. Thus, the area beyond the well screen becomes more hydraulically conductive. The suction also removes soil particles that may be trapped within the screen slots. Finally, the scraping action of the leather collars along the walls of the well tends to remove encrustation buildup and iron bacteria, if these are present. In the more contaminated wells at this site, iron bacteria appeared to be present in small quantities.

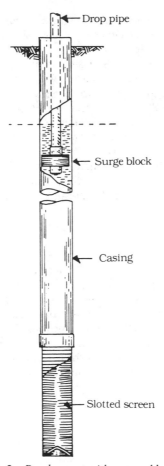

FIG. 2—*Development with a surge block.*

Results of the Surging

The rates of water level recovery prior to well rehabilitation were measured for all wells at the site. The surging results tended to increase slightly the rate of recovery of these wells. As described above, this is consistent with the removal of fine soil particles from the screen slots and the aquifer immediately surrounding the well screen.

The surging of most of the wells produced a reddish-brown-colored water during the early portions of the surging. The intensity of the coloration generally decreased with surging. The water retrieved from the wells was allowed to stand and settle. Soil particles settled out within 1 min of standing time. The coloration decreased with standing time as a flocculent formed and settled from the water. Since most of the wells were constructed with stainless steel tips and black iron pipe risers, the authors have concluded that the reddish-brown staining was the result of iron corrosion within the well. Based on the authors' personal observations, it is not uncommon for iron wells containing ground water high in trichloroethylene to show signs of severe corrosion.

Measurement of the apparent well bottom before and after surging indicated that the surging also removed significant quantities of fine sand, silt, and clay from the interior base of most wells even though the wells had been developed by pumping and flushing (purging) immediately after installation. In addition, these wells have been purged routinely by pumping on approximately a quarterly sampling cycle since 1983. It appears as though the previous purging had allowed fines to enter the wells and settle at the bases of the wells, where it is inaccessible to the purging pump unless there is significant agitation.

After surging one of the wells, it was observed that the amount of soil being pumped from the well could not be reduced. Measurement of the level of the apparent well bottom suggested that it was about 1 m higher than the installation records indicated. Surging would temporarily reduce this level, but the bottom would then increase to the presurging level. This suggested that soils were moving into the well screen through a rupture in the well casing. This was an unanticipated finding. The rupture appeared to be a result of installation rather than of the surging. This resulted in the recommendation of abandonment of this particular well.

Evaluation

The results of the surging at the southern Maine site indicated that the purging of wells by pumping prior to sampling may be ineffective in removing soil particles that have entered and settled at the bottom of the well. Typically, the surging pump or bailers are not lowered all the way to the base of the well and, consequently, any fine soil particles that settle to the bottom of the well are not removed. This is felt to be important in that the organic chemicals in the ground water at this site may partition strongly onto the soil particles [2]. During sampling of the wells the equilibrium of partitioned chemicals may be upset, causing the organic chemicals to enter into the well water. Furthermore, if pure

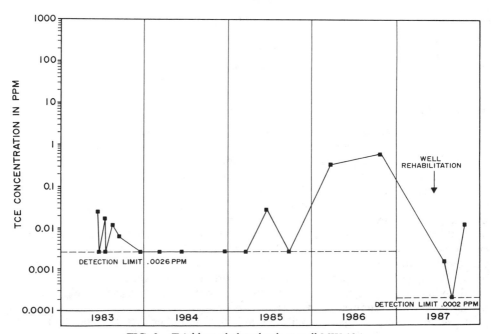

FIG. 3—*Trichloroethylene levels at well MW-19A.*

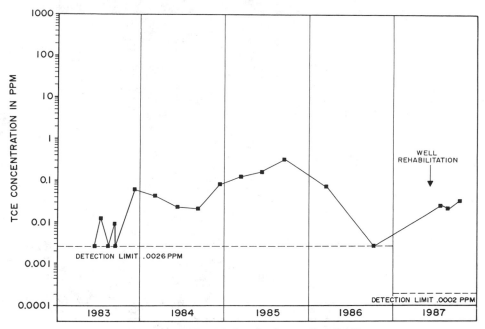

FIG. 4—*Trichloroethylene levels at well MW-25B.*

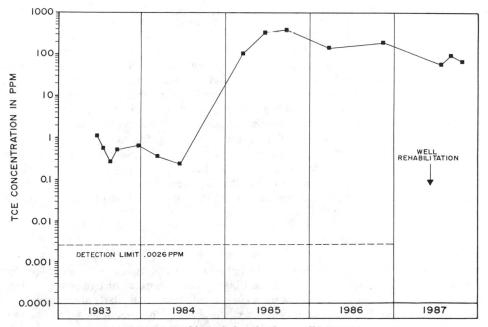

FIG. 5—*Trichloroethylene levels at well MW-26A.*

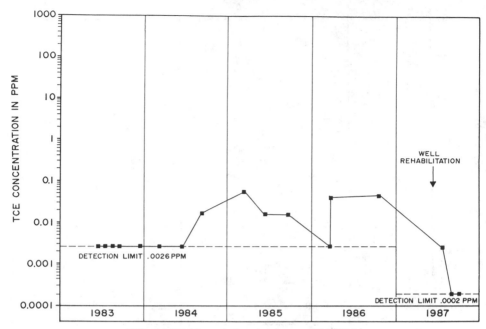

FIG. 6—*Trichloroethylene levels at well MW-28A.*

solvent is present, it may settle to the base of the well and go unnoticed. These conditions may lead to erratic and erroneous chemical analysis results.

It also appears reasonable that the surging technique may be appropriate subsequent to well installation as a means of development of the well. Surging appears to have several advantages over the typical pumping and flushing methodology used to develop wells. Surging would create a greater likelihood of removing any fine soil particles, which may, over a period of time, migrate into the well. These migrating soil particles may have an effect on the observed chemistry of the ground water within a well because of partitioning of the chemical with the soil particle. Surging also increases the hydraulic efficiency of the well screen by removing soil particles from the soil or rock fractures immediately surrounding the well. This has the effect of increasing the effective diameter of the well.

Three sets of sampling results were obtained after the wells had been surged at the southern Maine site. Initial indications are that samples from the wells show lower levels of TCE and other compounds than was the case prior to surging. The cause for the decrease in TCE concentration is uncertain; it is not known whether it was physical (e.g., volatilization) or geochemical (e.g., partitioning). The chemical results for four monitoring wells are presented in Figs. 3 through 6. The results will continue to be evaluated to see if the trends continue during future sampling.

Conclusions

Surge blocking is a useful technique for rehabilitating wells. It is a useful technique for removing fine soil particles from the surrounding aquifer materials which may migrate into the monitoring wells over a period of time after installation, particularly after many sampling events. These fine particles may cause partitioning of the chemicals within the well and lead to erratic and erroneous chemical analysis test results. Caution is urged when using this method on low-strength plastic well casings, since the suction pressures developed may cause the well to rupture or collapse.

References

[1] Gass, T. E., Bennett, T. W., Miller, J., and Miller, R., *Manual of Water Well Maintenance and Rehabilitation Technology,* National Environmental Research Center, U.S. Environmental Protection Agency, Ada, OK.
[2] *Proceedings,* Seminar on Transport and Fate of Contaminants in the Subsurface, U.S. Environmental Protection Agency, Boston, MA, 3–4 Feb. 1988.

Duane L. Winegardner[1]

Monitoring Wells: Maintenance, Rehabilitation, and Abandonment

REFERENCE: Winegardner, D. L., **"Monitoring Wells: Maintenance, Rehabilitation, and Abandonment,"** *Ground Water and Vadose Zone Monitoring, ASTM STP 1053,* D. M. Nielsen and A. I. Johnson, Eds., American Society for Testing and Materials, Philadelphia, 1990, pp. 98-107.

ABSTRACT: This paper presents an overview of the technical considerations and field procedures involved in monitoring well maintenance, rehabilitation, and abandonment. It includes criteria that should be considered throughout the life span of the well to ensure that the well serves its purpose optimally and that, when removed from service, it is no threat to the environment. The primary purpose of this paper is to provide guidance in selecting the most acceptable field procedures.

Even though monitoring wells are intended to be sources of reliable data, each well introduces a number of limited control variables that may affect the reliability of the analytical data. Regular maintenance of monitoring wells, both structurally and hydrologically, is required to preserve optimal performance of the well. Well maintenance principles are discussed in this paper, and recommendations for procedures are provided.

Major changes to a well may be required when it no longer serves its original purpose. The benefits of rehabilitation procedures are discussed and related to the risk of accepting changes in the established data base.

Abandonment (or decommissioning) of monitoring wells is necessary when they no longer serve their intended purpose. At that time, the well must be brought to a neutral condition which is compatible with the soil matrix, ground-water quality, and the chemicals monitored and which also preserves the hydrogeological integrity of the specific location. Examples of successful closures are presented to illustrate these general principles.

KEY WORDS: ground water, monitoring wells, well maintenance, well rehabilitation, well abandonment, well decommissioning, well redevelopment, aquifers, and well plugging

During the last few decades, as environmental awareness has increased in the public, so also has the number of ground-water quality investigations. In the 1960s and early 1970s, the rate of monitoring well installations began to accelerate. The advent of U.S. Environmental Protection Agency (EPA) and state regulatory initiatives has necessitated an increase in the number of monitoring wells.

In the early period of monitoring, investigations were conducted utilizing the limited knowledge and technical guidelines available at that time. Many monitoring wells were constructed in ways that would not meet today's performance or data quality requirements. Many wells constructed with hand-slotted screens or glued casing, or without proper borehole seals are still in existance.

Fortunately, as our understanding of hydrogeology and ground-water quality has developed, the technical and mechanical skills needed to measure pertinent parameters have become much more sophisticated and more standardized. Regulations require that

---

[1] Senior project coordinator, Engineering Enterprises, Inc., Norman, OK 73069.

ground-water investigation systems provide relevant, statistically valid data. Numerous authors have provided guidance for standardization of procedures for design, sampling, and data manipulation of monitoring well networks. This paper addresses a supporting phase of the work related directly to the individual monitoring well unit, which is to ensure that the well functions as it should. The paper is intended to provide general guidance in selecting the most appropriate field procedures.

Three primary topics of major concern for individual monitoring wells are maintenance, rehabilitation, and abandonment. Definitions of these terms follow:

Maintenance—The routine, continuing tasks that are intended to ensure that the well is a representative sampling point. This includes the minor structural repairs necessary to keep the well properly functioning.

Rehabilitation—An advanced effort beyond normal routine maintenance, intended to restore the well's original performance, or to alter the well to serve other purposes.

Abandonment—Permanent removal of the well from service, leaving it in a neutral status; decommissioning.

Maintenance

The purpose of most monitoring wells is to document physical and hydrogeological conditions (permeability, water level, and other conditions) and the discrete water quality. Most monitoring wells are routinely sampled and are considered fairly long-term investments. When changes in the water quality are noted in samples collected from a monitoring well, it is very important to be able to distinguish between those changes resulting from well problems and those which represent true variations in the water quality within the aquifer.

A primary indicator of a monitoring well's effectiveness is the record of the well's performance. A detailed data base describing the construction details, development procedures, hydraulic testing data, silting, and analytical quality records is needed to assess whether the water quality variations observed represent actual aquifer conditions.

Depending on the importance of the well, a wide variety of maintenance programs can be considered. At a minimum, the records should include the following data:

- the original design criteria (and purpose),
- the actual construction records,
- the original designation of the well by number or name,
- the periodic specific capacity test records,
- a listing of the sampler's observations at the well site,
- fluctuations of the water level,
- variations in depth due to silting or the presence of foreign objects,
- confirmation of casing elevations (especially in the frost belt),
- records of repairs or rework, and
- procedures and equipment used during sampling and testing.

Routine mechanical maintenance may be combined with scheduled sampling events. The minimal additional effort expended during these routine trips can often prevent much larger expenditures, which result from erroneous data.

Activities that should be considered routine maintenance include the following:

- bail testing of the well during normal purging to determine the specific capacity (and between, when sampling events are relatively infrequent);
- measurement of the depth before purging;
- preservation of exposed sections—repair of protective casing, covers, hinges, etc.; and
- occasional redevelopment by bailing, surging, or bottom pumping.

Any activities that would alter the character of the original data base should be limited. Examples of procedures which might be harmful include these:

- air-lifting to purge or redevelop without the use of eductor pipe,
- use of non-native water for jetting of screens or other purposes,
- addition of oxidants to remove bacterial growth,
- use of any material that interferes or reacts with the analyte, and
- failure to preserve well identification—the numbers should be clearly visible, and records should indicate the exact location with reference to permanent reference points.

Maintenance of a monitoring well should be an ongoing program which minimizes irregularities caused by the well.

Rehabilitation (Versus Replacement)

The use of the term "rehabilitation" indicates that the well is in need of significant alteration if it is to continue to serve its original purpose [1]. It is necessary to determine what is preventing the well from functioning and then to determine if the existing well is repairable. Specifically, will the reconstructed (or repaired) well provide data that are consistent or compatible with the existing data base? The alternative problem is that the postrepair analysis or water level data may skew a continuing data base and cause reviewers to make erroneous conclusions regarding the aquifer quality. Often it is prudent to abandon and decommission a nonfunctional well and to replace it with a properly constructed new well.

When renewing or altering the performance of an existing well, several factors must be considered. First, what was the original purpose of the well? If the well only served a limited purpose, such as water level measurement or chloride monitoring, alterations to its structure may not be detrimental to the data base. However, if the well is part of a sophisticated network monitoring the potential migration of low concentrations of toxic chemicals, rehabilitation should be carefully considered. Other questions that should be asked prior to any major reworking of a well include the following:

- How reliable is the original well construction record?
- Is the geological log accurate?
- What is the local geological setting?
- What caused the well to fail to serve its purpose?
- What are the expected results of rehabilitation?
- What is the remedial cost in comparison with the replacement cost?

If the answers to these questions, along with experienced professional judgment, indicate that the well should be rehabilitated, the next step is to plan and document the work effort thoroughly. Some common rehabilitation techniques include these:

- deepening because of lowering of the water table,
- sleeving to repair a physical problem,
- changing the basic purpose of the well (i.e., altering a monitoring well to become a recovery well), and
- treatment of screens to reduce plugging or encrustation.

Prior to performing any of these rehabilitation techniques, it is important that the present physical condition of the well be fully understood. Some possible inspection procedures include television logs, cement bond logs, and caliper logs. Each of these will provide additional insight as to the benefits of major rework on the well. Many interesting discoveries have been made by these procedures. The actual type and condition of the well screen may not be as expected. Breaks in plastic casing are not uncommon; also, older wells may not be of standard construction.

As a general rule, rehabilitation of a monitor well is a procedure that should be considered very carefully. In some situations, such as those where very limited access is available or where large-diameter deep monitor wells are involved, rehabilitation may be practical. However, in many cases, the alterations of the data base, along with the associated costs, dictate that the most reliable procedure is to abandon the well, move over a short distance, and construct a new well which is properly designed to accomplish the intended task. The following case histories provide descriptions of several well rehabilitation projects.

Case History No. 1

A very critical monitoring well was drilled in an industrial facility during aquifer remediation efforts (Fig. 1). When the well was constructed, it was screened throughout what appeared to be the uppermost aquifer. Later, questions about the water quality prompted a review of the well's performance, which indicated that a thin clay layer in the middle of the screened zone separated two distinct aquifers. The well was important to the monitoring system, and it was not possible to drill a new well at this location. Therefore, as a rehabilitative measure, a packer was placed at the confining layer. After the packer had been installed, the well was redeveloped, and the new specific capacity was measured to verify calculations of the flow through the sand pack around the packer. Analytical samples continue to be collected regularly.

For this particular well the most acceptable remedial solution was to tolerate leakage around the packer during purging and sampling. Hydraulic testing indicated that approximately 10% of the water being pumped came from water that had bypassed the packer. Calculations of dilution based on previous analysis are being made for each sampling event. The results are sufficient to determine the presence or absence of significant contamination.

Case History No. 2

A similar physical situation occurred at a large refinery where fairly deep wells are used to monitor a free oil plume (Fig. 2). Several of these wells are located where it is not pos-

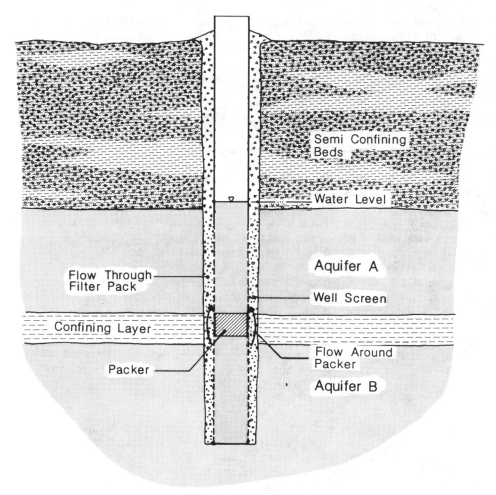

FIG. 1—*Case History No. 1.*

sible to operate a drilling rig. When the wells were installed, the floating oil layer was at the middle of the screen. Later, the water table rose to a level above the top of the screen. The choice of feasible remedial procedures was either to perforate the casing at the new oil level and accept the inefficiency of a less screened area with some silting or to extrapolate data gathered from other on-site wells. In this particular situation, the conclusion was reached that the data point was sufficiently important to justify the cost of perforating the casing.

Abandonment (Decommissioning)

When a monitor well no longer serves its intended purpose, the best practice is to convert it to a neutral status to prevent migration of any contaminants. A properly abandoned well preserves the hydraulic integrity by containing all formation fluids in their proper

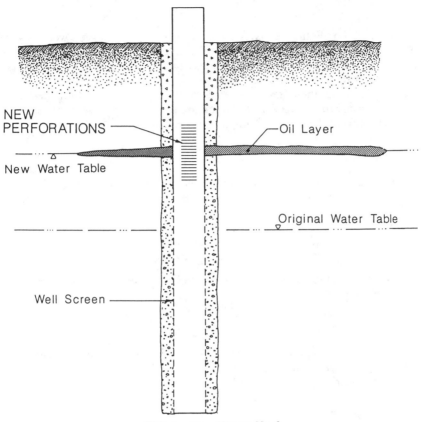

NEW
PERFORATIONS

Oil Layer

New Water Table

Original Water Table

Well Screen

FIG. 2—*Case History No. 2.*

location and not reacting with the environment. A wide variety of procedures has been used to decommission normal water and petroleum production wells [2–5]. Some of these procedures are directly applicable to monitor wells. However, because monitor wells are often associated with hazardous chemicals, special care must be taken. A decommissioned well must be considered a permanent fixture and the plugging must be completed accordingly.

The first necessary step is to plan the abandonment activity carefully, taking into account a variety of factors which will influence the material selection, ultimate performance, and economics. Some of the general considerations are discussed in the following paragraphs.

Well Construction

Plugging techniques must be compatible with the materials of the original well. It is very important to know the actual construction (as differentiated from the reported construction) of the well to be abandoned. This knowledge can be used to determine whether it will be necessary to pull the casing, perforate the casing, or overdrill and remove all existing materials, or whether simple grouting may be adequate.

General Geology

Some consideration should be given to the location of the plugged well, especially in relation to the level the abandonment that will be completed. The alternatives include finishing at grade level, below grade at a shallow depth, or at a specific depth relating to confining layers or other geological features.

After evaluation of all the criteria, a specific plan can be developed for each individual well which will ensure that its closure will maintain the formation fluid pressures at acceptable levels and prevent the spread of any remaining contaminants.

Local Geology

Development of a plugging plan is highly dependent on the variety and distribution of the subsurface materials encountered. Some of the factors to be considered are the following:

- Is the monitored aquifer artesian or a water table?
- What is the relationship between vertical and horizontal permeability?
- What are the fluid types and pressures? Are they all water, or oil and water, or do they include vapors?

Quality of Fluids

Containment of fluids requires a knowledge of the chemical and physical quality of the fluids. The design of the plugging activities must consider factors that will affect the performance of the materials used. Some items to be considered are these:

(*a*) chemicals and their concentrations in the water—i.e., volatile organics, chlorinated organics, organic and other acids, heavy metals, high or low pH, chlorides, or other corrosives; and (*b*) pathogenic bacteria or viruses, for which special precautions should be observed if such bacteria or viruses are found to be present.

Examples of Recommended Procedures for Abandonment

The following sections provide some general guidance for various simplified aquifer situations. The procedures presented here are intended to be guidelines that may be helpful in designing abandonment programs. The techniques and materials should be modified as needed for each actual case.

Case History No. 3

A shallow (7.6-m-deep) 50-mm-diameter monitor well in an unconfined sand aquifer was originally used to determine the concentration of very low levels of volatile organic compounds [i.e., benzene, toluene, and xylene (BTX)]. The well casing is polyvinyl chloride (PVC) and no protective casing is involved. The well is located in an agricultural area.

Several acceptable procedures can be used to plug this well. Because the depth is shallow, it is probable that the casing and screen can be pulled without breaking. If that is successful, the hole can then be re-drilled with a hollow stem auger. A bentonite-sand mixture can be pumped through the auger as it is withdrawn. The benefits of this procedures are that it is inexpensive and effective. Because the well is in a water table, total prevention of vertical fluid migration is not necessary, but it should not be more rapid than in the natural material. The chemicals present should have no effect on either the materials used or the aquifer. Because this well is in an agricultural setting, it is desirable that no grout or hardened material be left at a shallow depth where it may interfere with farming activities.

Variations of this procedure include drilling out the entire casing and filter pack (or guiding hollow stem augers over the casing) to a depth of a few feet below the bottom of the well and grouting, as described above. A second alternative procedure is to pressure grout through the existing casing and screen, then terminate the casing approximately 1 m below ground.

For this simple classic situation, any of the procedures just described would be sufficient to accomplish the overall goal.

Case History No. 4

A more complicated situation occurred near an old landfill in the Midwest (Fig. 3). This typical mixed-disposal municipal landfill was constructed in glacial till, which appeared, during initial excavations, to be situated in "impermeable clay." Unfortunately, at a depth of only a few metres below the landfill bottom, a reasonably permeable sand aquifer was

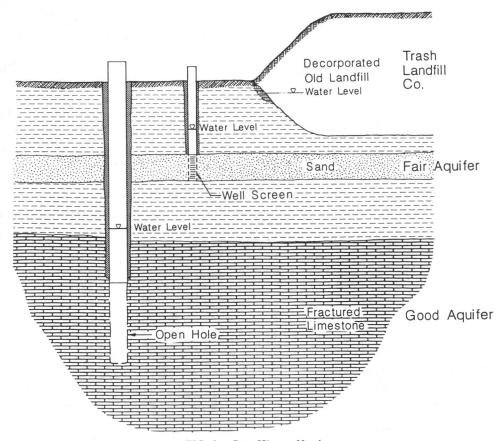

FIG. 3—*Case History No. 4.*

found. Glacial till occurs under the sand to an additional depth of 10 m, where fractured limestone has been found. On a regional scale, the potential gradient is downward. While vertical movement is slow, it is strongly suspected of being present at the site, based on the presence of continually increasing chlorides and iron concentrations in the sand aquifer with each sampling event.

Boreholes made for installation of monitoring wells adjacent to the landfill confirmed that the sand aquifer was present. A nearby steel-cased 30-m-deep water well extends downward into the limestone aquifer. As soon as contamination of the shallow aquifer was confirmed, concern was expressed that, ultimately, the deeper well may become a conduit for migration of fluids to the lower aquifer. Therefore, the decision was made to plug and abandon the deeper aquifer well.

Two general options for plugging this well were considered. The most effective (and expensive) method would be to clean the borehole to its full depth, fill the hole with drilling mud, and pull both the casing and the grout. A cement bentonite grout would be pumped through a tremie pipe to grout from the bottom to the top. The second option involved flushing and cleaning the well to its full depth and plugging it with the casing in place. A cement bentonite grout would be pumped into the bottom of the borehole to fill the uncased portion of the borehole and connect with the casing grout to form a solid bottom seal. The remainder of the casing, to within 2 m of the surface, would be filled with alternating layers of cement grout and short bentonite sections. At the surface, the casing would be cut off 2 m below the surface and the hole backfilled with native material.

After careful review, the second option was selected. The reasons for this selection were as follows:

- The well had been constructed by a reputable contractor and good well log documentation was available.

- Hazardous wastes in small quantities may have been placed in the landfill, but volatile organic compound concentrations in the shallow monitoring wells were either below confirmation levels or not present.

- It was reasoned that if the steel casing were to rust through, the alternating plastic bentonite layers would squeeze outward to fill any void.

- The cost of pulling the casing and grout would be very high.

- No municipal or residential water use was planned for this area in the near future.

In both of these cases, the plugging procedure was adapted to the site-specific situation. All of the factors (the geology, chemicals involved, well construction, local water uses, and other pertinent considerations) were carefully evaluated before designing and implementing the closure. Each change in situation, such as an increased chemical concentration, different geology, or other variable, would have been considered in the decision-making process.

General Steps to Consider when Abandoning a Monitoring Well

Preparations to abandon a monitoring well should be made with the utmost care to ensure that the most appropriate procedures and sealant materials are employed. The following instructions are presented as a guide to designing a monitoring well closure:

1. Evaluate the contaminants involved.
2. Consider the geological formations and aquifer involved.

3. Determine the fluid pressures and in-borehole flow.

4. Consider the local geography and land use.

5. Determine the borehole and casing extent, materials, and condition. (Clean them if necessary).

6. Prepare a plugging plan and select the materials and procedures (pulling the casing, cutting the casing, grouting in place, etc.)

7. Select an experienced qualified contractor who has the necessary hardware to complete the task.

8. Maintain full records to document the procedures, well location, and other pertinent information.

References

[1] Driscoll, F. G., *Groundwater and Wells,* Johnson Division, St. Paul, MN, 1986.

[2] Perazzo, J. A., Dorrier, R. C. and Mack, J. P., "Long-Term Confidence in Ground Water Monitoring Systems," *Ground Water Monitoring Review,* Vol. 4, No. 4, Fall 1984.

[3] Herndon, J. and Smith, D. K., "Plugging Wells for Abandonment," Union Carbide Corp., Nuclear Division, Office of Waste Isolation, Oak Ridge, TN, September 1976.

[4] Lutton, R. J., Strohm, W. E., Jr., and Strong, A. B., "Subsurface Monitoring Programs at Sites for Disposal of Low-level Radioactive Waste," NTIS NUREG/CR-3164, U.S. Nuclear Regulatory Commission, Washington, DC, April 1983.

[5] *Technical Manual: Injection Well Abandonment,* USEPA Office of Drinking Water Contract No. 68-01-5971, U.S. Environmental Protection Agency, Washington, DC, 1968.

Robert W. Gillham[1] and Stephanie F. O'Hannesin[1]

Sorption of Aromatic Hydrocarbons by Materials Used in Construction of Ground-Water Sampling Wells

REFERENCE: Gillham, R. W. and O'Hannesin, S. F., **"Sorption of Aromatic Hydrocarbons by Materials Used in Construction of Ground-Water Sampling Wells,"** *Ground Water and Vadose Zone Monitoring, ASTM STP 1053,* D. M. Nielsen and A. I. Johnson, Eds., American Society for Testing and Materials, Philadelphia, 1990, pp. 108–122.

ABSTRACT: Sorption of dissolved constituents from water within a sampling well onto well-casing materials or sampling equipment is frequently cited as a potential source of bias in ground-water sampling programs. This study examined the sorption of six monoaromatic hydrocarbons (benzene, toluene, ethylbenzene, and *m*-, *o*-, and *p*-xylene) onto seven sampling well materials. The materials included stainless steel, rigid polyvinyl chloride (PVC), polytetrafluoroethylene (PTFE), polyvinylidene fluoride, epoxy-impregnated fiberglass, flexible polyvinyl chloride (PVC), and polyethylene. Samples of the test materials were exposed to aqueous solutions containing the six organic compounds for time periods ranging from 5 min to 8 weeks. The concentration remaining in solution, at the end of the exposure period relative to the initial concentration, was taken as a measure of the degree of sorption.

No uptake of any compound onto stainless steel was noted, while some degree of sorption was observed for all compounds on all polymer well-casing materials. Sorption increased with increasing hydrophobicity of the organic compounds. The sorption increased in this order: benzene, toluene, *o*-xylene, *m*-xylene, ethylbenzene, and *p*-xylene. Of the polymer materials, rigid PVC showed the least sorption, followed by fiberglass and polyvinylidene fluoride (which were similar) and PTFE. The flexible tubing, polyethylene, and flexible PVC showed the highest rates of uptake, with significant losses from solution by the first sampling time (5 min).

The experimental data were shown to follow a diffusion model closely, with effective diffusion coefficients corresponding to different rates of uptake from solution. The model was used to extend the experimental results to monitoring wells of different diameter.

Stainless steel and PTFE are commonly recommended as preferred materials for constructing monitoring wells, particularly if organic contamination is expected. Under the conditions of this study, and based only on their relative sorption characteristics, several polymer materials, including rigid PVC, are as advantageous and possibly more advantageous than PTFE.

KEY WORDS: ground water, ground-water monitoring, sampling wells, organic contaminants, sampling bias, sorption

A significant decision in the design and implementation of a ground-water monitoring program is the choice of materials for construction of sampling wells. The choice can be influenced by regulatory stipulation and by cost, but also by technical factors, including the mechanical and chemical characteristics of the various materials available for well construction.

[1] Director and research coordinator, respectively, Waterloo Centre for Groundwater Research, University of Waterloo, Waterloo, Ontario, Canada N2L 3G1.

Materials used in the construction of sampling wells can affect the chemical characteristics of water samples through leaching of chemicals from the well material, chemical attack and dissolution of the material, or sorption of chemicals from the water in the well onto the well material. Because of the widespread occurrence of organic contaminants and the fact that many of these are of environmental importance at concentrations as low as a few micrograms per liter, considerable attention has been focused on the chemical characteristics of well materials and the potential for various materials to cause sample bias.

Based primarily on their resistance to leaching and chemical attack, glass, Type 316 stainless steel, and polytetrafluoroethylene (PTFE) have frequently been cited as the preferred materials when sampling for organics. Indeed, the Resource Conservation and Recovery Act (RCRA) Ground-Water Monitoring Technical Enforcement Guidance Document (TEGD) issued by the U.S. Environmental Protection Agency (EPA) [1] stipulates Type 316 stainless steel and PTFE as preferred materials for the construction of sampling wells. However, as more information becomes available concerning the sorptive characteristics of various materials, the preferred choice for well construction has become less clear and the subject somewhat controversial.

Reynolds and Gillham examined the rate of uptake of five halogenated organic compounds by several polymer materials [2]. Using exposure times of up to five weeks, the study concluded that sorption onto well casing materials could indeed be a significant source of sample bias. Nylon, polypropylene, and polyethylene showed the highest rates of sorption, while polyvinyl chloride (PVC) and PTFE showed rates of uptake that were generally similar to each other, but much lower than those of the other polymers. An exception was tetrachloroethylene, which was noted to sorb rapidly onto PTFE. It was further shown that the rate of uptake was consistent with a diffusion model. In similar studies, using stainless steel, PVC, and PTFE, several common organic contaminants of ground-water, and exposure times of 24 hours, Sykes et al. [3] reported that no statistically significant uptake of any of the materials was noted. Thus, when considering only sorptive processes and organic contaminants, the available data provide no basis for preference of PTFE over PVC as a sampling well material; indeed, the data suggest that in some situations PVC may be preferable.

This study examines the sorption of six monoaromatic constituents of petroleum products on seven materials that could be used in the construction of sampling wells and monitoring devices. The particular objectives were the following:

(a) To provide a qualitative ranking of the seven materials with respect to their potential to adsorb monoaromatic hydrocarbons,

(b) To test the diffusion model of the sorption process introduced by Reynolds and Gillham [2], and

(c) to extend the experimental results to other situations through application of the diffusion model.

Materials and Methods

The organic compounds considered in this study included six soluble constituents of petroleum products; benzene, toluene, ethylbenzene, and para-, meta-, and ortho-xylene. The constituents were obtained as analytical-grade compounds and are listed in Table 1 [4,5], along with selected chemical characteristics. The seven casing materials included Type 316 stainless steel, polytetrafluoroethylene (PTFE), rigid polyvinyl chloride (rigid PVC), flexible polyvinyl chloride (flexible PVC), polyvinylidene fluoride (PF), flexible polyethylene (PE), and an epoxy-impregnated fiberglass material (FG). The stainless steel, pol-

TABLE 1—*Organic compounds and selected chemical characteristics.*

Compound	Chemical Formula	Aqueous Solubility, mg/L[a]	Octanol-Water Partitioning Coefficient, K_{ow}[b]
Benzene	C_6H_6	1740	140
Toluene	C_7H_8	554	380
Ethylbenzene	C_8H_{10}	131	1410
m-Xylene	$C_6H_4(CH_3)_2$	134	1580
o-Xylene	$C_6H_4(CH_3)_2$	167	1318
p-Xylene	$C_6H_4(CH_3)_2$	157	1410

[a] The values are from Kebe et al. [4].
[b] Averages of selected values from Hansch and Leo [5].

yethylene, and flexible PVC were all obtained as tubing of approximately 6.5-mm outside diameter (OD) and 4.0-mm inside diameter (ID), while the rigid PVC was obtained as pipe of 14-mm OD and 8-mm ID. Virgin PTFE tubing (8 mm in OD and 4.8 mm in ID) was used, and polyvinylidene fluoride was acquired as rectangular wire having a circumference of 12.5 mm. The fiberglass material was obtained as a length of 5-cm-diameter tubing and was subsequently cut into strips 1.0 cm wide. All materials were cut into lengths 6.35 cm long and were cleaned by being washed in a strong organic-free detergent solution, followed by a rinse sequence using organic-free water, methanol, and organic-free water.

Several lengths of a particular material were placed into 160-mL glass hypovials. After being filled with a solution containing the six organics of interest, the vials were sealed with aluminum crimp caps lined with PTFE-faced silicon septa. Because of the larger diameter of the pipe, 250-mL hypovials were used for the rigid PVC. The surface area of the tubing in relation to the volume of solution in the hypovials was similar for all materials, ranging from 2.59 cm²/mL (for flexible PVC) to 2.94 cm²/mL (for rigid PVC). This ratio would be similar to that experienced in a monitoring well having an internal diameter of about 1.4 cm.

The test materials were exposed to an aqueous solution containing all the organic constituents for various lengths of time. The equilibrating solution was prepared by spiking an 18-L glass carboy of buffered organic-free water with an aliquot of a concentrated stock solution. The stock solution contained all six organic constituents dissolved in methanol, each at a concentration of approximately 13 000 mg/L. The concentrations of the organics in the equilibrating solution ranged from 1.0 to 1.4 mg/L, while the concentration of methanol was 118 mg/L. To prevent bacterial activity, sodium azide was added to the equilibrating solution at a concentration of 0.05%.

Fourteen exposure times were used, ranging from 5 min to 8 weeks. Five hypovials were prepared for each sampling material and for each exposure time, giving a total of 490 hypovials. Each set of five included triplicate vials containing the sampling material of interest and two blanks. The hypovials were filled by gravity flow from the carboy in a sequence of one blank, the three vials containing the sampling material, and the second blank. The second blank was used to check for losses by volatilization during the filling procedure. Efforts were made to minimize volatilization losses by filling the vials quickly, and all the vials were capped, leaving no head space, within a few seconds of being filled.

The samples were stored in the dark on a "tipping" shaker having a speed of two oscillations per minute. This did not agitate the samples, but the tipping motion ensured complete exposure of all surfaces to the equilibrating solution. The storage temperature was 22 ± 1°C.

At the appropriate time, the hypovials were removed from the shaker and shaken gently by hand to ensure that the solution was homogeneous; and when the hypovials were opened, duplicate sample bottles (10 mL) were filled for analysis.

The organic constituents were extracted from the aqueous phase using hexane at a water-to-hexane ratio of 9:1. The samples were placed on a rotary shaker for 15 min to allow equilibration between the water and hexane phases. For analysis, a 2.5-μL aliquot of hexane was removed and injected directly into a Shimadzu GC-9A gas chromatograph. The chromatograph was equipped with a flame ionization detector (FID) and capillary column. The column was a 0.32 mm by 60-m fused silica type with a 0.50-μL bonded Supelcowax 10 stationary phase.

Results and Discussion

Figure 1 shows the results for sorption of benzene on the seven test materials. In the upper graph, the mean of the two control samples at a particular time (C_T) is normalized by dividing by the initial concentration in the 18-L supply reservoir (C_0). The control data are very consistent up to a time of approximately three weeks, with somewhat greater scatter and a slight downward trend appearing at longer times. The data are generally centered on a relative concentration of 0.95. The apparent 5% loss could be due to volatilization during filling and sampling of the control hypovials, adsorption onto the glass walls of the hypovials, adsorption onto the PTFE liners of the caps, or losses around the seals between the caps and the hypovials. Considering the long duration of the tests, the losses are viewed as being small and the control data as being particularly consistent. The controls for the other five organic constituents are not shown but gave very similar results.

The concentrations determined from the duplicate analyses of the triplicate tests at each time were averaged and were normalized by dividing by the mean concentration of the controls for that particular material and at the corresponding time. The resulting relative concentration (C/C_T) is plotted versus time for each material. Referring to the results for benzene (Fig. 1), both flexible PVC and polyethylene show significant losses at the first sampling time (5 min), and by one day, approximately 80% and 90% of the benzene initially present in the hypovials had been lost to the polyethylene and flexible PVC, respectively. The concentrations remained fairly constant at later times, which suggests that some form of equilibrium had been established during the first day of exposure. At the other extreme, the relative concentrations determined for the hypovials containing stainless steel remained relatively constant and at a value close to 1.0 over the duration of the tests. Data for the remaining four polymer materials were similar for the first 24 h and did not differ substantially from the stainless steel data. At later times, however, all four materials showed a distinct downward trend in concentration, indicating sorption of benzene onto these materials. Of the four polymer materials, not including polyethylene or flexible PVC, rigid PVC showed the least uptake (25% at 8 weeks), followed by FG, PF, and PTFE. In the PTFE samples, by 8 weeks approximately 75% of the benzene initially present was lost from solution, apparently by sorption onto the PTFE.

Results for the other five organic compounds are given in Figs. 2a through 2c. In all cases, polyethylene and flexible PVC show very high losses, even after brief exposure, while there appeared to be little or no loss in the stainless steel tests. For the remaining four polymers, the order of loss remained the same in all cases (rigid PVC showing the least and PTFE the greatest), with the amount lost from solution varying from one organic constituent to another. Considering all six compounds, the losses from solution followed the

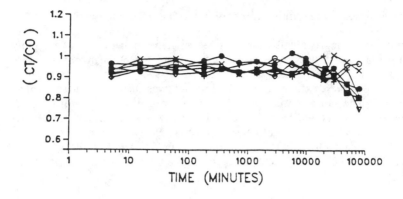

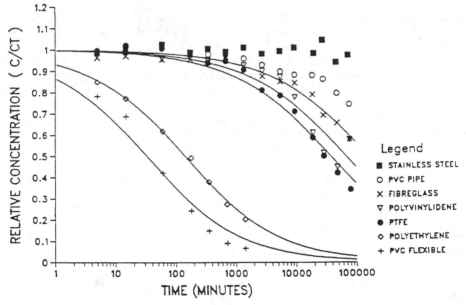

FIG. 1—*Sorption of benzene on the seven test materials. The solid lines are the fitted diffusion model.*

order of least for benzene, greater for toluene and *o*-xylene (which were similar), still greater for *m*-xylene and ethylbenzene (which were also similar), and greatest for *p*-xylene. Though there is not complete agreement, the rate of uptake tends to increase with decreasing water solubility or increasing octanol-water partitioning coefficient (K_{ow}) of the compounds (Table 1). It should also be noted that for compounds that showed the greatest loss from solution, losses were obvious well before 24 h had elapsed.

A statistical analysis was performed to determine the time at which significant uptake occurred for each compound-polymer combination. A statistical model was first fitted to the data to give a measure of the overall experimental uncertainty. This was required to establish the confidence limits on the data. A two-tailed Student's *t*-test was then used to determine the time at which the C/C_T values became significantly different from 1.0. Using

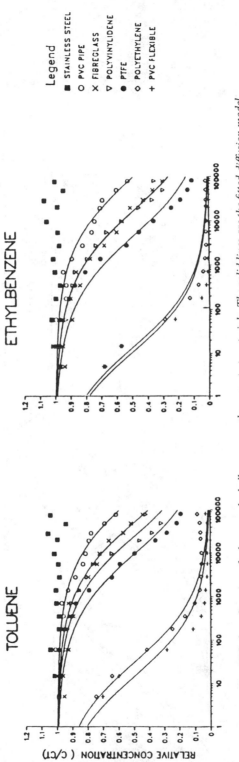

FIG. 2a—Sorption of toluene and ethylbenzene on the seven test materials. The solid lines are the fitted diffusion model.

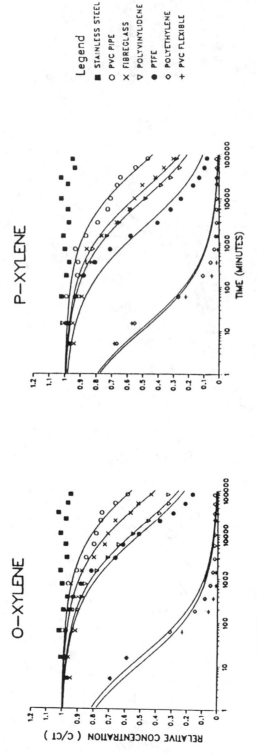

FIG. 2b—Sorption of o-xylene and p-xylene on the seven test materials. The solid lines are the fitted diffusion model.

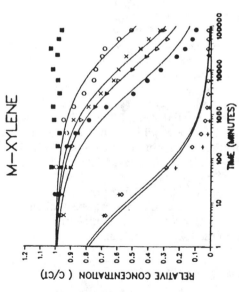

FIG. 2c—*Sorption of* m-*xylene on the seven test materials. The solid lines are the fitted diffusion model.*

TABLE 2—*Time interval within which the concentration phase for the compound and casing material became significantly different from 1.0.*

Material	Time, h					
	Benzene	Toluene	Ethylbenzene	m-Xylene	o-Xylene	p-Xylene
Stainless steel	>1344					
PVC (rigid)	48 to 96	24 to 48	12 to 24	12 to 24	12 to 24	12 to 24
FG	24 to 48	3 to 6	0.1 to 1.0	3 to 6	3 to 6	3 to 6
PF	24 to 48	6 to 12	1 to 3	1 to 3	0.1 to 1.0	1 to 3
PTFE	24 to 48	3 to 6	1 to 3	3 to 6	6 to 12	1 to 3
PE	0 to 0.1	0 to 0.1	0 to 0.1	0 to 0.1	0 to 0.1	0 to 0.1
PVC (flexible)	0 to 0.1	0 to 0.1	0 to 0.1	0 to 0.1	0 to 0.1	0 to 0.1

a 99% confidence level, C/C_T values of about 0.9 were determined to be significantly different from 1.0. Thus, the time at which there was significant uptake of the organic was identified as the time at which about 10% of the organic had been removed from the solution.

The results of the t-tests are summarized in Table 2, which shows the time corresponding to the first set of samples that showed significant uptake. For example, considering the combination of PVC and toluene, there was significant uptake by 48 h. The model did not allow interpolations between times. As a result, we know only that uptake became significant between the previous exposure time (24 h) and 48 h. For the polymer-compound combinations in the upper and left-hand regions of the table, sorption is less significant than for those in the lower and right-hand regions.

No significant sorption of any compound was observed for stainless steel, while significant sorption occurred for all compounds on all other materials. For PVC, significant sorption did not occur for any compound until an exposure time of 12 h had been exceeded (between 12 and 24 h). This suggests that, provided a PVC monitoring well is sampled within 12 h of purging, the sample will not be significantly affected by sorption processes. The other three rigid polymer materials (FG, PF, and PTFE) all show significant uptake of at least one compound by 3 h. The flexible tubings, on the other hand, show significant uptake of all compounds by the first sampling time (about 0.1 h). It is difficult to envision a purging and sampling procedure for these materials in which sorption could be eliminated as a potential source of sample bias.

The trends observed in Figs. 1 and 2 and Table 2 must be applied to field situations with a substantial degree of caution. In particular, though the qualitative trends should remain similar, they may differ quantitatively as a result of different surface area to volume ratios. This question is addressed in the following section.

Application of the Diffusion Model

If the experimental data can be represented by a physically or chemically based mathematical model, then it may be possible to use the model to predict the rate of uptake under conditions other than those of the experiments. As noted previously, Reynolds and Gillham showed that the time history of sorption of chlorinated organics onto sampling well materials could be represented reasonably well by a diffusion model [2].

The experimental procedures gave measured values of concentrations in the hypovials.

More specifically, the procedure involved diffusion from a reservoir of fixed volume (the hypovial) and initial concentration (C_0) into a polymer material of constant surface area. Assuming that the solution in the reservoir is thoroughly mixed at all times, a solution of the appropriate form of Fick's second law gives

$$\frac{C}{C_0} = \exp\left[\frac{RDt}{A^2}\right] erfc\left[\frac{(RD)^{1/2}t^{1/2}}{A}\right] \tag{1}$$

The parameter RD combines the effects of chemical sorption (R) and the physical diffusion characteristics of the polymer (D). The parameter A is the ratio of the solution volume to polymer surface area (L), t is time, C_0 is the initial concentration in the reservoir (M/L^3), and C is the concentration in the reservoir at times greater than zero (M/L^3). In deriving Eq 1, it is assumed that the medium is semi-infinite. In the case of a tubing material, the organic compounds diffuse towards the middle of the wall from both the outside and the inside. Once the concentration profiles meet, presumably near the middle of the wall, the medium would no longer act as semi-infinite and Eq 1 would no longer apply. Thus, the applicability of Eq 1 may be limited to a relatively early time, or to materials that show relatively low rates of uptake, or to both.

A computer model was used to fit Eq 1 to the experimental data. This involved an adjustment of RD to give the least-squares best fit to the data. The solid lines of Figs. 1 and 2 were calculated in this manner, and the best-fit RD values are given in Table 3.

Referring to Figs. 1 and 2, the forms of the experimental and calculated curves are very similar. Though not conclusive, the results provide strong evidence that the mechanism of uptake is indeed molecular diffusion.

Two exceptions to the above trend should be noted. The diffusion equation did not match the data for the flexible tubings (PE and flexible PVC) particularly well. Uptake of the organics in these materials was very rapid, and thus, the assumption of a semi-infinite medium would be violated at a very early time. Therefore, for most of the experiment the model is not appropriate and, thus, the relatively poor agreement is not surprising. In the second case, with the exception of benzene, the diffusion model tended to predict lower concentrations in solution at intermediate times (between about 1.0 and 24 h) for PTFE than were actually measured in solution. It is unclear at this time whether the differences represent a particular characteristic of PTFE or whether they are again the result of the semi-infinite medium assumption. The model would, of course, not fit the stainless steel

TABLE 3—*Values of* RD *determined for each compound and each sampling material.*[a]

Material	Benzene	Toluene	o-Xylene	m-Xylene	Ethylbenzene	p-Xylene
			RD Values, cm^2/s			
PVC (rigid)	...	5.9×10^{-9}	7.3×10^{-9}	1.4×10^{-8}	1.2×10^{-8}	1.8×10^{-8}
FG	8.5×10^{-9}	2.8×10^{-9}	2.6×10^{-8}	7.0×10^{-8}	8.7×10^{-8}	9.3×10^{-8}
PF	1.9×10^{-8}	5.7×10^{-8}	8.9×10^{-8}	1.3×10^{-7}	8.3×10^{-8}	1.5×10^{-7}
PTFE	3.3×10^{-8}	1.4×10^{-7}	1.3×10^{-7}	3.7×10^{-7}	3.0×10^{-7}	6.1×10^{-7}
PE	7.5×10^{-6}	4.5×10^{-5}	8.9×10^{-5}	1.1×10^{-4}	1.0×10^{-4}	1.2×10^{-4}
PVC (flexible)	4.6×10^{-5}	1.0×10^{-4}	1.3×10^{-4}	1.3×10^{-4}	1.4×10^{-4}	1.4×10^{-4}

[a] Note that, because there was no measurable uptake, RD values could not be calculated for stainless steel.

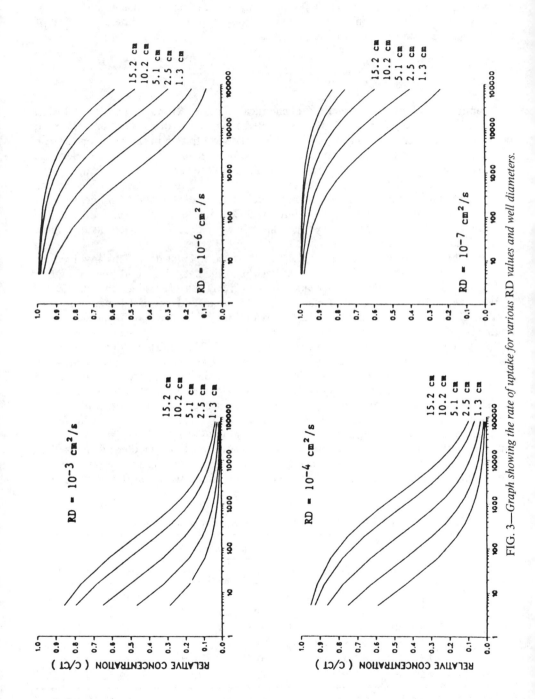

FIG. 3.—*Graph showing the rate of uptake for various RD values and well diameters.*

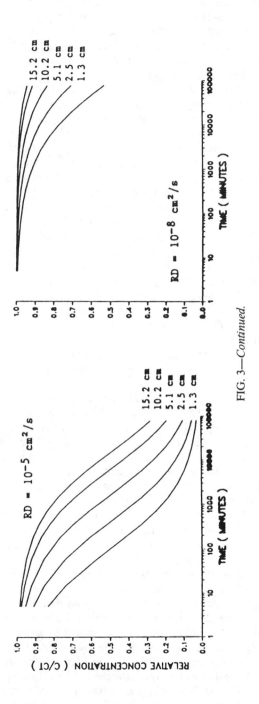

FIG. 3—*Continued.*

data since there was no significant decline in concentration. The model could also not fit the benzene data for PVC for the same reason.

Referring to Table 3, the best-fit RD values range over about four orders of magnitude, from 8×10^{-9} cm²/s for benzene in fiberglass to 1×10^{-4} cm²/s for p-xylene in flexible PVC. The low values correspond to cases in which uptake was relatively slow, while the highest values correspond to high rates of uptake (flexible tubing). If the RD parameter is the combined effect of both diffusive and chemical partitioning processes, then the wide range in the values of Table 3 indicates a very wide range in either the diffusive characteristics or the chemical partitioning characteristics of the materials included in the study. Unfortunately, from the sorption experiments, there is no means of quantitatively separating the diffusive process from the chemical partitioning process.

The RD values increase down the table, indicating that the inertness of the materials is in the following order: stainless steel $>$ PVC (rigid) $>$ FG $>$ PF $>$ PTFE $>$ PE $>$ PVC (flexible). This is reasonably consistent with both the qualitative and statistical evaluation of the data. Furthermore, the RD values generally increase from left to right, which suggest that the inertness of the organic compounds is in the order: benzene $>$ toluene $>$ o-xylene $>$ m-xylene $>$ ethylbenzene $>$ p-xylene. Though there are exceptions, the order of decreasing inertness is in the order of decreasing water solubility and increasing octanol-water partitioning coefficient (Table 1). It should also be noted that the RD values tend to be more sensitive to the sampling material than to the organic compound. Over the compounds studied, the RD values generally vary by less than a factor of ten, while over the sampling materials, the RD values vary by three to four orders of magnitude.

Based on the foregoing results, the authors suggest that the diffusion model gives a reasonably accurate representation of the data and can therefore be used to extend the results beyond the condition of the experiment. In particular, the diffusion model was used to examine the effect of well diameter on the rate of uptake. As the well diameter increases, the solution volume to surface area ratio increases, and thus lower rates of uptake would be expected.

Simulations were performed using RD values ranging from 10^{-3} to 10^{-8} cm²/s and well diameters ranging from 1.3 to 15.2 cm. The results are given in the graphs of Fig. 3. In these graphs, C_T refers to the initial concentration of the contaminant in the water entering the well. In addition to increased uptake with increasing RD values, as observed previously, the rate of uptake declines significantly with increasing well diameter. For example, if a particular contaminant and well material combine to give a RD value of 10^{-6} cm²/s, then in a 1.3-cm-diameter well, 10% of the contaminant would be lost by sorption onto the well material in about 15 min. On the other hand, if a 15.2-cm well were used, it would take almost 3000 min (50 h) before 10% of the contaminant would be lost. Clearly, the potential effects of sorption on sample bias are reduced with increasing well diameter.

If it is agreed that a 10% decline in concentration as a result of sorption processes is tolerable, then Fig. 3 can be further reduced. The time at which a diffusion curve declines to a relative concentration of 0.9 represents the tolerable time of exposure. Thus, from Fig. 3, the maximum time of exposure (the maximum time that the sample should be exposed to the well material), for a particular RD value, can be determined for the various well diameters. This can then be plotted on a graph of time (maximum exposure time) versus well diameter for various RD values. This procedure was followed in generating Fig. 4.

Figure 4 is applicable to the design of monitoring wells. For example, if it has been determined that wells having a diameter of 5 cm will be used in a particular project, and that the wells will be purged one day and sampled the next (24-h exposure time), then a material having an RD of about 3×10^{-7} cm²/s for the contaminants of interest should be selected. If the contaminant of concern is p-xylene, then all materials from Table 3, with the excep-

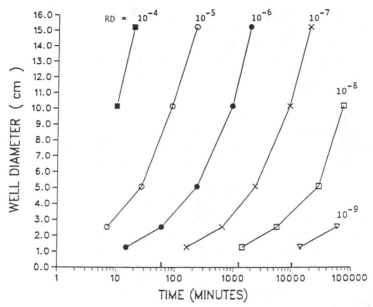

FIG. 4—*Time versus well diameter for various* RD *values, in centimetres squared per second.*

tion of polyethylene and flexible PVC, would be suitable for construction of the sampling wells. Had 1.3-cm wells been used ($RD = 10^{-8}$ cm^2/s), then, of the polymer materials, only PVC and fiberglass would be suitable. Alternatively, if the material is selected and the RD for the material and the anticipated contaminants are known, then Fig. 4 could be used to select the well diameter and sampling schedule (the time between purging and sampling).

Conclusions

Considering only the uptake of contaminant from solution and under the limited range of conditions studied, stainless steel was clearly the most favorable of the well materials examined, showing no significant uptake over the eight-week period of the tests. On the other hand, all polymer materials adsorbed all compounds to some extent, with the order of sorption (more favorable to less favorable) being as follows: rigid PVC < fiberglass < polyvinylidene fluoride < polytetrafluoroethylene < polyethylene < flexible PVC. The flexible tubings (polyethylene and flexible PVC) showed substantial uptake after only 5 min of exposure. Of the four remaining materials, rigid PVC showed substantially lower rates of sorption than PTFE, and fiberglass and polyvinylidene fluoride showed intermediate characteristics.

A diffusion model provided a good fit to the data, suggesting, though not proving, that the sorption mechanism is diffusion into the polymer material. The trends in the values of the effective diffusion coefficients (RD) indicate that the rate of uptake increases with increasing hydrophobicity of the organic compounds and with the physical characteristics of the polymer materials. For the compounds studied, RD values increased by about an order of magnitude with increasing hydrophobicity, while, for the polymers studied, RD values ranged over about four orders of magnitude. At least for the aromatic compounds of this study, the rates of uptake appear to be much more sensitive to the polymer characteristics than to the chemical characteristics of the organic compounds.

Though all polymer materials sorbed all organics, with the exception of flexible PVC and polyethylene, all might still be suitable for construction of ground-water monitoring wells. Based on simulations using the diffusion model, through selection of an appropriate well diameter and an appropriate interval between well flushing and sample collection, significant bias as a result of sorption processes could be avoided. Clearly, however, rigid PVC is the most favorable of the rigid polymer materials.

With rigid PVC being the most commonly used material for the construction of ground-water monitoring wells, with many monitoring networks having already been established using PVC, and considering the much lower cost of PVC, there is considerable reluctance to accept the EPA recommendation that stainless steel or PTFE materials be used. The results of this study, as well as the work of others [2,3], support the resistance to the use of PTFE.

There are situations for which PVC is not suitable. Generally, these include sites where there are high concentrations (or free product) of solvents, such as acetone, that will dissolve PVC. In these particular cases, stainless steel, or PTFE would indeed be preferable. In monitoring for petroleum products and many other industrial solvents, PVC would appear to be technically satisfactory, while offering a considerable price advantage over both stainless steel and PTFE.

Acknowledgments

The authors are grateful for the efforts of Greg Friday, who performed much of the laboratory work and to Michel Robin, who performed the statistical analysis. The study was funded through the Health and Environmental Sciences Department of the American Petroleum Institute (API). We appreciate the efforts of Bob Hockman, of the Groundwater Technical Task Force of API for technical input and interest during the course of the study.

References

[1] "RCRA Groundwater Monitoring Technical Enforcement Guidance Document," U.S. Environmental Protection Agency, Washington, DC, 1986.
[2] Reynolds, G. W. and Gillham, R. W., "Absorption of Halogenated Organic Compounds by Polymer Materials Commonly Used in Ground Water Monitors," *Proceedings,* Second Canadian/American Conference on Hydrogeology and Hazardous Wastes in Ground Water: A Soluble Dilemma, Banff, Alberta, Canada, 25–29 June 1985, National Water Well Association, Worthington, OH, pp. 125–132.
[3] Sykes, A. L., McAllister, R. A., and Homolya, J. B., "Sorption of Organics by Monitoring Well Construction Materials," *Ground Water Monitoring Review,* Vol. 6, No. 4, 1986, pp. 44–47.
[4] Kebe, J. O., Brookman, G. T., Atkins, R. S., and Unites, D. F., "Report to the American Petroleum Institute for an Evaluation of the Natural Fate of Aqueous Gasoline Components and Treatment Alternatives, Task 1A—Literature Survey: Hydrocarbons Solubilities and Attenuation Mechanisms," TRC Project No. 2134-N31, American Petroleum Institute, Washington, DC, 1984.
[5] Hansch, C. and Leo, A., *Substituent Constants for Correlation Analysis in Chemistry and Biology,* Wiley, New York, 1979.

Aquifer Hydraulic Properties and Water-Level Data Collection

Curtis A. Kraemer,[1] John B. Hankins,[2] and Carl J. Mohrbacher[2]

Selection of Single-Well Hydraulic Test Methods for Monitoring Wells

REFERENCE: Kraemer, C. A., Hankins, J. B., and Mohrbacher, C. J., "**Selection of Single-Well Hydraulic Test Methods for Monitoring Wells,**" *Ground Water and Vadose Zone Monitoring, ASTM STP 1053,* D. M. Nielsen and A. I. Johnson, Eds., American Society for Testing and Materials, Philadelphia, 1990, pp. 125–137.

ABSTRACT: Single-well hydraulic tests yield order-of-magnitude estimates of the hydraulic conductivity of aquifer materials around a single well. Although the single-well tests are less accurate than multiple-well pumping tests, they are often an attractive option when a quick, inexpensive estimate of hydraulic conductivity is required. There are three basic methods for determining the hydraulic conductivity from a single well:

- *Slug test*—This method is also commonly referred to as a falling-head test or bailer test. It involves the instantaneous addition or removal of a given volume (slug) of water from a monitoring well with measurements of the recovery collected over a period of time.
- *Constant-head test*—This method involves a measured discharge of water into a monitoring well to maintain a constant water level within the well. The method is more commonly used where the aquifer materials have moderate to high hydraulic conductivity values.
- *Single-well pumping test*—This method is no different from an aquifer pumping test except that water level measurements are collected from just the pumped (monitoring) well. The pumping rate is kept constant throughout the test, and water levels are measured over a period of time.

The choice of a single-well test method in a specific situation depends on a variety of factors. A decision-making flowchart has been developed to help select a hydraulic test method that incorporates both geologic conditions and monitoring well construction details. This flowchart will allow a hydrogeologist or ground-water engineer to assess quickly which test method is the most appropriate for a specific situation.

KEY WORDS: slug injection, slug withdrawal, pumping test, constant-head test, falling-head test, bailer test, hydraulic test, hydraulic conductivity

Most ground-water monitoring programs include an assessment of the rate of ground-water flow and a determination of the area of potential contamination. To make this assessment, the hydraulic conductivity of the aquifer must be estimated. The hydraulic conductivity of the aquifer can be determined by several methods, including multiple-well pumping tests, laboratory analysis of soil samples, and single-well hydraulic tests. Multiple-well pumping tests yield the greatest amount of data about the aquifer (its transmissivity, storativity, leakage, boundary conditions, and other kinds of data), but these tests are costly and time-consuming and, if the aquifer is contaminated, pose a problem of dealing

---

[1] Senior hydrogeologist, Atlantic Environmental Services, Inc., Colchester, CT 06415.

[2] Hydrogeologists, TRC Environmental Consultants, Inc., East Hartford, CT 06108.

with the contaminated discharge. In multiple-well pumping tests, the heterogeneities of the aquifer are averaged so that the hydraulic properties of the aquifer as a whole can be accurately determined. Papers by Freeze and Cherry [1], Walton [2], Driscoll [3], Mandel and Shiftan [4], and Todd [5] are just five of the several references that discuss the procedures and analyses of multiple-well pumping tests.

Laboratory analyses of relatively undisturbed core samples were developed to evaluate hydraulic properties of earth materials. Lambe describes the laboratory tests that can be made on both cohesive and cohesionless soils using a variety of different procedures [6]. Soil laboratory analyses are particularly useful on soils with a very low hydraulic conductivity, where *in situ* test methods either are impractical or may take a very long time to complete. The two main types of laboratory tests are falling-head and constant-head tests. A Shelby tube or some other sampling device is used to collect an undisturbed cohesive soil sample. In the laboratory, the sample can be oriented so that either the horizontal or the vertical hydraulic conductivity is measured. A laboratory analysis of a cohesionless soil is not practical for a ground-water monitoring program because of the difficulty of obtaining representative undisturbed samples of the aquifer.

A particle-size distribution analysis is another laboratory technique that can be used to estimate hydraulic conductivity. Based on hydraulic conductivity work on uniform sands, Hazen derived the empirical equation

$$K = d_{10}^2$$

where K equals the hydraulic conductivity, in centimetres per second, and d_{10} is the particle-size diameter, in millimetres, at which 10% by weight of the soil is finer. Freeze and Cherry indicate that this method of estimating hydraulic conductivity is useful for most soils in the fine sand to gravel range [1].

Hvorslev developed a series of empirical formulas for the *in situ* determination of hydraulic conductivity in individual monitoring wells (piezometers) [7]. Empirical formulas based on laboratory experiments were developed for both slug and constant-head tests. These single-well hydraulic tests have been found to be an inexpensive method of quickly estimating the hydraulic conductivity in the vicinity of a monitoring well. The use of these tests has increased significantly over the last 30 years.

There are a number of assumptions which Hvorslev made when developing the formulas for the single-well hydraulic tests [7]. These assumptions, which also apply to single-well pumping tests, are as follows:

- the well screens are placed in a porous material (soil);
- the soil has infinite directional isotropy;
- there is no disturbance, segregation, swelling, or consolidation of the soil;
- there is no sedimentation or leakage;
- there is no air or gas entrained in the soil, well screen, or riser; and
- the hydraulic losses in the riser, well screen, or filter are negligible.

The assumptions of Hvorslev also apply to single-well pumping tests.

Single-well pumping tests were developed simultaneously with multiple-well pumping tests but tended to be overlooked because they yielded fewer data on the aquifer characteristics. Analysis and interpretation of single-well pumping tests is more complicated than for slug and constant-head tests. A detailed description of each of the three types of single-well tests (slug tests, constant-head tests, and pumping tests) is given in the following sections of this paper. The three methods of testing are considered together in a decision-making flowchart in the final section. This flowchart will allow the reader to assess which test method is most appropriate for a specific situation.

Slug Tests

Slug tests are often favored over multiple-well pumping tests because of their relatively low cost and the fact that they can be run on single wells. Up to several dozen of these tests can be performed in a single day on wells screened in materials of moderate permeability, whereas multiple-well pumping tests can take hours, days, or even weeks to complete. Unfortunately, the data provided by slug tests are less reliable and less useful than those provided by multiple-well pumping tests. In general, slug tests produce data which are order-of-magnitude estimates of hydraulic properties at best. In addition, the tests do not provide data on important aquifer parameters such as storativity. The tests should not, therefore, be considered a replacement of multiple-well pumping tests but, instead, should be used as a tool to gain a quick order-of-magnitude estimate of the hydraulic conditions in the immediate vicinity of a single well.

Slug tests, and to some extent any type of single-well test, should be considered hydraulic tests rather than true aquifer tests because they generally influence only the zone within a metre, or so, of the well. The tests are thus highly sensitive to drilling methods and well construction techniques. Well construction techniques which disturb as little of the aquifer as possible are favored and include the following:

- hollow stem continuous auger,
- cable tool,
- air rotary,
- water/mud rotary, and
- solid stem continuous auger.

Final selection of a drilling technique must be based on site-specific conditions [8].

The high sensitivity of slug tests to well construction techniques can make them a valuable tool in assessing the effectiveness of well development. If a well has not been developed adequately, the calculated hydraulic conductivity from a slug test will be lower than the true value for the aquifer because of the presence of fine material in the area around the well screen. Faust and Mercer suggest that these "skin effects" are due to the entrainment of gas within the aquifer or well screen or to the invasion of the aquifer by drilling fluids [9]. Successive slug tests during well development can, therefore, help to determine whether additional development is warranted. Black and Kipp suggest the use of slug tests in conjunction with multiple-well pumping tests to evaluate observation well response time [10].

In general, slug tests should only be performed in materials of moderate to low hydraulic conductivities (10^{-7} to 10^{-2} cm/s). Hydraulic conductivities higher than this will cause water levels to recover too quickly for accurate measurement. Slug tests in aquifers with conductivities near the high end of the range (10^{-2} cm/s) require a system of rapid water level measurement such as a pressure transducer and data logger. If the hydraulic conductivities are too low, the recovery of the water levels will be too slow to be practical.

Slug Test Methodology

Injection/Withdrawal of Slug—An assumption common to all the techniques for analyzing slug test data is that the injection or withdrawal of the slug of water is instantaneous.

The solution to the problem is the same whether water is added or removed from the well. In wells where part of the screen extends into the unsaturated zone above the water table, only slug withdrawal techniques are applicable. If water is added to a well screened into the unsaturated zone, the water will flow out of the well into the unsaturated materials at a rate proportional to a value between the saturated and unsaturated conductivity.

Slug injection consists of adding a known volume of water into the well as rapidly as possible. For small-diameter wells (less than 10-cm (4-in.) in diameter), this can be done easily by one person with a bucket. For larger wells, more elaborate injection schemes may be necessary. Bredehoeft and Papadopulos, and Neuzil suggest that injection tests with shut-in pressure may be used to speed the test in tight formations [11,12].

There are a variety of techniques used for slug withdrawal. If a pump is to be used, the pumping rate must be considerably greater than the maximum inflow rate and the pumping time must be kept short so that an instantaneous withdrawal can be approximated. The favored method for slug withdrawal is the weighted float method. A long piece of weighted pipe with a diameter slightly less than the well casing is sealed at both ends and lowered into the well until it reaches a level at which it floats. The presence of the weighted float displaces the water level in the well upward to some nonequilibrium level. The water level is then allowed to return to its static level at which time the weighted float is rapidly pulled from the well. The water level in the well then drops instantaneously, the volume of the drop being equal to the volume displaced by the weighted float. The upward recovery of water levels are then measured versus elapsed time. An obvious advantage of the weighted float method is that no water is actually withdrawn from the well. In cases where the ground water is contaminated and discharge from the well must be contained, this may be a major advantage.

Water Level Measurement During Slug Tests—As with pump tests, the rate of water level recovery after a slug of water is added or removed from a well decreases logarithmically. Measurements should thus be made rapidly at first and then less rapidly as the test progresses. It is important to know the initial head difference caused by the slug. This can be measured directly at time zero or determined indirectly by calculating the rise or fall of water level which a slug of known volume should have produced. Several references, including papers by Hvorslev and the U.S. Environmental Protection Agency (EPA), suggest that the tests should be run until recovery is 85 to 90% completed [7,8]. Recovery information beyond these levels offers little additional information and may be quite time-consuming to collect; the EPA indicates that 99% recovery takes twice as long as 90% recovery [8]. Wells screened in low-permeability materials may take too long to recover to make the measurement of 90% recovery practical. In these cases, it may be found, after plotting the data during the test, that a solution is possible when as little as 50% of the recovery has been achieved.

Water level measurements taken during slug tests can be made using the same techniques as those used for pumping test measurements. If recovery is expected to be rapid, a pressure transducer should be used. An electronic data logger attached to this pressure transducer allows the best data collection. For wells that recover more slowly, an electric sounder or other type of manual measuring device may be sufficient. Mechanical strip-chart recorders with weighted floats offer an additional option for wells that recover slowly.

Analytical Techniques to the Solution of Slug Test Data

Analytical techniques for the solution of slug test data were first developed by Hvorslev [7]. Hvorslev states that if ground water flows to a well according to Darcy's law, then the rate of flow must be proportional to the hydraulic conductivity of the material around the well, the amount by which the water level has been displaced from equilibrium, and the surface area of the well. The basic Hvorslev equation for the solution of slug test data is

$$K = \frac{A}{F} \frac{1}{(T_2 - T_1)} \ln \frac{H_1}{H_2}$$

where K = the hydraulic conductivity, A = the cross-sectional area of the well, F = a factor representing the shape of the well intake, T_x = time x, and $H_x = S_x/S_o$, where S_o is the initial drawdown at the start of the test. Recovery data are plotted on a graph with the log of H_x/H_o on the y-axis and linear time on the x-axis. A best fit line is drawn through the data and two arbitrary points on the line are chosen for the calculation (T_1, H_1; T_2, H_2). Shape factors have been determined empirically for a variety of field situations and are presented by Hvorslev [7], the Naval Facilities Engineering Command (NAVFAC) [13], Bouwer and Rice [14], Cedergren [15], Lambe and Whitman [16], and the U.S. Department of Interior [17].

The list of field situations covered by these references includes both confined and unconfined aquifers. Although Cooper, Bredehoeft, and Papadopulos have shown that the results of this method are not precise [18], the method is still considered valid for determining order-of-magnitude estimates of hydraulic conductivity. A computer program which solves for hydraulic conductivity using the technique described above has been developed by Thompson [19].

Cooper et al. recognized several problems with the Hvorslev technique and presented an alternate method using a curve matching for the analysis of slug test data [18]. The technique is applicable only to confined aquifers and produces values for both transmissivity and storativity. The authors point out, however, that the value of storativity determined with this method has questionable reliability. Bredehoeft and Papadopulos extended the method of Cooper et al. to tight formations [11]. Their method involves the use of shut-in pressure to speed a return to equilibrium water levels after a slug of water is injected.

Constant-Head Tests

Constant-head tests of monitoring wells were developed concurrently with slug tests and are conducted by adding a known rate of flow of water into a well to maintain a constant piezometric head. Hydraulic conductivity values can be calculated from the known discharge and constant piezometric head values by using an empirically derived formula.

Constant-head tests have the same advantages of low cost and convenience as slug tests. As many as ten tests can be made daily on individual wells. The practical range of soil hydraulic conductivities which can be tested by the constant-head method is 10^{-4} to 10^1 cm/s. This range covers a wide variety of soil types, more than either the slug or single-well pumping tests. A constant head can be maintained for an indefinite amount of time, allowing a large volume of water to be introduced into the aquifer. Because of this, a constant head test can average the hydraulic conductivity of a larger portion of the aquifer than a slug test can.

The information provided by constant-head tests, like slug tests, is less reliable and useful than information provided by either multiple-well pumping tests or laboratory tests. The hydraulic conductivity values generated by constant-head tests are order-of-magnitude estimates. The tests do not provide values of other important aquifer characteristics, such as storativity or leakage.

There are several potential disadvantages to performing constant-head tests. Constant-head tests, particularly those done in permeable materials, require large volumes of water. Often ground-water monitoring wells are located far from sources of suitable water (sediment free and contaminant free). Without a nearby source of water, the ability to conduct the test becomes logistically difficult. If the wells are part of an ongoing ground-water monitoring program, it may not be desirable to discharge water into the well. In many cases, federal or state regulations will prohibit the injection of water (even clean water) or require a ground-water discharge permit. The discharge will certainly affect the water quality of the aquifer immediately surrounding the well for some period of time. Perhaps the greatest disadvantage of the constant-head test is that it cannot be conducted in wells where the well screen extends above the water table.

Constant-Head Test Methodology

The constant-head test procedure is the most simple of the three single-well hydraulic tests. Prior to discharging water into the well, an accurate static water level is measured. Water is then discharged into the well until a constant piezometric head is maintained by a constant discharge.

The piezometric head can be measured a number of ways, including by transducers, water level indicators, or a chalked steel tape. Transducers are likely to give the most reliable readings. They can handle rapid small-scale fluctuations better than the other two measurement instruments. Fluctuations occur when the flow into the well is great enough to cause turbulence. The discharge rate should be accurately measured and the test should be run until the piezometric head and the discharge have stabilized.

Analytical Techniques to the Solution of Constant-Head Test Data

Analytical techniques for the solution of constant-head test data from monitoring wells were developed by Hvorslev [7]. The basic Hvorslev equation for the solution of constant head test data is

$$K = \frac{q}{FH_c}$$

where K = the hydraulic conductivity, q = the rate of flow into the well, F = a factor accounting for the shape and soil conditions surrounding the well intake, and H_c = the constant piezometric head. Hvorslev empirically developed shape factors for a variety of field situations which are presented in papers by Hvorslev [7], NAVFAC [13], Bouwer and Rice, [14], Cedergren [15], Lambe and Whitman [16], and the U.S. Department of the Interior [17], as well as in many other geotechnical or soil mechanics textbooks. The factors consider confined and unconfined conditions as well as a variety of geometric configurations for well intakes. Many modifications of Hvorslev's original techniques have been developed by others, such as the U.S. Department of the Interior [17].

Single-Well Pump Tests

Although much has been written about multiple-well pumping tests, very little has been written about using single-well pumping tests. Under certain conditions, however, single-well pumping tests may be the only feasible means of determining hydraulic conductivity in the vicinity of a well.

As stated previously, slug tests cannot be effectively applied in situations where the hydraulic conductivity of the aquifer materials approaches 10^{-2} cm/s. Constant-head tests also have a practical range of hydraulic conductivities which can be measured. Very permeable formations will require large volumes of water, applied rapidly. Unless a fire hydrant and fire hose are available, a constant-head test may not be possible. Furthermore, large flows of water into a small-diameter monitoring well will produce rapid head fluctuations due to turbulence.

Single-well pumping tests share an advantage with constant-head tests over slug tests. Pumping tests and constant-head tests can be run indefinitely, which allows increasing amounts of water to be withdrawn or added, thus increasing the volume of aquifer that is hydraulically tested. The typical slug test, which would involve about three linear metres (10 ft) displacement of water in a 5-cm (2-in.) monitoring well with a 3-m (10-ft) screen, displaces water in only a 8.1-cm (3.2 in.) radius around the well, if the effective porosity of the aquifer is 0.1. A 2-h pumping test or constant-head test at 18.9 L/min (5 gal/min) displaces water in a 152-cm (60-in.) radius around the well under the same conditions. (This example ignores the subtleties of elastic deformation of the water and aquifer, but it is sufficiently precise to make the point.) As stated in the previous section, slug tests are sensitive to the degree of well development and skin effects. Pumping tests and constant-head tests of sufficient duration begin to overcome these liabilities.

Pumping tests have their own set of liabilities, however. At hazardous waste sites, release of pump discharge water on the ground may be disallowed by regulators or may pose potential liability risks. Containing the water for later disposal may not be cost-effective.

In moderately permeable formations, it may not be possible to maintain a constant pumping rate as the test progresses. As the drawdown increases, discharge will decrease, especially if a peristaltic or centrifugal pump is being used. (Usually, these types of pumps are preferred over submersible pumps because of the ease of decontamination between wells and their flexibility with regard to the well diameter.) As long as the total height that the water must be lifted does not exceed about 7.6 m (25 ft), the pump discharge valve can be adjusted to maintain a constant discharge rate. Although, theoretically, suction can lift a water column equal to 1 atm of pressure [~10.4 m (34 ft)], experience has shown that as the lift approaches 7.6 m (25 ft) the discharge cannot be maintained at a constant rate.

Single-Well Pumping Test Methodology

Mandel and Shiftan [4] and Strausberg [20] present brief discussions of single-well pumping tests as an alternate hydraulic testing technique. However, the theory behind this method is really just a limited application of the more general theory of pumping tests with observation wells. This more general theory was first presented by Theis for a confined aquifer [21]. Subsequent workers developed the theory for a variety of conditions, such as an aquifer confined by a leaky aquitard, an unconfined aquifer, partially penetrating wells, and recharging and impermeable aquifer boundaries. A comprehensive treatment of well hydraulics can be found in Walton [2]. Widely available textbooks, such as those by Freeze and Cherry [1] and Todd [5] give excellent introductions to the topic.

For hydraulic testing of a single well, the semilog method developed by Cooper and Jacob is the best suited [22]. Although curve-matching techniques could be employed, the semilog method allows calculations to be made in the field, even during the test. Such real-time results allow for maximum flexibility.

The Theis equation is

$$s(r,t) = \frac{Q}{4\pi T} \int_u^\infty \frac{e^{-u}du}{u}$$

where $s(r,t)$ is the drawdown (at any radius from the pumped well and any time since the start of pumping) and

$$u = \frac{r^2 S}{4Tt}$$

This equation can be represented by the infinite series

$$s = \frac{Q}{4\pi T} \left(-0.5772 - \ln u + u - \frac{u^2}{2 \cdot 2!} + \frac{u^3}{3 \cdot 3!} + \cdots \right)$$

After sufficient time has elapsed since the start of pumping, the sum of the series beyond the second term can be ignored. Substituting the equation for u and changing to the base-10 logarithm yields

$$s = \frac{2.3Q}{4\pi T} \log \frac{2.25Tt}{r^2 S}$$

The radius (r) in this equation is the distance to the observation well in a multiple-well pumping test, but is considered the effective well radius in a single-well test. This effective well radius is not known. Plotting drawdown (s) versus time (t) on a semilog graph (time on the log scale) yields a straight line after sufficient time has elapsed. Todd has stated the if $u < 0.01$, the Cooper-Jacob approximation is valid [5]. The above equation can be solved for transmissivity (T)

$$T = \frac{2.3Q}{4\pi \Delta s}$$

where Δs is the drawdown over one log (t) cycle and Q is the constant discharge rate.

As stated previously, constant discharge is often difficult to maintain. If the discharge port valve is partially closed at the start of the test, it can be opened incrementally to maintain a constant discharge. When the maximum depth that a centrifugal or peristaltic pump can lift has been exceeded, the discharge will sharply decrease and the water level in the well will begin to fluctuate. If this occurs, the pump should be shut off and the residual drawdown versus time measured. If the diameter of the well being tested in 10.2 cm (4 in.) or more, submersible pumps designed for domestic wells can be used. Air-lift pumps are another feasible option for small-diameter wells [wells with diameters less than 10.2 cm (4 in.)] with greater than 7.6 m (25 ft) of lift.

Recovery data from single-well pumping tests can be used to calculate transmissivity. In fact, recovery tests have the advantage over pumping tests in that they are not as sensitive to fluctuations in discharge rate, well losses, and nonhorizontal flow. Late-stage data (the portion used in the Cooper-Jacob approximation) for a recovery test cover a period when flow to the well is slow and therefore laminar, minimizing well losses. In the case of uncon-fined aquifers, late-stage data are collected when the aquifer has recovered most of its saturated thickness, thereby more closely conforming to the Theis equation assumptions. One possible liability of using recovery data in an unconfined aquifer is that air bubbles may become entrained in the dewatered portions of the aquifer, thereby decreasing the effective transmissivity of the materials during recovery. Therefore, Strausberg suggests that the recovery data should be analyzed in conjunction with pumping data [20].

Recovery test analysis is based on the fact that the drawdown of multiple pumping wells is additive. Thus, an imaginary recharging well is superimposed on the pumping well at the beginning of the recovery period and is considered to be recharging water at the same rate that the pumping well had been discharging. The effects of the imaginary recharging well are added to the effects of the imaginary extension of pumping. Residual drawdown is plotted against the ratio of the total elapsed time versus the time since recovery began. Analysis proceeds according to the Cooper-Jacob method outlined above. More in-depth treatment of recovery tests is given in the general references cited above.

Analysis of pumping test data can be complex. Under suitable conditions pumping tests yield more information about the aquifer than constant-head or slug tests. Recharging or impermeable boundaries will cause a change in slope of the linear portion of the drawdown versus log time plot. Pumping an unconfined aquifer may cause a phenomenon called delayed yield, which would manifest itself in the drawdown versus log time plot as a change in slope similar to a recharging boundary. This phenomenon, which is related to vertical flow components, is treated by Neuman [23–25] in depth and in introductory form in the text books just referenced [23–25]. In the authors' experience and in the small amount of literature available concerning single-well pumping tests, delayed water table response does not occur early enough in the test to interfere with calculation of T by the Cooper-Jacob method.

Often monitoring wells are constructed primarily to collect water quality samples and often they do not fully penetrate the aquifer. Todd gives an introduction on how to correct drawdown values for partially penetrating wells [5]. The well can be considered fully penetrating if it penetrates at least 85% of the aquifer. Also, if the geologic log for the well indicates that the vertical hydraulic conductivity is likely to be at least an order of magnitude less than the horizontal conductivity (e.g., if there are frequent stringers of lower permeability material), partial penetration can be ignored. Similarly, if the aquifer is subdivided by lower permeability layers and the well screen almost fully penetrates one of the subaquifers, partial penetration can be ignored.

Hydraulic conductivity is defined as T/b. In the case of a partially penetrating well, selection of a value for b would proceed according to the logic used to consider partial penetration. In aquifers where the vertical conductivity is less than the horizontal, b is equal to the screen length. Todd suggests that where the screen penetrates at least 85% of a subpart of the aquifer, b equals the thickness of that subpart [5]. If the screen extends through all of the aquifer and some of the confining layer, b equals only the aquifer thickness. If drawdowns have been corrected for partial penetration, b equals the total aquifer thickness. Drawdowns in a pumping test of an unconfined aquifer should be less than 10% of the aquifer thickness, b, so that the Theis assumption of constant T is not seriously violated.

Determining the Best Test Method

Figure 1 shows the decision flowchart for selecting the correct type of hydraulic test. The decision process moves left to right through the flowchart. The decision process leads to the selection of one of five possible options for the hydraulic test: a pumping test with discharge water released to the environment, a pumping test with discharge water con-

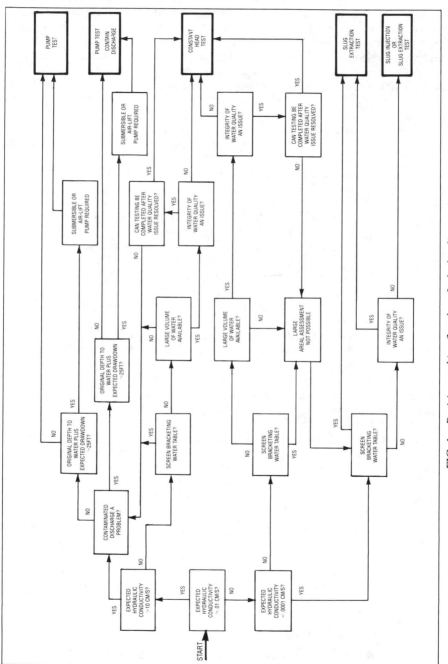

FIG. 1—*Decision-making flowchart for hydraulic tests.*

tained, a constant-head test, a slug extraction test, and a slug injection or extraction test. Under favorable conditions, the flowchart may yield more than one possible hydraulic test method. In that case, the most convenient method may be selected. If the possible choices are between a slug test and a constant-head test, the constant-head test should be selected because it yields results more representative of the aquifer. A step-by-step description of the flowchart follows.

The first two decisions which must be made concern the expected hydraulic conductivity. An estimate of the hydraulic conductivity can be made from the geologic log of the well. Particle size analysis is not necessary since only an order-of-magnitude estimate is needed. The textbook by Freeze and Cherry gives a useful chart showing hydraulic conductivity ranges for various geologic materials [1].

If the hydraulic conductivity is less than 10^{-2} cm/s, a pumping test is not possible and the decision process moves down to consider whether the conductivity is greater than 10^{-4} cm/s. Strictly speaking, pumping tests are not restricted to conductivities greater than 10^{-2} cm/s. Strausberg successfully conducted a low-discharge pumping test on materials which had a measured hydraulic conductivity of 2.4×10^{-4} cm/s. [20]. However, somewhat specialized equipment is needed, and some step testing is needed to select the proper discharge rate.

If the hydraulic conductivity is greater than 10^{-4} cm/s and less than 10^{-2} cm/s, a slug test or constant-head test can be performed. Let us assume that the conductivity is expected to be less than 10^{-4} cm/s. In that case, only a slug test can be performed, and the decision process moves downward and to the right in the flowchart to the question of well construction. If the well screen extends above the water table, testing methods which involve adding water to the well should not be used.

If the well screen does bracket the water table, a slug extraction test should be conducted and the decision process is complete. If the screen is entirely below the water table the decision process moves down and to the right to the question of the integrity of the water quality. If the tested well is going to be sampled for water chemistry, adding water to the well may dilute (or increase!) contaminant concentrations. In reality, this is not a serious consideration for slug tests because purging the well prior to sampling should remove all the water added. However, in sensitive situations, where the results of sampling may have to stand up in court, it would be better to use the slug extraction technique. If the water quality is not an issue or hydraulic testing can be delayed until sampling is complete, the decision process moves forward to the lower right corner of the flowchart where either an extraction or injection test can be used.

We are now finished with the lowest branch of the flowchart (dealing with conductivity less than 10^{-4} cm/s). If the expected hydraulic conductivity is less than 10^{-7} cm/s, a normal slug test is not practical because of the long time required to observe head decline. In this case, various laboratory methods, such as hydraulic testing of undisturbed core samples, may be used.

Returning to the lowest box in the first column of the flowchart; if the expected hydraulic conductivity is greater than 10^{-4} cm/s, a constant-head test can be considered. The decision process moves forward to consideration of whether the screen is bracketing the water table. If it is, a constant-head test cannot be used, and the decision process moves down and to the right. In this case, a slug extraction test is the only option.

If the screen is not bracketing the water table, the decision process moves upward and to the right to the question of availability of water. Recall that to arrive at this point, the hydraulic conductivity is expected to be between 10^{-4} and 10^{-2} cm/s. In this case, a sufficient supply of water can be delivered to the well with a garden hose (assuming a 5.1-cm (2 in.) well and 3 m (10 ft) of screen or less). If sufficient water is not available, the decision process moves down and to the right to a slug extraction test. Otherwise, the decision process moves forward to the water quality issue. Constant-head tests require adding significant amounts of water to the well, and dilution of subsequent water quality samples becomes a real issue. Well purging may not be sufficient to remove all the water added. If hydraulic testing can be delayed until water sampling is complete, a constant-head test can still be performed.

We have now completed discussing the lower half of the flowchart and will turn to the situation of hydraulic conductivities greater than or equal to 10^{-2} cm/s. In this case the decision process moves straight up in the flowchart to consideration of whether the expected conductivity is greater than or less than 10 cm/s. This value of hydraulic conductivity is equal to the approximate upper limit for constant-head tests. If the expected conductivity is less than this upper limit, the decision process moves forward towards the constant-head option, and the issues of screen positioning, water availability, and integrity of water quality are addressed. If any of these issues preclude the use of a constant-head test, the decision process moves up and back to the left in the flowchart to consideration of a pumping test. A slug test is not possible because we are considering the situation of the hydraulic conductivity being greater than 10^{-2} cm/s, the upper limit for slug tests.

Before a pumping test can be conducted, the issue of disposal of the discharge water must be addressed. This question is represented by the third box up from the start of the flowchart. If contaminated discharge water is a problem, then it must be contained. Containing discharge water is usually expensive and should be avoided if possible. However, some regulators require that all well discharge be contained. In situations where the types of contaminants and their concentrations are such that released water could cause a surface soil contamination problem, prudence dictates that the discharge water be contained.

Whether or not the contaminated discharge presents a problem, the next issue to be addressed is the expected depth to water towards the end of pumping. If a centrifugal or peristaltic pump is used, the distance from the water surface in the well to the top of the highest casing should not be greater than 7.6 m (25 ft) anytime during the test. Obviously, if the static water level is more than 7.6 m (25 ft) down, the decision is predetermined. Drawdowns are not easy to predict beforehand, and trial and error often is "par for the course." If the drawdown creates too great a lift for the centrifugal or peristaltic pump, the test must be redone (after water levels have re-equilibrated) with a lower discharge rate or with a submersible or air-lift pump.

Conclusions

The decision flowchart presented does not always indicate that only one testing method can be used. Materials with hydraulic conductivities at either end of the spectrum will limit the choice to one method; i.e., high conductivity limits the choice to the pumping test, low conductivity to the slug test or laboratory test. Constant-head tests cover the middle range, which characterizes most geologic materials. The conductivity ranges covered by pumping tests and slug tests overlap with the range for constant-head tests without overlapping with each other. Thus, if a choice exists, it is between conducting a slug test or a constant-head test or it is between conducting a pumping test or a constant-head test.

If a choice exists between a constant-head test and a slug test, the constant-head test is preferred because it tests a larger portion of the aquifer and therefore should give more representative results. When the choice is between a constant-head test and a pumping test, the pumping test is preferred because of the potential for determining more about the aquifer (i.e., leakage, boundary conditions) than is currently possible with constant-head tests.

References

[1] Freeze, R. A. and Cherry, J. A., *Groundwater,* Prentice-Hall, Englewood Cliffs, NJ, 1979, p. 604.
[2] Walton, W. C., *Groundwater Resource Evaluation,* McGraw-Hill, New York, 1970, p. 664.
[3] Driscoll, F. G., *Groundwater and Wells,* Johnson Division, Universal Oil Products Co., St. Paul, MN, 1986, p. 1089.
[4] Mandel, S. and Shiftan, Z. L., *Groundwater Resources,* Academic Press, New York, 1981, p. 269.
[5] Todd, D. K., *Groundwater Hydrology,* Wiley, New York, 1980, p. 535.
[6] Lambe, T. W., *Soil Testing for Engineers,* Wiley, New York, 1951, p. 165.
[7] Hvorslev, M. J., Waterways Experiment Station Bulletin 36, U.S. Army Corps of Engineers, Vicksburg, MS, 1951.
[8] *RCRA Ground-Water Monitoring Technical Enforcement Guidance Document,* U.S. Environmental Protection Agency, Washington, DC, 1986.
[9] Faust, C. R. and Mercer, J. W., *Water Resources Research,* Vol. 20, 1984. pp. 504–506.
[10] Black, J. H. and Kipp, K. L., Jr., *Journal of Hydrology,* Vol. 34, 1977, pp. 297–306.
[11] Bredehoeft, J. D., and Papadopulos, S. S., *Water Resources Research,* Vol. 16, 1980, pp. 233–238.
[12] Neuzil, C. E., *Water Resources Research,* Vol. 18, 1982, pp. 439–441.
[13] *Design Manual—Soil Mechanics, Foundations, and Earth Structures,* Naval Facilities Engineering Command (NAVFAC), Alexandria, VA, 1971, Chapter 4.
[14] Bouwer, H. and Rice, R. C., *Water Resources Research,* Vol. 12, 1976, pp. 423–428.
[15] Cedergren, H. R., *Seepage, Drainage and Flow Nets,* Wiley, New York, 1967.
[16] Lambe, T. W. and Whitman, R. V., *Soil Mechanics,* Wiley, New York, 1969, p. 553.
[17] U.S. Department of Interior, *Earth Manual,* U.S. Government Printing Service, Washington, DC, 1974, p. 810.
[18] Cooper, H. H., Bredehoeft, J. D., Papadopulos, I. S., and Bennett, R. R., *Water Resources Research,* Vol. 3, 1967, pp. 263–269.
[19] Thompson, D. B., *Ground Water,* Vol. 25, 1987, pp. 212–218.
[20] Strausberg, S. I., *Ground Water Monitoring Review,* Vol. 2, 1982, pp. 23–26.
[21] Theis, C. V., *Transactions of the American Geophysical Union,* Vol. 2, 1935, pp. 519–524.
[22] Cooper, H. H., Jr., and Jacob, C. E., *Transactions of the American Geophysical Union,* Vol. 27, 1946, pp. 526–634.
[23] Neuman, S. P., *Water Resources Research,* Vol. 8, 1972, pp. 1006–1021.
[24] Newman, S. P., *Water Resources Research,* Vol. 9, 1973, pp. 1102–1103.
[25] Newman, S. P., *Water Resources Research,* Vol. 11, 1975, pp. 329–342.

Kenneth R. Bradbury[1] and Maureen A. Muldoon[1]

Hydraulic Conductivity Determinations in Unlithified Glacial and Fluvial Materials

REFERENCE; Bradbury, K. R. and Muldoon, M. A., **"Hydraulic Conductivity Determinations in Unlithified Glacial and Fluvial Materials,"** *Ground Water and Vadose Zone Monitoring, ASTM STP 1053,* D. M. Nielsen and A. I. Johnson, Eds., American Society for Testing and Materials, Philadelphia, 1990, pp. 138-151.

ABSTRACT: Experiences with many measurements of the hydraulic conductivity of unlithified glacial and fluvial materials in Wisconsin suggest that hydraulic conductivity must be viewed in terms of the operational scale of measurement, based on the scale of the problem at hand and the volume of the materials of interest. Frequently, the hydraulic conductivity of a given lithostratigraphic unit appears to increase as the operational scale of measurement increases. In particular, laboratory methods can yield hydraulic conductivities one to two orders of magnitude lower than conductivities determined in field tests on the same materials. The operational scale of most laboratory methods is much smaller than the operational scale of most field problems, and laboratory tests, although often logistically and financially attractive, may be of little value in characterizing the hydraulic conductivity of Pleistocene and recent deposits at working field scales.

KEY WORDS: hydraulic conductivity, ground water, glacial deposits, pumping tests, permeability, laboratory tests, particle size, field permeability tests

Saturated hydraulic conductivity is one of the most basic and important parameters in hydrogeology, and almost all analyses of ground-water movement, from simple back-of-the-envelope calculations to sophisticated numerical models, require selecting representative values for hydraulic conductivity. In the past, hydrogeologists and engineers concerned themselves mainly with issues of water supply, and studies of ground-water flow and drawdown at production wells required hydraulic conductivity or transmissivity estimates determined for large volumes of aquifer material. Today, in the rapidly evolving science and technology associated with contaminant movement and ground-water monitoring, such issues as assessing site suitability for waste disposal facilities, predicting rates and directions of contaminant movement, selecting locations for monitoring equipment, and planning remedial actions all require more detailed hydraulic conductivity information than was generally needed in the past. In particular, there have recently been two major changes in the type of hydraulic conductivity information required by hydrogeologists. First, there is a growing interest in ground-water movement in "tight," clayey materials, which must be understood when planning for waste disposal and ground-water quality protection. Second, studies of contaminant hydrogeology often require detailed information about the distribution and spatial variability of hydraulic conductivity of particular materials over small areas. Classical pumping tests, long the accepted means of determining aquifer parameters in the field, generally work best for measuring global averages of parameters in relatively conductive materials but may not always be appropriate

---

[1] Wisconsin Geological and Natural History Survey, Madison, WI 53705.

or suitable for the detailed hydraulic conductivity measurements needed in contaminant or waste disposal studies.

The purpose of this paper is to report the authors' recent experience with measurement of saturated hydraulic conductivity in unlithified glacial and fluvial materials in Wisconsin. Techniques currently used include pumping tests, specific capacity tests, piezometer "slug" tests, laboratory permeameter tests, and estimates based on particle-size distributions. In addition, the Wisconsin Geological and Natural History Survey is developing a data base containing hydraulic conductivity determinations taken from geotechnical investigations at waste disposal facilities. This data base associates each hydraulic conductivity measurement with a spatial location and with the geologist's rigorously defined lithostratigraphic unit. A lithostratigraphic unit consists of a group of geologic materials having characteristic physical properties which can be identified and mapped in the field. These recent experiences lead to several observations about the measurement and interpretation of hydraulic conductivity in such materials. Although the following observations may not hold in all cases, they may be general enough to have wide application:

1. Measured values of hydraulic conductivity tend to increase as the scale of the measurement increases.
2. Results of laboratory permeameter tests are almost invariably lower in values than results of field tests on the same materials.
3. Hydraulic conductivity estimates based on particle-size distributions give widely varying results depending on the estimation method used and the type of material being evaluated.

The following discussion attempts to illustrate and support these observations.

Measurement Scale Effects

Although commonly considered a characteristic physical property, the "true" hydraulic conductivity of saturated near-surface materials is difficult to define. The physical characteristics (particle size, roundness, sorting, and other characteristics) of geologic materials vary spatially, and this spatial variation can be large in many glacial and fluvial settings, where significant changes in the depositional environment often occur over short distances. Even in relatively homogeneous materials, detailed field determinations have shown that the spatial variability in hydraulic conductivity can be large. For example, Sudicky [1] reported that laboratory values of hydraulic conductivity varied by a factor of between 5 and 30 over 2-m cores along a transect in fine-to-medium sand at the Bordon site in Ontario, Canada. Likewise, Smith [2] documented significant variations in hydraulic conductivity over short distances in an apparently uniform stratified sand. Any experiment that tests the hydraulic conductivity of such a variable material necessarily produces a value which represents some type of average conductivity for the volume of material tested. The hydrogeologist is usually forced to assume that this "average" value is the appropriate value to use for the problem at hand. However, as greater or smaller volumes of material are tested, the test encounters a greater or smaller number of aquifer inhomogeneities, which might be thought of as more or less conductive zones.

Hydraulic conductivity tests commonly used today generally fall into four operational scales based on the volume of material sampled during the test. These methods might be classified as follows:

(a) Small-scale tests (<1 m^3): permeameter tests, particle-size estimates;

(*b*) Site-specific tests (1 m^3 to hundreds of cubic metres): specific capacity tests, single-well pumping tests, slug tests;

(*c*) Local tests (hundreds to thousands of cubic metres): multiple-well pumping tests; and

(*d*) Regional tests (thousands of cubic metres): regional aquifer analysis, inverse models.

As used here, *small-scale tests* refer to laboratory determinations on small samples obtained from the field. Such tests include flexible-wall and solid-wall permeameter determinations [3] and empirical estimates of hydraulic conductivity based on various measures of the particle-size distribution [4–10]. *Site specific* determinations include piezometer tests (also referred to as slug tests or bail tests) [11], estimates based on the specific capacity of single production wells [12], and single-well pumping tests, in which drawdown is measured in the pumped well. *Local* hydraulic conductivity measurements cover larger areas and volumes of material than site-specific determinations and include the aquifer pumping test, using multiple observation wells [13]. *Regional* evaluations establish the hydraulic conductivity over large areas (many square kilometres) using such techniques as forward ground-water model calibration [14], inverse ground-water modeling [15], or flow net analyses [16]. Other techniques for determining hydraulic conductivity, such as tracer experiments [17,18], are not considered here.

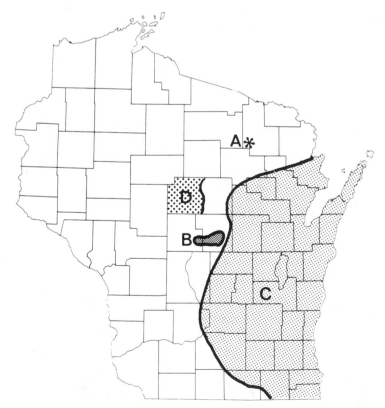

FIG. 1—*Locations of areas in Wisconsin where the hydraulic conductivity of glacial material was evaluated: (a) Crandon site, (b) Buena Vista basin, (c) region of eastern Wisconsin where data from 40 waste disposal sites were evaluated, (d) western Marathon County.*

The following three case studies include measurements of the hydraulic conductivity of unlithified Pleistocene sediment at locations in eastern, central, and northern Wisconsin (Fig. 1). The examples include data from ongoing or recently completed hydrogeologic investigations and are selected to demonstrate a variety of methods and a variety of measurement scales. In each case at least two methods are evaluated.

Example I: Horizontal Hydraulic Conductivity in Coarse-Grained Fluvial Sediment

Hydraulic conductivity measurements on coarse-grained fluvial deposits in two parts of central and northern Wisconsin illustrate the effects of various measurement scales. The Buena Vista basin (Fig. 1) is an agricultural area in Wisconsin's central sand plain which has been the subject of recent hydrogeologic research related to ground-water contamination by agricultural chemicals [19–21]. The aquifer in the Buena Vista basin consists of approximately 30 m of fluvial and lake sediment composed of well-sorted, coarse-to-medium-grained quartz sand.

At the Crandon site in north-central Wisconsin (Fig. 1), the Exxon Minerals Co. recently completed an extensive hydrogeologic evaluation for the purpose of predicting the environmental impacts of a proposed subsurface mine [22,23]. The materials at Crandon can be generally grouped into till, lake sediment, and fluvial sediment. The fluvial sediment, classified by Exxon as coarse-grained stratified drift, consists of fine-to-coarse-grained sand and gravel and ranges from poorly sorted to well sorted [23].

The sediment in both areas is stratified but is so noncohesive that obtaining undisturbed samples for laboratory analysis is almost impossible. Therefore, even though the ratio of horizontal to vertical hydraulic conductivities in this material can range from about 2 to 20 [24], all values reported here are considered to represent horizontal conductivities.

Workers at both the Buena Vista and Crandon areas determined hydraulic conductivity by methods ranging from laboratory analyses to major pumping tests. In addition, both areas have been the subject of extensive numerical computer simulations. At the Crandon site, Exxon used a two-dimensional finite-element model to simulate aquifer drawdowns caused by mine dewatering [25]. Through the model calibration process Exxon arrived at a "best" hydraulic conductivity value to describe ground-water movement in the fluvial sediment over the large area (40 km^2) of a regional model. In the Buena Vista basin, Stoertz [26] has used a two-dimensional parameter estimation model [15] to arrive at hydraulic conductivity estimates which allow the model to best fit an extensive set of hydraulic-head data. Table 1 summarizes the hydraulic conductivity data available at these two sites and shows the approximate volume of aquifer material evaluated in each type of analysis. In general, the hydraulic conductivity data are log-normally distributed; therefore, Table 1 presents geometric means rather than arithmetic means to describe the data. In Table 1 and subsequent tables and graphs, sample volumes for particle-size analyses and permeameter test are calculated as the actual volume of material tested. Thus, for permeameter tests, this is the volume of material placed in the permeameter; for particle-size tests it is the volume of sample used in the sieve or hydrometer analysis.

At both sites, the mean hydraulic conductivity appears to increase as the volumetric scale of the test increases. When the test volume is plotted versus hydraulic conductivity (Fig. 2), this trend is very evident, with a more rapid increase at smaller scales than at larger scales. In the Buena Vista data, the geometric mean hydraulic conductivities vary by a factor of about 9 among the various methods. The Crandon data show variation by a factor of about 60. The cause of this difference in range of variation between the two areas is not completely clear, and it may be partially due to the greater number of samples at the Crandon site. However, this variation probably also occurs because the Crandon sediment

TABLE 1—*Summary of the hydraulic conductivity data, test method, and volume of material tested at two sites in north-central Wisconsin.*

Method	Approximate Volume Tested, m³	Number of Tests	Hydraulic Conductivity (Geometric Mean), m/s
Crandon Area [22,23,25][a]			
Particle-size estimates (GS)	2×10^{-5}	~100	2.1×10^{-6}
Slug tests (ST)	5	15	3.0×10^{-5}
Pumping test (PT)	1.9×10^{8}	1	1.3×10^{-4}
Model (M)	1.2×10^{9}	1	1.22×10^{-4}
Buena Vista Area [12,19,20,26][b]			
Particle-size estimates (GS)	2×10^{-5}	32	1.0×10^{-4}
Permeameter tests (P)	6×10^{-4}	7	1.5×10^{-4}
Slug tests (ST)	1	44	2.9×10^{-4}
Specific capacity tests (SC)	1000	266	6.4×10^{-4}
Pumping tests (PT)	9×10^{6}	10	7.3×10^{-4}
Inverse model (M)	4×10^{9}	3	7×10^{-4}

[a] Particle-size estimates were performed using the method of Hazen [6].
[b] Particle-size estimates were performed using the method of Masch and Denny [4].

is much more heterogeneous than sediment in the Buena Vista area. The sandy sediment at the Crandon site includes both till and fluvial sediment, often interbedded in a complex fashion [22,25]. At each site, regional models produce the highest values of hydraulic conductivity, while estimates based on laboratory particle-size data yield the lowest values. Both areas show strong and similar trends, which suggests that larger scales of measurement in both areas result in larger conductivity values.

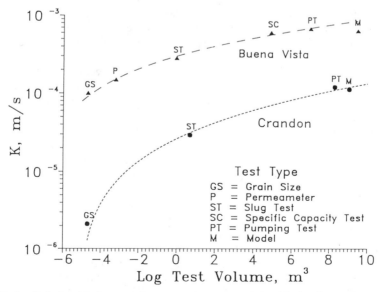

FIG. 2—*Relationship between the geometric mean measured hydraulic conductivity, K, and the test volume for the Crandon site and Buena Vista basin.*

Example II: Field Versus Laboratory Determinations of Hydraulic Conductivity for Tills in Eastern Wisconsin

Comparisons between field and laboratory values of hydraulic conductivity for tills in eastern Wisconsin show widely varying results for similar materials. Pleistocene materials in eastern Wisconsin consist of a series of till sheets locally or regionally separated by layers of stream or lake sediment. Mickelson et al. [27] classified these materials into a series of lithostratigraphic units which can be mapped over large areas and identified in subsurface borings. As a result, there is a well-understood geologic and stratigraphic framework for Pleistocene materials in eastern Wisconsin, and it is possible to study hydrogeologic properties of the various till units within this classification scheme. The availability of much data from numerous proposed or operating solid waste disposal facilities in the area provides an opportunity for comparisons of field and laboratory results on similar materials over wide areas.

Rodenbeck et al. [28] summarize hydrogeologic and geotechnical properties of tills from 40 proposed or existing landfill sites in eastern Wisconsin (Fig. 1). Only hydraulic conductivity results are discussed here, with the goal of comparing field and laboratory measurements. All data were obtained from consultants' reports and other documents available at the Bureau of Solid Waste Management of the Wisconsin Department of Natural Resources. The methodology is not entirely consistent between sites, but in general, field conductivities are determined by piezometer "slug" tests, while laboratory conductivities are obtained from solid-wall or flexible-wall permeameters using either "undisturbed" or remolded samples. The orientation of the laboratory samples is frequently unknown; however, till units are generally unstratified, and orientation may not be as important for till as for other glacial sediments.

Table 2 (modified from Rodenbeck et al. [28]) lists the geometric mean field and laboratory hydraulic conductivities for the till of the Kewaunee, Horicon, and Oak Creek Formations and till of six members of the Kewaunee Formation. The number of samples for each unit varies from 15 to 133 for formations and from 4 to 35 for formation members. In general, till of the Horicon Formation shows the highest hydraulic conductivities, while till of the Oak Creek Formation shows the lowest.

A plot of these data, grouped by lithostratigraphic unit (Fig. 3) highlights major discrepancies between field and laboratory measurements of hydraulic conductivity. In general, laboratory measurements are 1.2 to 2.6 orders of magnitude lower than field measurements. It is tempting to explain this discrepancy as an apparent anisotropy related to a bias of field tests toward horizontal hydraulic conductivity and a bias of laboratory tests toward

TABLE 2—*Hydraulic conductivity data for eastern Wisconsin tills.*

| Unit | Code | Hydraulic Conductivity (Geometric Mean), m/s | | | |
		Laboratory	N	Field	N
Kewaunee Formation	KF	6.3×10^{-10}	133	1.4×10^{-8}	48
Middle Inlet Member	MI	6.3×10^{-10}	16	1.0×10^{-7}	17
Kirby Lake Member	KL	5.0×10^{-10}	26	1.6×10^{-8}	35
Glenmore Member	G	1.6×10^{-10}	17	1.0×10^{-8}	14
Chilton Member	C	2.0×10^{-10}	10	6.3×10^{-9}	4
Valders Member	V	4.0×10^{-9}	19	3.2×10^{-7}	12
Haven Member	H	5.0×10^{-10}	27	5.0×10^{-8}	9
Horicon Formation	HF	7.9×10^{-8}	15	6.3×10^{-7}	25
Oak Creek Formation	OCF	3.2×10^{-10}	57	1.0×10^{-8}	72

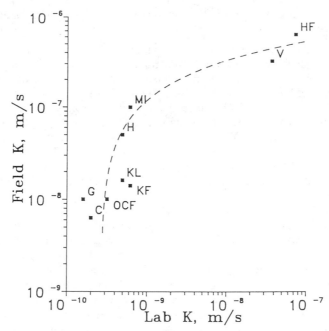

FIG. 3—*Relationship between laboratory and field determinations of hydraulic conductivity at waste disposal sites in eastern Wisconsin. The letters refer to the abbreviations of lithostratigraphic units in Table 2.*

vertical hydraulic conductivity. However, three lines of reasoning oppose this hypothesis. First, till units are generally not well stratified, and high anisotropy ratios would thus not be expected. Second, calibration of a numerical model of ground-water flow near Green Bay, Wisconsin [29], produced vertical to horizontal anisotropy ratios of only 1 to 10 for tills of the Kewaunee Formation, much lower than the ratios of 30 to 150 suggested by Table 2. Third, many of the eastern Wisconsin till units contain horizontal and vertical joints [30]. These joints may add significant secondary permeability to the tills, and field tests on such materials may yield results several orders of magnitude greater than results of laboratory tests on unfractured samples [31].

Example III: Field and Laboratory Investigations of the Hydraulic Conductivity of Till Units in Central Wisconsin

One weakness of the study of Rodenbeck et al. [28] reported above is a potential lack of consistency in data collection from site to site. However, Muldoon [32] reports very similar results in a study of the hydrogeologic properties of four till units in western Marathon County in central Wisconsin (Fig. 1). Muldoon investigated the hydraulic conductivity of till of the Medford and Edgar Members of the Marathon Formation and the Bakerville and Merrill Members of the Lincoln Formation using field piezometer tests, flexible-wall permeameter tests on cores, and empirical estimates based on particle-size distributions. Till of the Marathon Formation tends to contain more clay and less sand than does till of the Lincoln Formation. In addition, till of the Marathon Formation shows more variation in particle size than does till of the Lincoln Formation. These materials contain significant amounts of clay and are generally cohesive enough to permit the recovery of undisturbed cores for laboratory analyses.

Muldoon's [32] methods were as follows. Each lithostratigraphic unit was identified and sampled in the field through a series of 59 stratigraphic borings, and particle-size distributions of the resulting disturbed grab samples were determined through standard sieve and hydrometer analyses. In addition, several continuous undisturbed cores of each unit were obtained at selected characteristic sites using hollow-stem augers with a 0.08-m (3-in.)-diameter overshot split-spoon sampler. Using these core holes, 27 standpipe piezometers were installed at 11 locations throughout the study area. The piezometers consisted of 0.03-m (1.25-in.) polyvinyl chloride (PVC) pipe with slotted PVC screens about 1 m in length. Each screen was surrounded by a silica sand pack, and the remaining annular space was sealed using granular bentonite. The piezometers were used to conduct falling-head slug tests, which were interpreted using the methods of Hvorslev for a well point in an infinite medium [11].

Muldoon [32] also performed laboratory tests of hydraulic conductivity using a triaxial cell as a flexible-wall permeameter. Flexible-wall permeameters confine the sample within a flexible but impermeable membrane, thus allowing field consolidation to be duplicated and minimizing piping along the sides of the sample. Samples obtained during the coring operation were trimmed to a diameter of 0.035 m (1.4 in.), placed in the triaxial cell, and encased in two latex membranes. The samples were saturated by applying back pressure to both the inflow and outflow burettes and then were consolidated to calculated overburden pressure. After application of a gradient of about 13.8 kPa (2 psi) across the samples, the inflow and outflow burettes were monitored for about two days or until steady-state conditions were established.

Permeameter and Piezometer Tests

Comparing laboratory and field results for the Marathon County till units (Table 3 and Fig. 4) shows that the geometric mean hydraulic conductivities determined in the laboratory are 2.5 to 3 orders of magnitude smaller than the field results. In addition, the laboratory results show less variation between the two formations than do the field results. The difference between the laboratory- and field-measured hydraulic conductivity values is comparable to that observed by Herzog and Morse [33] for similar tests in the fine-grained Vandalia till of Illinois. Many till units display only weakly developed horizontal fabric, and vertical anisotropy alone probably cannot explain such large discrepancies between the field and laboratory results. The difference in scale between the two tests probably

TABLE 3—*Summary of field and laboratory measurements in tills in Marathon County, Wisconsin* [32].

Lithostratigraphic Unit	Code	Hydraulic Conductivity (Geometric Mean), m/s			
		Laboratory	N	Field	N
Marathon Formation	MA	3.4×10^{-10}	7	5.8×10^{-8}	20
Medford Member	Mf	6.4×10^{-10}	5	3.6×10^{-7}	4
Edgar Member	Ed	1.8×10^{-10}	2	9.5×10^{-8}	16
Lincoln Formation	LN	1.1×10^{-9}	6	2.3×10^{-6}	5
Bakerville Member	Bk	1.4×10^{-9}	2	2.5×10^{-6}	2
Merrill Member	Mr	9.2×10^{-10}	4	2.2×10^{-6}	3

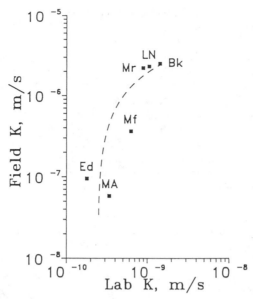

FIG. 4—*Relationship between laboratory and field determinations of hydraulic conductivity in western Marathon County, Wisconsin. The letters refer to the abbreviations of lithostratigraphic units in Table 3.*

causes a large part of the discrepancy. The larger scale field tests are likely to include the effects of small sand layers or fractures, which would increase hydraulic conductivity. Till of the Edgar Member contains fractures in places, and the Medford Member displays subhorizontal fissility where exposed.

Hydraulic Conductivity Estimates Based on Particle-Size Distributions

Many workers have attempted to establish empirical relationships between hydraulic conductivity and more easily measured physical properties of unconsolidated materials [4–10]. Muldoon [32] used five empirical relationships to estimate the hydraulic conductivity of four till units. The equations used included those developed by Bedinger [8] and Hazen [6], which utilize some measure of "effective particle size" to predict hydraulic conductivity, as well as the method of Krumbein and Monk [5], which expands the effective diameter relationships by including a measure of sorting of the sample. The final two equations used [9,10] estimate hydraulic conductivity from the percentage of the sample that falls within a given size class.

Detailed particle-size analyses were performed on 111 samples. For each sample, a cumulative particle-size distribution curve was plotted, and the grain diameters needed for the hydraulic conductivity calculations were then determined from these curves. A sample particle-size curve is shown in Fig. 5, along with the five equations used to calculate hydraulic conductivity. In order to compare the results of the different methods, all permeability values were converted to hydraulic conductivity values and all hydraulic conductivity values were converted into units of metres per second.

The five geometric mean hydraulic conductivity values calculated for each lithostratigraphic unit are presented in Table 4 and shown graphically in Fig. 6.

The equations developed by Bedinger [8] and Hazen [6] are very similar in form and

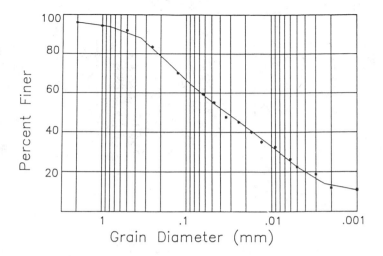

Bedinger: $K_{(gal/day/ft^2)} = 2000 * D50^2$

Hazen: $K_{(cm/sec)} = D10^2$

Krumbein & Monk: $k_{(darcies)} = 760 * Dm^2 * e^{(-1.31 * \sigma_\phi)}$

Cosby et al: $\log K_{(in/hr)} = (.0153 * \%sa) - .884$

Puckett et al: $K_{(m/sec)} = 4.36 \times 10^{-5} * e^{(-.1975 * \%cl)}$

FIG. 5—*Sample particle-size distribution curve and five empirical equations used to estimate hydraulic conductivity: D50 = median diameter, in millimetres; D10 = diameter, in millimetres, at which 10% of the sample is finer; Dm = mean diameter, in millimetres; σ_ϕ = phi standard deviation; %sa = percentage of the sample coarser than 0.05 mm; %cl = percentage of the total sample finer than 0.002 mm (from Muldoon [32]).*

were both derived using results from sand samples. These two methods, however, give dissimilar results. The equation developed by Bedinger [8] tends to overestimate hydraulic conductivity in comparison with values determined using field tests. Hazen's approximation [6], which was developed for clean filter sands, consistently underestimates the hydraulic conductivity, in comparison with field results, by one to two orders of magnitude. This method, however, provides the closest agreement with the laboratory-measured values although consistently overestimating K by approximately one order of magnitude. The equation of Krumbein and Monk [5], developed using laboratory-prepared sand samples, is the only one used which takes into account the sorting of a sample. This method predicts hydraulic conductivity values for all of the till units that are within one order of magnitude of field-derived values. The equation by Cosby et al. [9], developed using soil samples from throughout the United States, overestimates the hydraulic conductivity, in comparison with both field and laboratory results, by several orders of magnitude. The equation by Puckett et al. [10], derived using samples from each horizon or six ultisols, uses the percentage of clay in the less-than-2-mm fraction of a sample to estimate K. This equation provides good estimates of field hydraulic conductivity for most of the till units.

TABLE 4—Summary of hydraulic conductivity values estimated from particle-size distribution curves using various empirical equations.

Lithostratigraphic Unit	Code	Number of Samples	Hydraulic Conductivity (Geometric Mean), m/s				
			Bedinger	Hazen	Krumbein and Monk	Cosby	Puckett
Marathon formation	MA	98	7.8×10^{-7}	6.0×10^{-9}	3.6×10^{-8}	3.2×10^{-5}	1.3×10^{-6}
Medford member	Mf	13	2.7×10^{-9}	3.8×10^{-9}	1.5×10^{-8}	2.6×10^{-5}	7.1×10^{-7}
Edgar member	Ed	85	2.3×10^{-6}	9.5×10^{-9}	8.4×10^{-8}	4.0×10^{-5}	2.4×10^{-6}
Lincoln formation	LN	13	7.1×10^{-6}	1.7×10^{-8}	3.2×10^{-7}	6.0×10^{-5}	4.0×10^{-6}
Bakerville member	Bk	7	1.2×10^{-5}	1.2×10^{-8}	4.1×10^{-7}	7.4×10^{-5}	4.0×10^{-6}
Merrill member	Mr	6	4.1×10^{-6}	2.5×10^{-8}	2.5×10^{-7}	4.8×10^{-5}	4.1×10^{-6}

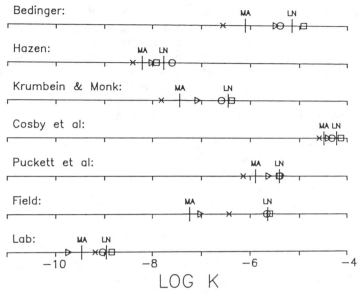

FIG. 6—*Plot of geometric mean hydraulic conductivities estimated from particle-size analyses for tills in western Marathon County, Wisconsin, using various empirical equations. Field and laboratory results are shown for comparison. MA = Marathon Formation; LN = Lincoln Formation; X = Medford Member; ▷ = Edgar Member; □ = Bakerville Member; O = Merrill Member (from Muldoon [32]).*

Comparison of the results from all five empirical methods suggests that particle-size estimates of hydraulic conductivity should be used with caution. The five methods used in this study provide estimates of hydraulic conductivity that range over three to four orders of magnitude for any given lithostratigraphic unit. Each method is most applicable for the type of sediment used to derive it and should not be extended to other types of material without performing some field tests to verify the results. The methods of Krumbein and Monk [5] and Puckett et al. [10] most closely predict the field hydraulic conductivities of the units studied. The sorting parameter included in the method of Krumbein and Monk may account for the wide applicability of this equation. The method of Puckett et al. was derived for fine-grained materials that may be similar, in terms of particle-size distribution, to the till of the Lincoln and Marathon Formations. Similar particle-size distributions could explain the good predictive capability of this equation.

Discussion

Although attractive for financial and logistical reasons, laboratory tests, at the scale commonly run, appear frequently to underestimate the "operational" hydraulic conductivity of most field situations involving undisturbed Pleistocene material. At least four effects may be partly responsible for this discrepancy. First, *sample bias* can occur during sample collection. More permeable materials tend, in general, to be less cohesive and more difficult to sample than less permeable materials. For example, in obtaining a core of a clay till with interbedded sand seams, the core may tend to break or crumble at the sand seams. These sandy materials might not be recovered at all or might be accidentally discarded in the field, yet might be very important in controlling ground-water movement at the site.

Coarse gravel materials might not be recovered at all. Second, *preferential flow* undoubtedly occurs in many Pleistocene materials along joints, fractures, interbeds, and macropores. Small laboratory samples cannot contain all of these features, and the resulting measurement of hydraulic conductivity will be too low. Third, *sample disturbance,* particularly compaction but also chemical and biological change, can occur during sampling, transport, and storage of samples. Most sample disturbance tends to reduce hydraulic conductivity. Fourth, *directionality* can be an important consideration in anisotropic materials. For example, most common core sampling tends to be vertical, but ground-water flow is often horizontal. In particular, vertical and horizontal hydraulic conductivities must be tested in a similar manner so that any apparent anisotropy is not due to differences between the test methods used for the vertical and horizontal directions.

While frequently criticized, single-well hydraulic conductivity estimates (slug tests, specific capacity tests, single-well pumping tests) are useful methods for characterizing the hydraulic conductivity of shallow glacial and fluvial deposits because most common contamination and monitoring problems occur on this operational scale, and the measurements are relatively inexpensive and repeatable. Classical multiple-well pumping tests yield parameters for large volumes of material but can be logistically difficult and expensive in unconfined aquifers. Regional conceptual and numerical (inverse) models yield values of hydraulic conductivity on the regional scale, but such values may be too general for many current hydrogeologic investigations.

Conclusions

Because of the relatively high spatial variability of Pleistocene materials, the value and meaning of their hydraulic conductivity can be difficult to define. For most field problems in hydrogeology, the operational hydraulic conductivity varies with the scale of the problem and with the scale of measurement. The most significant conclusion from the data presented here is that field measurements should be conducted on the same scale as the field problem. For example, regional studies may require regional information in the form of large-scale pumping tests or flow net analyses, while very localized studies require small-scale tests. In particular, the utility of laboratory measurements of hydraulic conductivity appears doubtful. Such measurements appear to yield conductivity values significantly lower than those observed in larger scale field tests. Hydraulic conductivity measurements based on particle-size distributions must be used with caution because significant variation occurs with the various methods of interpretation.

References

[1] Sudicky, E. A., *Water Resources Research,* Vol. 22, No. 13, 1986, pp. 2069–2082.
[2] Smith, L., *Mathematical Geology,* Vol. 13, No. 1, 1981, pp. 1–21.
[3] Olson, R. E. and Daniel, D. E., in *Permeability and Groundwater Contaminant Transport, ASTM STP 746,* T. F. Zimme and C. O. Riggs, Eds., American Society for Testing and Materials, Philadelphia, 1981, pp. 18–64.
[4] Masch, F. D. and Denny, K. J., *Water Resources Research,* Vol. 2, No. 4, 1966, pp. 665–677.
[5] Krumbein, W. C. and Monk, G. D., *Transactions of the Petroleum Division, American Institute of Mining and Metallurgical Engineering,* Vol. 151, 1943, pp. 153–163.
[6] Hazen, A., "Some Physical Properties of Sands and Gravels with Special Reference to Their Use in Filtration," 24th Annual Report, Massachusetts State Board of Health, Boston, 1893.
[7] Rose, H. G. and Smith, H. F., *Water Well Journal,* Vol. 3, No. 3, March 1957, pp. 30–32.
[8] Bedinger, M. S., "Relation Between Median Grain Size and Permeability in the Arkansas River Valley," *U.S. Geological Survey Professional Papers,* No. 292, pp. C31–C32.
[9] Cosby, B. J., Hornberger, G. M., Clapp, R. B., and Ginn, T. R., *Water Resources Research,* Vol. 20, No. 6, 1984, pp. 682–690.

[10] Puckett, W. E., Dane, J. H., and Hajek, B. F., *Soil Science Society of America Journal*, Vol. 49, 1985, pp. 831–836.

[11] Hvorslev, M. J., "Time-Lag and Soil Permeability in Groundwater Observations," Bulletin 36, U.S. Army Corps of Engineers Waterways Experiment Station, Vicksburg, MS, 1951.

[12] Bradbury, K.R. and Rothschild, E. R., *Ground Water*, Vol. 23, No. 2, 1985, pp. 240–246.

[13] Walton, W. C., "Selected Analytical Techniques for Well and Aquifer Evaluation," Bulletin 49, Illinois State Water Survey, Urbana, IL, 1962.

[14] Wang, H. F. and Anderson, M. P., *Introduction to Groundwater Modeling*, Freeman, San Francisco, 1982, pp. 109–110.

[15] Cooley, R. L., *Water Resources Research*, Vol. 13, No. 2, 1977, pp. 318–324.

[16] Lohman, S. W., "Ground-Water Hydraulics," *U.S. Geological Survey Professional Papers*, No. 708, 1922, pp. 45–49.

[17] Taylor, S. R., Moltyana, G. L., Howard, K. W. F., and Killey, R. W. D., *Ground Water*, Vol. 25, No. 13, 1987, pp. 321–330.

[18] Anderson, M. P., *Reviews of Geophysics*, Vol. 25, No. 2, 1987, pp. 141–147.

[19] Stoertz, M. W., "Evaluation of Groundwater Recharge in the Central Sand Plain of Wisconsin," M.S. thesis (Geology), University of Wisconsin—Madison, Madison, WI, 1985.

[20] Faustini, J. M., "Delineation of Groundwater Flow Patterns in a Portion of the Central Sand Plain of Wisconsin," M.S. thesis (Geology), University of Wisconsin—Madison, Madison, WI, 1985.

[21] Zheng, C., Wang, H. F., Anderson, M. P., and Bradbury, K. R., *Journal of Hydrology*, Vol. 98, 1988, pp. 61–78.

[22] Golder and Associates, "Geohydrologic Characterization, Crandon Project," report prepared for Exxon Minerals Corp., Atlanta, GA, 1982.

[23] STS Consulting Engineers, "Hydrogeologic Study Update," report prepared for Exxon Minerals Corp., Green Bay, WI, 1984.

[24] Weeks, E. P., *Water Resources Research*, Vol. 5, No. 1, 1969, pp. 196–214.

[25] D'Appolonia Consulting Engineers, "Hydrologic Impact Assessment," Appendix 4.1A, prepared for Exxon Minerals Corp., Pittsburgh, PA, 1984.

[26] Stoertz, M. W., "Parametric Estimation Groundwater Modeling of the Central Wisconsin Sand Plain," Ph.D. thesis (Geology), University of Wisconsin—Madison, Madison, WI, 1989.

[27] Mickelson, D. M., Clayton, L., Baker, R. W., Mode, W. N., and Schneider, A. F., "Pleistocene Stratigraphic Units of Wisconsin," Miscellaneous Paper 84-1, Wisconsin Geological and Natural History Survey, Madison, WI, 1984.

[28] Rodenbeck, S. A., Simpkins, W. W., Bradbury, K. R., and Mickelson, D. M., "Merging Geotechnical Data with Pleistocene Lithostratigraphy—Examples from Eastern Wisconsin," *Proceedings*, Tenth Annual Madison Waste Conference, Department of Engineering Professional Development, University of Wisconsin—Madison, Madison, WI, 1987, pp. 454–476.

[29] Feinstein, D. T. and Anderson, M. P., "Recharge to and Potential for Contamination of an Aquifer System in Northeastern Wisconsin," Technical Report WIS WRC 87-01, Wisconsin Water Resources Center, Madison, WI, 1987, pp. 70–71.

[30] Connell, D. E., "Distribution, Characteristics, and Genesis of Joints in Fine-Grained Till and Lacustrine Sediment, Eastern and Northwestern Wisconsin," M.S. thesis (Geology), University of Wisconsin—Madison, Madison, WI, 1984.

[31] Freeze, R. A. and Cherry, J. A., *Groundwater*, Prentice-Hall, Englewood Cliffs, NJ, 1979, pp. 149–152.

[32] Muldoon, M. A., "Hydrogeologic and Geotechnical Properties of Pre-Late Wisconsin Till Units in Western Marathon County, Wisconsin," M.S. thesis (Geology), University of Wisconsin—Madison, Madison, WI, 1987.

[33] Herzog, B. L. and Morse, W. J., "A Comparison of Laboratory and Field Determined Values of Hydraulic Conductivity at a Waste Disposal Site," *Proceedings*, Seventh Annual Madison Waste Conference, Department of Engineering Professional Development, University of Wisconsin—Madison, Madison, WI, 1984, pp. 30–52.

Beverly L. Herzog[1] and Walter J. Morse[2]

Comparison of Slug Test Methodologies for Determination of Hydraulic Conductivity in Fine-Grained Sediments

REFERENCE: Herzog, B. L. and Morse, W. J., "**Comparison of Slug Test Methodologies for Determination of Hydraulic Conductivity in Fine-Grained Sediments,**" *Ground Water and Vadose Zone Monitoring, ASTM STP 1053,* D. M. Nielsen and A. I. Johnson, Eds., American Society for Testing and Materials, Philadelphia, 1990, pp. 152–164.

ABSTRACT: Slug tests were conducted on four lithologic units at a waste disposal site in southwestern Illinois. All tested units were fine-grained glacial tills that had laboratory-determined hydraulic conductivity values between 10^{-9} and 10^{-6} cm/s. A total of 37 field tests of hydraulic conductivity were run in 29 open-hole piezometers, with some tests lasting up to a year. The piezometers were oriented both vertically and at a 45° angle to test the effects of possible vertical fractures on hydraulic conductivity.

All slug test results were analyzed using three different methods. The methods selected were those of Cooper, Bredehoeft, and Papadopulos; Hvorslev; and Nguyen and Pinder. The geometric mean values for each geologic unit determined by all three methods were within an order of magnitude and well within one standard deviation.

All three methods could not be applied to all data with the same degree of reliance. This is to be expected since each method has different underlying assumptions. However, all data did fit at least one theoretical curve very well. The importance of these results is twofold: (1) any of the three methods can produce reliable results if properly applied, and (2) data that appear to be unusable by one method may be usable by changing to another method of analysis.

KEY WORDS: hydraulic conductivity, slug tests, fine-grained sediments, low permeability

Laboratory tests for hydraulic conductivity are relatively inexpensive and fast, and tend to produce consistent results; however, they must be corroborated by field tests. In the field a larger volume of material can be tested *in situ,* thus minimizing the influence of scale effects. While laboratory-determined values can provide upper and lower boundaries of hydraulic conductivity estimates for moderately sorted to well-sorted sand [1,2], discrepancies between laboratory and field values of more than two orders of magnitude have been found for fine-grained sediments [3–5].

In a recent study by the Illinois State Geological Survey to determine why solute migration at a chemical waste disposal site exceeded predicted rates by 100 to 1000 times, researchers determined that the low predictions had resulted from using only laboratory-determined hydraulic conductivity values in the calculations [4,6]. The geology of the site consisted primarily of fine-grained glacial tills [7], as shown in Fig. 1. Because the tills were expected to exhibit low hydraulic conductivities, making conventional pumping tests impractical, slug tests were primarily used to collect field data.

---

[1] Associate hydrogeologist, Illinois State Geological Survey, Champaign, Il 61820.
[2] Associate staff geologist, Illinois State Geological Survey, Champaign, IL 61820.

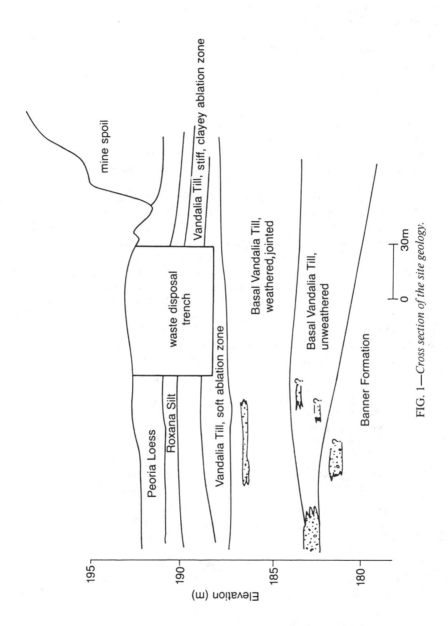

FIG. 1—Cross section of the site geology.

Field hydraulic conductivity data were analyzed using three different slug test analytical techniques. The slug tests were originally designed to be analyzed by the method proposed by Cooper, Bredehoeft, and Papadopulos [8]. This method was chosen because the theory assumed boundary conditions that could reasonably be met in the field. The data were reanalyzed using Hvorslev's method [9], because this method is commonly used in the engineering community. A method of analyzing hydraulic conductivity, proposed more recently by Nguyen and Pinder [10] and which uses data primarily collected during the early portion of the slug test, was selected to provide a comparison with the other methods.

Procedure

The 37 slug tests were conducted in 29 open-hole piezometers. Angle holes, in addition to the standard vertical holes, were drilled in an attempt to include vertical fractures in the test conditions. The piezometers were constructed by boring a hole to a selected depth with a hollow-stem auger drill rig. Samples of geologic materials collected with a split spoon were used only to verify the geologic horizon. A polyvinyl chloride (PVC) glue-joint casing with a 6.4-cm inner diameter (ID) was lowered through the hollow-stem auger to the bottom of the hole. The augers were then withdrawn from the hole; the bottom of the hole was sealed with 60 to 150 cm of an expanding cement plug; and the hole was backfilled with a mixture containing 70% (by volume) clean silica sand and 30% granular bentonite to within approximately 1.2 m of the surface. Expanding cement was also used as a surface seal. Bentonite was not chosen for the plug to preclude the possibility of cracking due to the presence of organic solvents. Construction of the angle holes was similar to that of the vertical holes, except that the entire annular space was filled with cement grout pumped from near the bottom to the surface to prevent bridging. Construction details for the vertical and angle piezometers are shown in Fig. 2. These piezometers were not intended to provide water-quality samples; a separate set of wells was constructed for that purpose.

After the surface seal had set, a Shelby tube 5 cm in diameter was lowered through the

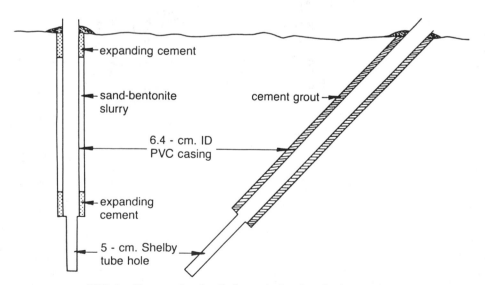

FIG. 2—*Construction details for vertical and angle piezometers.*

casing and pushed 60 cm below the bottom of the casing to collect a sample and to create the open hole for the slug test. The hole and casing were immediately filled to the top with water, initiating the slug test. The immediate introduction of water, which applied a positive pressure to the soil, kept the hole from caving in. The water levels were recorded as a function of time until they stabilized. A second set of slug tests was conducted on 8 of the piezometers after the water levels stabilized, making a total of 37 slug tests in 29 piezometers.

Cooper, Bredehoeft, and Papadopulos Method

The first method used to analyze data from the slug tests was the curve-matching method of Cooper, Bredehoeft, and Papadopulos [8,11]. Figure 3 shows the parameters to be measured in the use of this method: stabilized water level, head at time zero (H_0), head at any time ($H(t)$), radius of the casing (r_c), radius of the open hole (r_s), and length of the open hole ($z_2 - z_1$). The method assumes that the aquifer is infinite, horizontal, homogeneous, and isotropic, and that the open hole completely penetrates the aquifer to be tested. Compressibility of both the formation and the water are included in the analysis. The piezometers used in this study did not completely penetrate the unit tested, because testing more than a small interval of the hard, fine-grained materials is not practical.

Water levels in the piezometers were recorded until they stabilized. This stabilized (or initial) water level, $H(t)$ for $t = \infty$ or $t < 0$, is used as the datum from which the heads in the piezometer are measured. The ratio of the measured head at any time to the initial head, $H(t)/H_0$, is plotted with the logarithm of time (Fig. 4). This curve is then matched to one of the type curves presented by Papadopulos, Bredehoeft, and Cooper [11] (Fig. 4). From this curve match, the time (t) is selected for which $Tt/r_c^2 = 1$. T is the transmissivity of the tested material and r_c is the radius of the casing in which the head change occurs. The transmissivity was then determined from the equation

$$T = \frac{1.0 r_c^2}{t} \tag{1}$$

For the vertical holes, r_c is the actual radius of the casing; however, for the angle holes, r_c must be corrected to correspond to the radius of a round vertical casing penetrating the same vertical interval and containing the same volume of water as the angled casing. In the horizontal plane, the cross section of the angled casing forms an ellipse with a major radius of 1.414 r_c and a minor radius of r_c. A circle with the same area of this ellipse has a radius of 1.19 r_c, which has the value used for r_c for the angle-hole calculations. The length of vertical penetration of the open hole is used as the aquifer thickness (Fig. 3). These corrections for the angle holes are necessary because the casing and open hole enclose a larger volume than does a vertical hole of the same radius. Hydraulic conductivity is then calculated by dividing the transmissivity obtained in Eq 1 by the length of vertical penetration.

Of the 37 sets of data collected, 7 could not be reasonably analyzed by this method because they produced curves that did not match any of the type curves.

Hvorslev Method

Hvorslev's basic time lag method [9] is based on the assumption that inflow at any time is proportional to the hydraulic conductivity and to the unrecovered head difference. His

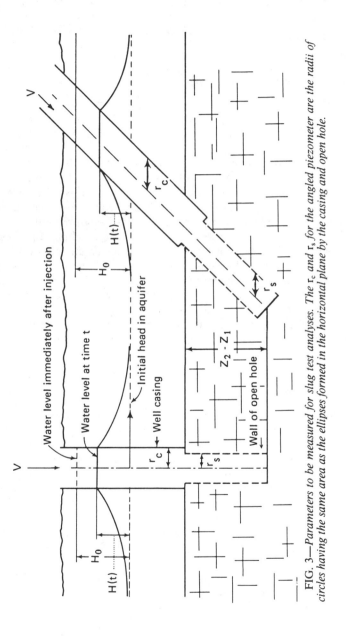

FIG. 3—*Parameters to be measured for slug test analyses. The r_c and r_s for the angled piezometer are the radii of circles having the same area as the ellipses formed in the horizontal plane by the casing and open hole.*

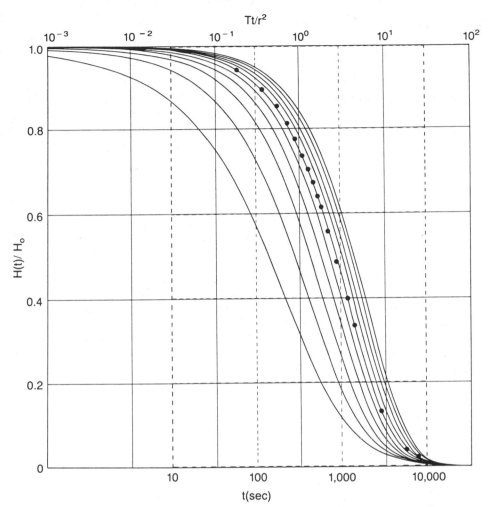

FIG. 4—*Semilog plot of field data for the head ratio versus time overlain on type curves using the Cooper, Bredehoeft, and Papadopulos method. The type curves are from Refs 8 and 11.*

assumptions, which include a homogeneous, isotropic medium and incompressibility of both formation and water, lead to the differential equation

$$q(t) = \pi r^2 - \frac{dh}{dt} = FKH(t) \qquad (2)$$

where

   q = rate of flow into the formation, cm^3/s,
   F = factor which depends on the shape and size of the intake area, dimensionless,
   $H(t)$ = head at any time, t, measured using the stabilized or initial head, H_0, as the datum, cm, and
   K = hydraulic conductivity, cm/s.

Hvorslev also defined a basic time lag, T_0, as

$$T_0 = \frac{\pi r^2}{FK} \tag{3}$$

For hydrostatic pressure changes, the time lag was defined as "the time for water to flow into or from the device until a desired degree of pressure equalization is obtained." The basic time lag, T_0, is the time that would be required for a complete pressure equalization if the original flow rate were maintained. When Eq 3 is substituted into Eq 2, and $H(t) = H_0$ is used as the initial condition, the solution to Eq 2 is

$$\frac{H(t)}{H_0} = e^{-t/T_0} \tag{4}$$

When the natural log of the left-hand side of the equation is plotted against time, as shown in Fig. 5, a straight line should result. The basic time lag is the time at which $t = T_0$ or when $H(t)/H_0 = 0.37$ that is, $\ln (H(t)/H_0) = -1$). The hydraulic conductivity is then determined from Eq 3. Hvorslev evaluated this equation for a laboratory permeameter and four geologic situations. The four field situations are shown in Fig. 6.

These slug tests were analyzed using Configurations c and d, depending on whether the piezometer was at the top or middle of a unit. A difficulty arises from using Configurations c and d in that the equation requires that the ratio of vertical to horizontal conductivity be known. For these analyses, the average ratio for the formation determined from the

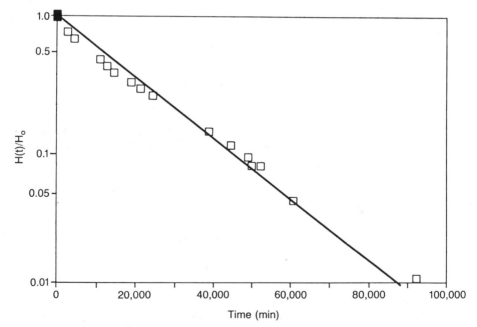

FIG. 5—*Semilog plot of field data for head ratio versus time for analysis by Hvorslev's method.*

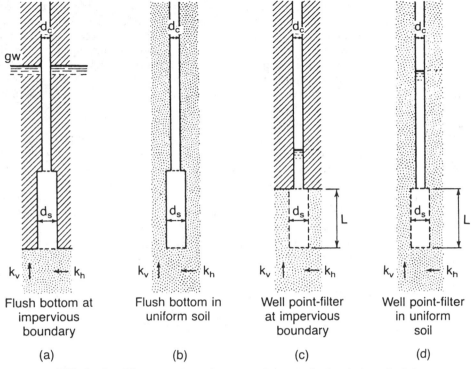

FIG. 6—*Possible geometric configurations for use with Hvorslev's method [9].*

method of Cooper, Bredehoeft, and Papadopulos was used because no other source for determining this ratio was available. Although the hydraulic conductivity values determined from the angle piezometers include both vertical and horizontal components, observation of vertical joints indicates a greater hydraulic conductivity in the vertical direction. Since the equation is not very sensitive to this ratio, this estimate should not be a significant factor. An advantage of using the geometry for Configurations c and d is that the borehole is open to more of the formation than it is in Configurations a and b. For Configurations a and b, the entire length of the borehole is cased, so water is in direct contact only at the open end of the casing.

Two of the data sets did not appear linear and, therefore, were not analyzed by this method.

Nguyen and Pinder Method

The method of Nguyen and Pinder [10] is designed to account for situations involving partially penetrating wells in aquifers in which the short-term effects of a water table or leakage from a confining layer can be ignored. According to its authors, this method appears to be most appropriate for dealing with slug tests in materials of moderate to low hydraulic conductivity and also can be used for determining storativity. The underlying assumptions (except for full aquifer penetration), boundary conditions, and parameters to be measured (Fig. 3) are the same as in the Cooper, Bredehoeft, and Papadopulos method.

The Nguyen and Pinder method is a direct calculation method that uses data obtained during the early part of the slug test. Because our slug tests were designed for analysis by the Cooper, Bredehoeft, and Papadopulos method, few early data were collected. In addition, error may be introduced by using discrete measurements obtained by steel tape or electronic water-level meter rather than a continuous record of water levels. Such measurement error may be more significant for this method than for the other two methods because small changes in water levels occur between measurements during the early time period, which is the critical time period for this method.

The method of Nguyen and Pinder [10] is illustrated in Figs. 7 and 8. Figure 7 is a log-log plot of head, $H(t)$, relative to time, t, for a typical piezometer. The slope of this line (C_1) is used in the calculation of hydraulic conductivity. When the data begin to deviate from a straight line on the log-log plot, the boundary conditions are no longer being met and the test is over. Figure 8 is a semilog plot of $1/t$ relative to $\ln(-\Delta H(t)/\Delta t)$. The earliest time data available are plotted on the time scale. For our data sets, the change in head between the first two or three measuring times was less than the measurement instrument could detect, causing the first one or two data points to vary significantly from the straight line and to be ignored. The slope of this line (C_2) was then used in the hydraulic conductivity calculation. The aquifer parameters were calculated as follows

$$K = \frac{r_c^2 C_1}{4 C_2 (z_2 - z_1)}$$

(5)

where

r_c = radius of the well casing, cm,
r_s = radius of the open hole, cm,
$z_2 - z_1$ = vertical length of the open hole, cm,
C_1 = slope from the log-log plot, cm, and
C_2 = slope from the semilog plot, s.

The relevant time span in each test was determined from the graphs, and the slopes were obtained from least-squares lines through the data. Four sets of data were not analyzed by the Nguyen and Pinder method because they did not appear linear on the semilog plot. Possibly the early data were not precise enough.

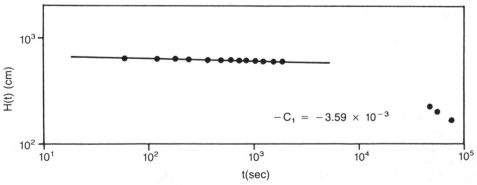

FIG. 7—*Plot of the head in the formation versus time on a log-log basis for analysis by the Nguyen and Pinder method.*

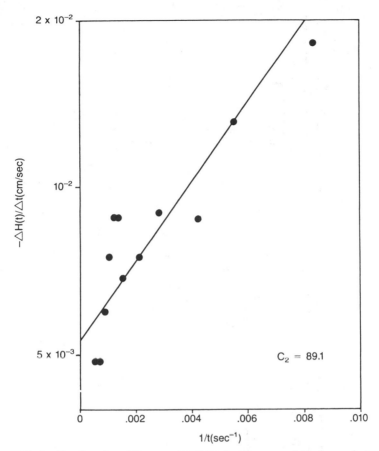

FIG. 8—*Semilog plot of l/t versus ΔH(t)/Δt for Nguyen and Pinder analysis.*

Results

Not all data could be analyzed by all three methods with the same degree of reliance. This is to be expected since each method has different underlying assumptions and uses data from different time periods. The Cooper, Bredehoeft, and Papadopulos method assumes a confined aquifer that is fully penetrated, and includes soil and water compressibility. Hvorslev's method does not require full formation penetration and ignores soil and water compressibility. The Nguyen and Pinder method allows for some leakage into the formation, partial penetration of the formation, and soil and water compressibility, but may be highly affected by piezometer construction because of the short time span of testing. However, all of our data did fit the theoretical curve of at least one method very well.

Results of these tests are summarized in Table 1. Of particular interest is a comparison of the geometric means of hydraulic conductivity values of each lithologic unit. Data are presented as the geometric means because hydraulic conductivity values in field soils are generally acknowledged to be log-normally distributed [*12,13*], so arithmetic averaging would not be appropriate [*14*]. For this computation, only one value from each boring was used. If one boring was tested twice, a geometric mean was taken of the two values, and

TABLE 1—Results of slug tests: the geometric means and log standard deviations for multiple tests with each method in each unit.

Material	Orientation	Method of Cooper, Bredehoeft, and Papadopulos			Hvorslev's Method			Method of Nguyen and Pinder		
		N	Geometric Mean, cm/s	Log Standard Deviation, cm/s	N	Geometric Mean, cm/s	Log Standard Deviation, cm/s	N	Mean, cm/s	Log Standard Deviation, cm/s
Vandalia Till, upper (stiff) ablation zone	vertical	2	1.3×10^{-7}	...	2	1.2×10^{-7}	...	1	1.2×10^{-7}	...
	45° angle	0	...	...	1	3.0×10^{-7}	...	1	4.5×10^{-7}	...
Vandalia Till, lower (soft) ablation zone	vertical	7	3.8×10^{-5}	0.89	10	1.2×10^{-5}	1.12	8	1.9×10^{-5}	0.88
	45° angle	2	5.3×10^{-5}	...	2	8.4×10^{-6}	...	2	1.7×10^{-5}	...
Weathered basal Vandalia Till	vertical	8	1.2×10^{-6}	1.39	8	8.4×10^{-7}	1.19	7	2.3×10^{-6}	1.38
	45° angle	3	6.0×10^{-6}	0.19	3	2.1×10^{-6}	0.35	4	8.2×10^{-6}	0.33
Unaltered basal Vandalia Till	vertical	5	8.4×10^{-8}	0.59	6	3.9×10^{-8}	1.13	5	6.5×10^{-8}	0.76
	45° angle	3	4.0×10^{-7}	0.68	3	5.7×10^{-7}	0.89	4	6.6×10^{-7}	0.32

that value was used to calculate the geometric mean for the unit. This procedure was followed to give equal weight to each boring.

For some formations, the data had a range of about three orders of magnitude. Because of the heterogeneity of glacial tills, this variability was not unexpected. However, the geometric means of hydraulic conductivity values calculated by the three analytical methods were within one order of magnitude and well within one standard deviation for all four lithologic units tested.

In all cases except for the soft portion of the ablation zone, the values calculated using the angle borings were greater than those from vertical borings; this difference ranged from less than a factor of two for the base of the ablation zone to more than an order of magnitude for the unaltered till (Table 1). The variability of the data was generally much less for the angle holes than for the vertical holes. These greater values suggest that the vertical structures do have a significant impact on the ground-water flow characteristics of some of the geologic materials at this site.

The importance of these results is twofold: (1) any of the three methods can produce reliable results if properly applied, and (2) data that appear to be unusable by one method may be usable by changing to another method of analysis with a different set of assumptions.

Acknowledgments

This research was carried out under Cooperative Agreement No. R810442-01 of the U.S. Environmental Protection Agency, Land Pollution Control Division, Hazardous Waste Engineering Research Laboratory, Cincinnati, Ohio. Mike H. Roulier was the project technical officer and Robert A. Griffin of the Illinois State Geological Survey was the principal investigator. Additional support was provided by SCA Chemical Services, Wilsonville, Illinois, and the Illinois Environmental Protection Agency.

Although the research described in the article has been funded in part by the U.S. Environmental Protection Agency (EPA), it has not been subjected to agency review and therefore does not necessarily reflect the views of the EPA and no official endorsement should be inferred.

References

[1] MacFarlane, D. S., Cherry, J. A., Gillham, R. W., and Sudicky, E. A., "Migration of Contaminants in Groundwater at a Landfill: A Case Study—1. Groundwater Flow and Plume Delineation," *Journal of Hydrology,* Vol. 63, 1983, pp. 1–29.

[2] Taylor, S. R., Moltyaner, G. L., Howard, K. W. F., and Killey, R. W. D., "A Comparison of Field and Laboratory Methods for Determining Contaminant Flow Parameters," *Ground Water,* Vol. 25, 1987, pp. 321–330.

[3] Olson, R. E. and Daniel, D. E., "Measurement of the Hydraulic Conductivity of Fine-Grained Soils," *Permeability and Groundwater Contaminant Transport, ASTM STP 746,* American Society for Testing and Materials, Philadelphia, 1981, pp. 18–64.

[4] Herzog, B. L. and Morse, W. J., "Hydraulic Conductivity at a Hazardous Waste Disposal Site: Comparison of Laboratory and Field-Determined Values," *Waste Management and Research,* Vol. 4, 1986, pp. 177–187.

[5] Muldoon, M. A., Attig, J. W., Bradbury, K. R., and Mickelson, D. M., "Hydraulic Conductivity of Pre-Late-Wisconsin Till Units in Central Wisconsin," *Abstracts with Programs,* Geological Society of America, Vol. 19, No. 4, 1987, p. 235.

[6] Griffin, R. A., Herzog, B. L., Johnson, T. M., Morse, W. J., Hughes, R. E., Chou, S. F. J., and Follmer, L. R., "Mechanisms of Contaminant Migration Through a Clay Barrier—Case Study, Wilsonville, Illinois," *Proceedings,* Eleventh Annual Research Symposium of the Solid and Hazardous Waste Research Division, EPA-600/9-85-013, U.S. Environmental Protection Agency, Cincinnati, OH, September 1985, pp. 27–38.

[7] Follmer, L. R., "Soil—An Uncertain Medium for Waste Disposal," *Proceedings,* Seventh Annual Madison Waste Conference, University of Wisconsin—Extension, Madison, WI, September 1984, pp. 296–311.

[8] Cooper, H. H., Bredehoeft, J. D., and Papadopulos, I. S., "Response of a Finite-Diameter Well to an Instantaneous Charge of Water," *Water Resources Research,* Vol. 3, 1967, pp. 263–269.

[9] Hvorslev, M. J., *Time Lag and Soil Permeability in Ground-Water Observation,* Bulletin 36, Waterways Experiment Station, U.S. Army Corps of Engineers, Vicksburg, MS, 1951.

[10] Nguyen, V. and Pinder, G. F., "Direct Calculation of Aquifer Parameters in Slug Test Analysis," *Groundwater Hydraulics,* J. Rosensheim and G. D. Bennet, Eds., Water Resources Monograph 9, American Geophysical Union, Washington, D.C., 1984, pp. 222–239.

[11] Papadopulos, S. S., Bredehoeft, J. D., and Cooper, H. H., "On the Analysis of Slug Test Data," *Water Resources Research,* Vol. 9, 1973, pp. 1087–1089.

[12] Rogowski, A. S., "Watershed Physics: Soil Variability Criteria," *Water Resources Research,* Vol. 9, 1972, pp. 1015–1023.

[13] Nielson, D. R., Biggar, J. W., and Ehr, K. R., "Spatial Variability in Field-Measured Soil-Water Properties," *Hilgardia,* Vol. 42, 1973, pp. 215–259.

[14] Keisling, T. C., Davidson, J. M., Weeks, D. L., and Morrison, R. D., "Precision with Which Selected Soil Parameters Can be Estimated," *Soil Science,* Vol. 124, 1977, pp. 241–248.

Steven P. Sayko,[1] *Karen L. Ekstrom,*[1] *and Rudolph M. Schuller*[1]

Methods for Evaluating Short-Term Fluctuations in Ground Water Levels

REFERENCE: Sayko, S. P., Ekstrom, K. L., and Schuller, R. M., **"Methods for Evaluating Short-Term Fluctuations in Ground Water Levels,"** *Ground Water and Vadose Zone Monitoring, ASTM STP 1053,* D. M. Nielsen and A. I. Johnson, Eds., American Society for Testing and Materials, Philadelphia, 1990, pp. 165–177.

ABSTRACT: During several investigations of ground water contamination, relatively large fluctuations in ground water levels—which occurred in a time interval shorter than that necessary to collect a round of water levels—were observed. However, because of their short-term nature, it was not possible to collect the data necessary to characterize fully the effects of these fluctuations when using standard water level measurement techniques. Without such a characterization, these short-term fluctuations in ground water levels can seriously impair interpretation of the ground water flow system.

In a case history presented, tidal and other environmental influences were large enough to make a normal round of water level data useless in interpreting ground water flow. Monitoring of water levels at the site for a period of eight days, using electronic water level recording instruments, allowed evaluation of the fluctuations and determination of an acceptable method of collecting site water levels. To evaluate ground water levels properly, multiple rounds of water levels had to be collected and averaged to filter the noise, or short-term fluctuations, from the water level data. Monitoring for these short-term fluctuations in ground water levels should be considered good engineering practice in any ground water investigation.

KEY WORDS: ground water, electronic date loggers, ground water levels, fluctuations, ground water monitoring

Most hydrogeologic investigations include the collection of water level data from wells to evaluate ground water flow conditions. Collection of a round of water level data by hand at a site may take several hours to complete. However, for data interpretation purposes, these data are considered to have been collected instantaneously. In several instances, the authors have observed fluctuations in ground water elevations at a site which were large in comparison with the total variation in the water head across the site, and these fluctuations were occurring in a time interval shorter than that necessary to collect a round of water levels. Tidal changes and intermittent ground water pumping were the most often observed influences. Because the normal method of collecting water levels is one measurement per well, these fluctuations are not immediately apparent to the investigator.

Given the conditions described above, a typical round of ground water levels cannot be validly interpreted as instantaneous data. Further, if the fluctuations are not uniform in time and space, a truly instantaneous set of water level measurements can also fail to yield good data for proper interpretation of ground water flow conditions. For the purposes of most ground water investigations, short-term water level fluctuations—due to an inter-

---

[1] Environmental Resources Management, Inc., Exton, PA 19341.

mittent pumping, tides, barometric pressure, or other environmental influences—are noise in the water level data and must be filtered out for proper interpretation of ground water flow conditions.

Discussion of influences on ground water levels—loading, barometric pressure, and tidal influences—is found widely in the literature. Good summaries of influences on ground water levels are given by McWhorter and Sunada [1] and Todd [2].

Purpose and Objectives

The intent of this paper is to present the implications of short-term fluctuations in ground water levels, methods for monitoring for fluctuations, and a method to compensate for (filter) those fluctuations. No attempt is made to explain the driving forces behind the water level fluctuations. For illustrative purposes, one site that was investigated is described. The difficulties encountered in interpreting ground water elevation data from that site, the methods used to monitor water levels, evaluation of the water level data, and the results of that evaluation are included in that description. The work described has been applied at several sites, ranging from fractured bedrock to sand and gravel aquifers, and the generalizations made in this paper are the results of that experience. The methods described are generic and should provide data useful for evaluation of the ground water flow conditions at any site. To the authors' knowledge, this investigative method is not currently applied as standard investigative practice.

Site Description

The study site is located in the Atlantic Coastal Plain Physiographic Province of southern New Jersey. It is approximately one mile south of the Delaware River. The site covers approximately 12.5 ha (31 acres) of former farmland (Fig. 1). It is bordered on the south and east by a swamp and on the north by a former borrow pit which has been excavated below the water table and is now a shallow lake. The swamp is bisected by a small stream. On the west, the site is bordered by a fallow field. One mile (1.6 km) beyond this field is an active sand and gravel pit. Approximately 3 000 000 L (800 000 gal) of ground water a day are pumped for processing operations at this pit and discharged as wastewater to the ground after use. An unknown amount of water is lost by evaporation during this process.

An active plant production well is located in the center of the site. Records indicate that the pump cycles on and off as necessary during the day to fill a 19 000-L (5000-gal) holding tank. The daily ground water production averages between 46 000 and 61 000 L (12 000 and 16 000 gal).

Site investigations to date have focused on the aquifer between the water table and a thick and regionally continuous clay stratum, approximately 46 m (150 ft) below grade (Fig. 2). The aquifer materials above the regional clay are sands and gravels with interspersed lenses of clay. The clay lenses, although discontinuous, generally separate the aquifer vertically into three definable levels: the shallow, intermediate, and deep levels. The variation in water level elevation across the site in each aquifer level is small, usually less than 0.3 m (1 ft).

There are 26 monitoring wells on, or in the immediate vicinity of, the site. Many of the wells are located in nests of two or three. All of the monitoring wells are constructed of 5-cm (2-in.)-diameter polyvinyl chloride (PVC) casing with 1.5 to 3.0-m (5 to 10-ft) lengths of screen. The plant production well is finished in the intermediate aquifer level.

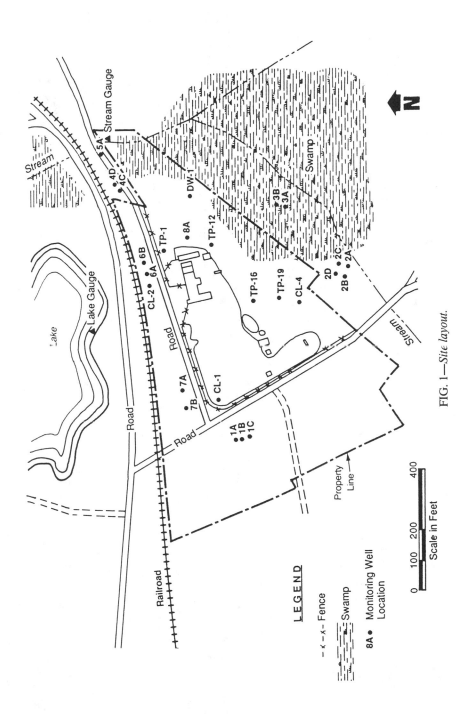

FIG. 1—*Site layout.*

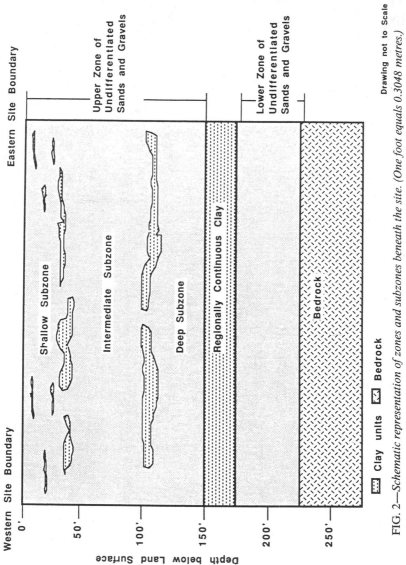

FIG. 2—Schematic representation of zones and subzones beneath the site. (One foot equals 0.3048 metres.)

Problem Description

Several rounds of ground water level data were collected during initial site investigations. The data were inconsistent in that ground water flow directions appeared to be different with each round of water level measurements collected. At least one data set completely defied any reasonable interpretation. In addition, the interpreted directions of flow were incompatible with the ground water quality isopleth maps, which were derived from repeated ground water sampling events and were consistent over time. In short, potentiometric surface maps generated from the water level data collected were of questionable validity, and the direction of ground water flow beneath the site was not known.

It was hypothesized that the influence of local aquifer stresses, such as pumping from the plant well, gravel pit, or municipal or agricultural wells, could cause fluctuations in ground water levels in less time than that necessary to collect a round of water level measurements. Such fluctuations in water levels, if of sufficient magnitude in relation to the hydraulic gradient at the site, could cause the inconsistencies observed in the data.

Instrumentation

To monitor for short-term water level fluctuations, electronic water level recording instrumentation was installed. In Situ Hermit SE-1000B electronic data loggers and In Situ water level (pressure) transducers were selected for the monitoring program. One SE-1000B can monitor two transducers. Pressure transducers [68.9 kPa (10 psi)] were installed in selected wells and monitored by the data loggers. This system is capable of resolving water level changes of 0.003 m (0.01 ft) with an accuracy of 0.5% (0.034 m (0.11 ft)) over its full range, which is 0 to 7 m (0 to 23 ft) of water. The transducers are vented to the atmosphere, so that water level fluctuations due to barometric pressure can be observed. Data from this instrument are transferred directly to an IBM personal computer for graphing and numerical evaluation.

Electronic water level recorders were chosen over Steven's continuous water level recording devices, primarily because the data loggers can transfer the collected data to a computer for analysis and graphing. In addition, in the authors' experiences, the data loggers have been more reliable, especially in small-diameter wells, where the floats of Steven's recorders often stick to the casing.

To minimize the possibility of vandalism and to protect the data loggers from wet weather, the instruments were locked into 200-L (55-gal) drums, and the transducer cables were fed out through the bung. Since the test was performed during hot summer temperatures, the drums were painted white to reduce thermal buildup.

Test 1

Monitoring Method Used in Test 1

In July 1986, the first extended water level test was conducted at the site. When planning the test, it was necessary to determine how long the test should run, how often the water levels should be measured, and what wells would provide the most useful information.

An eight-day test duration was selected so that a full work week and weekend could be observed with a day of overlap. This duration spans a typical industrial production cycle, which repeats weekly, pumping during the work week and off during nonworking hours (weekends). A sampling interval of 15 min (96 per day) was selected. This interval was small enough to resolve transient effects of the plant production well (which cycles on only for approximately 30 min every 2 h), tidal effects, or other phenomena occurring in a 15-

min or greater time frame. The data logger is capable of sampling intervals as frequent as 1 min; however, in prior experiences, this resolution was not necessary and generated an unwieldy, large data base.

The two well nests selected for water level monitoring are located on opposite sides of the site (Fig. 1). One data logger was placed at each well nest, to monitor the intermediate and deep wells in that nest. Those wells are screened in the intermediate and deep aquifer levels, which appear to be hydraulically connected to the plant production well, which was expected to be the single greatest influence on ground water levels at the site. The shallow level of the aquifer appeared to be the most hydraulically isolated from the production well. Therefore, since the intent of the investigation was to determine the maximum fluctuations, monitoring of water levels in the shallow portion of the aquifer was not considered necessary.

Results of Test 1

The data collected during the first test indicated the significance of the pumping effects from the plant production well. More surprisingly, because of the distance from the river, the data showed tidal effects. Figures 3 and 4 are graphs of data collected in the intermediate and deep levels, respectively, at one well nest. For simplicity of presentation, only the data from these two wells are presented, as they are generally representative of other wells on the site. All wells monitored during this test showed tidal influence, but only the intermediate-level wells showed the influence of the plant production well.

The effects of short-term pumping of the plant well were visible in the intermediate-level well at "spikes" superimposed upon the sinusoidal curve from tidal effects (Fig. 3). The pumping "spikes" were observed continuously during the work week, as the plant operates 24 h per day. When the plant shuts down on the weekend the spikes disappear (it is not a 100% shutdown).

In the deeper aquifer (Fig. 4), seven and one half full tidal cycles (1490 min per cycle) are evident over the duration of the test (each cycle consists of two highs and two lows). However, the effects of the plant production well are not observed. The magnitude of the tidal effects varied among the wells. Subsequent tests have demonstrated that the magnitude of the tidal effects decreases with distance from the lake.

Importance of the Results of Test 1

The maximum water level fluctuations due only to tidal effects were approximately 0.12 m (0.4 ft). This change in water level occurred in approximately 6 h. During the time required to collect a round of water level measurements, approximately 3 h, the water levels could vary as much as 0.06 m (0.2 ft) as a result of tidal effects. The water level fluctuations due to pumping by the plant production well were as large as 0.18 m (0.6 ft) in a 15-min interval. Because the water levels were contoured at 0.06-m (0.2-ft) intervals, these fluctuations are considered significant, and it is apparent that a round of water levels collected over a 3-h period could not be interpreted as an instantaneous set of data.

A natural criterion for determining whether short-term fluctuations in water levels are significant in the collection of water levels appears to be comparison with water level contour intervals. As a rule of thumb, the authors consider any fluctuation greater than one half of a contour interval to be an effect of significance, requiring consideration when collecting and evaluating water level data. Water levels on the site had been contoured at an interval of 0.06 m (0.2 ft), with a maximum change in water level elevation of less than 0.3 m (1 ft) across the site. Thus, tidal fluctuations at this site could influence the water

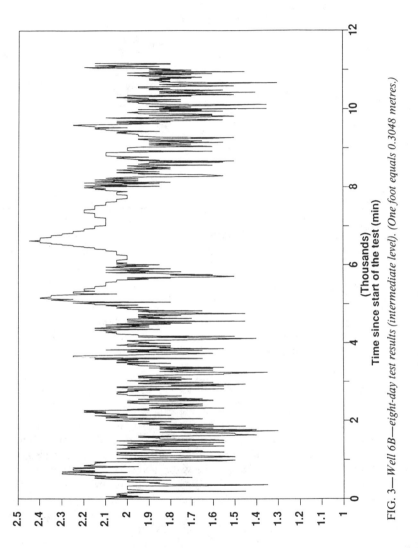

FIG. 3—*Well 6B—eight-day test results (intermediate level). (One foot equals 0.3048 metres.)*

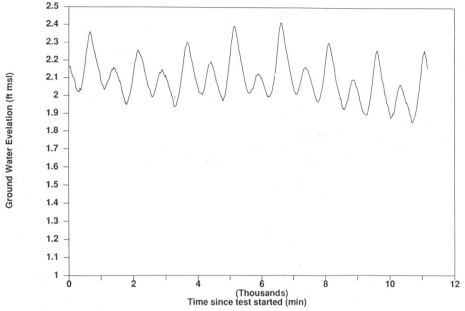

FIG. 4—*Well 6A—eight-day test results (deep level). (One foot equals 0.3048 metres.)*

levels one full contour interval over a measurement event, while pumping of the plant well changed water levels up to three contour intervals in 15 min.

Even at sites where the water level contour interval is much greater than the magnitude of short-term water level fluctuations, these fluctuations can be a concern. Rapid changes in water levels, especially those due to intermittent pumping, can interfere with data collection during a pump test.

Planning Additional Water Level Studies

Review of the data collected in Test 1 demonstrated that the water level fluctuations observed were uniform in neither time nor space. As a result, it was likely that a full round of truly instantaneous water level measurements would show different directions of ground water flow, depending on the time of day the water levels were collected. Ground water would be discharging to the lake during low tide and gaining from the lake at high tide. However, over the course of one tidal cycle, there must be either a net loss or gain of water. Definition of the net flow conditions is necessary for evaluation of contaminant migration in ground water. With that need in mind, the short-term fluctuations in water levels are noise in the data set. To evaluate net flow conditions properly, this noise must be filtered from the data.

The method selected to filter noise from the water level data was averaging of multiple measurements, collected at uniform intervals, over one full tidal cycle. The major question for planning additional investigation was, "How frequently should water levels be measured so as to provide a valid average water level?" All the wells could have been instrumented with electronic water level recorders; however, this would have been an expensive solution. The following discussions will demonstrate that electronic water level recorders are necessary only in those wells showing a transient influence of the plant production well.

In wells with tidal influence only, water levels collected by hand at intervals of less than 300 min provide an average water level with the accuracy necessary for water level contouring.

As tidal influence was the largest amplitude fluctuation in most wells, the duration of one tidal cycle (1490 min) was selected as the averaging period. To generate an unbiased average, the sampling interval should be an even interval of the tidal cycle (e.g., 5 or 10 min). However, the data in Test 1 were collected at 15-min intervals. Averaging 99 points would cover 1485 min, 100 points 1500 min. Evaluation of the error introduced by sampling at 15-min intervals demonstrated that the error introduced would be less than 0.001 m (0.0003 ft).

Evaluating the Error Associated with Sampling Frequency

Because of the large number of wells at the site it was not desirable to instrument every well with electronic data recorders. The data collected in Test 1 were evaluated to determine the frequency of water level measurements necessary to find an average water level during one tidal cycle with an acceptably small error associated with that average.

The data collected in Test 1 were loaded into an electronic spreadsheet, Lotus 123, and were analyzed. Tables 1 and 2 are excerpts from the spreadsheets created. Table 1 lists data from Fig. 3, an intermediate-level well which is influenced by intermittent pumping of the plant production well. Table 2 lists data from Fig. 4, a deep well which is not influenced by intermittent pumping of the plant production well.

The first and second columns of those tables are the elapsed test time and the water level elevation as recorded by the data logger, respectively. Columns 3 through 6 are moving averages of the water level data. Columns 7 through 9 are errors associated with the moving averages. The following discussion describes the calculated values in detail.

Average water levels values were calculated for several sampling frequencies: 15, 60, 135, and 300 min (99-point, 25-point, 11-point, and 5-point averages, respectively). The formula generated for each average was copied down each respective column in the spreadsheet. This method of generating "running" averages within the data set is referred to as moving averages. The electronic spreadsheet is perfectly suited for this task.

To determine the error associated with an average based on a limited number of data points, the true average value is needed for comparison. It was assumed that the 99-point moving average (Column 3) was the true moving average; that is, the values in column 3 would be identical to those generated if the sampling frequency had been infinitely small.

A 25-point moving average was generated, using data spaced 60 min apart (Column 4). Next, the 25-point error, the difference between the 99-point and the 25-point moving average value at a given time was calculated (Column 7). Finally, the mean and standard deviation of the error were calculated (at the bottom of the column). This was repeated for data spaced 135 min and 5 h apart (11-point and 5-point averages, respectively).

For the deep well without the influence of transient pumping, the average of the errors was less than 0.001 m (0.003 ft). The standard deviation of the error was less than 0.0015 m (0.005 ft) for all three averages (25-, 11-, and 5-point), and the largest error observed was −0.0043 m (−0.014 ft). This magnitude of error was determined to be acceptable. The criterion for an acceptable error was defined as follows: three standard deviations of the error should be less than one quarter of a contour interval, that is, 0.015 m (0.05 ft) for a 0.06-m (0.2-ft) contour interval. This criterion was more than sufficient and could be achieved without difficulty because the deep well did not show measurable effects of the intermittent pumping.

For the intermediate-level well, which was influenced by intermittent pumping, the

TABLE 1—Data for the monitoring well in the intermediate aquifer interval.

Elapsed Time (minutes)	Ground Water Elevation (ft msl)	15-min Moving Average (ft msl) 99-point	60-min Moving Average (ft msl) 25-point	135-min Moving Average (ft msl) 11-point	300-min Moving Average (ft msl) 5-point	60-min Error (ft)	135-min Error (ft)	300-min Error (ft)
10140	1.90	1.846	1.844	1.773	1.730	0.002	0.073	0.116
10155	1.95	1.846	1.800	1.886	1.720	0.046	-0.040	0.126
10170	1.95	1.846	1.906	1.941	1.940	-0.060	-0.095	-0.094
10185	2.00	1.842	1.828	1.914	1.870	0.014	-0.071	-0.028
10200	1.75	1.841	1.838	1.832	1.860	0.003	0.010	-0.019
10215	2.00	1.837	1.784	1.814	1.890	0.053	0.023	-0.053
10230	1.90	1.838	1.904	1.786	1.940	-0.066	0.052	-0.102
10245	1.75	1.837	1.830	1.755	1.910	0.007	0.082	-0.073
10260	2.00	1.830	1.814	1.782	1.880	0.016	0.048	-0.050
10275	1.90	1.828	1.776	1.732	1.850	0.052	0.096	-0.022
10290	2.05	1.825	1.894	1.895	1.970	-0.069	-0.070	-0.145
10305	2.05	1.821	1.812	1.955	1.800	0.009	-0.134	0.021
10320	1.90	1.820	1.804	1.873	1.890	0.016	-0.053	-0.070
10335	1.70	1.816	1.768	1.791	1.690	0.048	0.025	-0.126
10350	2.00	1.814	1.884	1.805	1.820	-0.070	0.009	-0.006
10365	1.50	1.813	1.812	1.777	1.780	0.001	0.036	0.033
10380	1.55	1.804	1.804	1.759	1.710	0.009	-0.054	0.103
10395	1.70	1.809	1.752	1.832	1.730	0.057	-0.023	0.079
10410	1.85	1.809	1.876	1.714	1.800	-0.067	0.095	0.009
10425	1.40	1.812	1.818	1.855	1.700	-0.006	-0.042	0.112
10440	1.85	1.812	1.810	1.905	1.680	0.002	-0.093	0.132
Average						0.000	0.001	0.001
Std Dev						0.027	0.043	0.064
Maximum						0.059	0.159	0.215
Minimum						-0.070	-0.134	-0.167

TABLE 2—*Data for the monitoring well in the deep aquifer interval.*

Elapsed Time (minutes)	Ground Water Elevation (ft msl)	15-min Moving Average (ft msl) 99-point	60-min Moving Average (ft msl) 25-point	135-min Moving Average (ft msl) 11-point	300-min Moving Average (ft msl) 5-point	60-min Error (ft)	135-min Error (ft)	300-min Error (ft)
10140	1.91	2.008	2.017	2.005	2.006	-0.009	0.003	0.002
10155	1.93	2.008	2.017	2.005	2.012	-0.009	0.002	-0.004
10170	1.95	2.007	2.015	2.007	2.011	-0.007	0.000	-0.004
10185	1.96	2.007	2.016	2.009	2.013	-0.009	-0.002	-0.006
10200	1.98	2.007	2.016	2.010	2.016	-0.009	-0.002	-0.008
10215	1.99	2.007	2.015	2.008	2.017	-0.008	-0.001	-0.010
10230	2.01	2.007	2.012	2.007	2.018	-0.004	-0.000	-0.010
10245	2.03	2.007	2.011	2.007	2.016	-0.004	-0.000	-0.008
10260	2.04	2.007	2.009	2.003	2.016	-0.004	0.004	-0.009
10275	2.05	2.007	2.005	2.005	2.013	-0.002	0.003	-0.006
10290	2.06	2.007	2.007	2.007	2.012	0.002	0.000	-0.004
10305	2.06	2.007	2.007	2.009	2.005	0.000	-0.002	0.002
10320	2.07	2.007	2.004	2.010	2.004	0.000	-0.003	0.003
10335	2.06	2.007	2.006	2.008	2.006	0.003	-0.001	0.001
10350	2.06	2.007	2.003	2.008	2.002	0.006	-0.001	0.005
10365	2.04	2.007	2.004	2.007	2.000	0.004	-0.000	0.007
10380	2.04	2.007	2.004	2.007	2.005	0.002	-0.000	0.002
10395	2.03	2.007	2.001	2.002	2.006	0.003	0.004	0.001
10410	2.03	2.007	2.002	2.003	2.006	0.005	0.003	0.000
10425	2.01	2.006	2.004	2.007	2.008	0.001	-0.002	0.001
10440		2.005						-0.003
				Average		-0.001	0.000	0.000
				Std Dev		0.004	0.001	0.005
				Maximum		0.007	0.004	0.012
				Minimum		-0.012	-0.004	-0.014

results were much different. As the number of points in the average decreased, the standard deviation of the error increased. None of the averages could meet the criterion previously defined (except the 99-point average, which was defined to be exact).

In summary, the preceding discussion demonstrates that water levels in wells not affected by the intermittent plant pumping could be measured by hand every 300 min or less (as long as the measuring interval was a uniform increment of one tidal cycle, 1490 min) and averaged with acceptable accuracy. This could easily be accomplished by several rounds of hand measurements. Wells showing effects of the plant well pumping had to be measured more frequently than at 60-min intervals. This could not be done practically by hand and required that those wells be instrumented with electronic water level recorders.

Test 2

Method Used in Test 2

The purpose of the second test was to collect multiple water level measurements from each on-site monitoring well so that average potentiometric surface maps could be constructed for each aquifer level.

This second water level test was run for 26 h. The water levels were measured in all wells, either by hand at 115-min intervals or by using an electronic water level recorder. The measurements collected at each monitoring well were arithmetically averaged and used to construct an average potentiometric surface map for each aquifer level. The water levels in the lake and the swamp were also monitored during the test.

Six monitoring wells, those closest to the plant production well, were monitored with pressure transducers. The water levels in the remaining 20 wells, the lake, and the swamp were measured by hand. As in the first test, the data loggers recorded at 15-min intervals. The task of measuring the water levels in the remaining 20 wells, the lake, and the stream was split between six persons in three shifts (two to a shift). In this manner, one round of water level measurements could be completed in approximately 1 h.

Results of Test 2

Data collected for all wells during the second test confirmed the presence of tidal fluctuations in most of the site wells. No significant tidal fluctuations were evident in the shallow-level wells. The water level variations in the 26 wells monitored ranged from 0.02 to 0.21 m (0.07 to 0.71 ft) over the 26-h period. The water level elevation variations in the lake and stream over the same period were 0.37 m and 0.25 m (1.21 ft and 0.81 ft), respectively. Tidal fluctuations in the lake and stream were approximately in phase with the fluctuations in the monitoring wells.

The potentiometric surface maps prepared for each aquifer level, using the averaged water level elevations, provided ground water flow patterns consistent with the distribution of contaminants at the site. The maps showed that the variations in water level across the site were relatively small: 0.32 m (1.06 ft) in the shallow level, 0.08 m (0.25 ft) in the intermediate level, and 0.17 m (0.54 ft) in the deep level. The short-term fluctuation in water levels observed in the 15-min measurements often exceeded the total water level variation across the site for the intermediate and deep aquifer levels. This, and the fact that the amplitude of the fluctuations varied in time and with the location, explains why the maps constructed previously from a single round of measurements were difficult to interpret and inconsistent with the established contaminant migration patterns at the site.

Conclusions

Fluctuations in ground water levels, occurring in a time period shorter than that necessary to collect one round of water levels, may invalidate the assumption that the round of water levels was collected instantaneously. However, without long-term (24-h or more) monitoring of ground water levels in one or more wells, as described herein, the investigator may be unaware of these short-term fluctuations. Routine monitoring of ground water levels with electronic water level recorders has proved to be invaluable at several sites in detecting, and subsequently compensating for, such fluctuations. The data collected with electronic recorders not only document fluctuations in ground water levels, but may also aid in associating water level fluctuations with a source or sources (e.g., ground water pumping and tides).

Summary

In the case history described, ambient fluctuations in ground water levels were large enough and rapid enough that a round of water level measurements, collected in the normal fashion, provided data that were inconsistent with the contaminant migration patterns observed at the site. However, collection of water levels at regular intervals, both electronically and by hand, allowed compensation for tidal and pumping effects and enabled an evaluation of the net ground water flow conditions at the site to be made.

Use of electronic water level recorders is recommended as a diagnostic tool in ground water investigations, particularly in those areas having low hydraulic gradients. Furthermore, electronic instruments offer the flexibility of programmable data collection intervals, extended duration monitoring, direct data transfer to the computer, and high reliability. Besides the site investigation described, data loggers have been used by the authors at several other sites. In each case, the method has detected disturbances caused by previously unknown influences. Monitoring ambient water levels can enable investigators to evaluate the ground water system better, to compensate for short-term water level fluctuations, and to plan future investigations. Monitoring for fluctuations in ground water levels during any investigation of ground water contamination should be considered good engineering practice.

References

[1] McWhorter, D. B. and Sunada, D. K., *Ground-Water Hydrogeology and Hydraulics,* Water Resources Publications, Fort Collins, CO, 1977.
[2] Todd, D. K., *Groundwater Hydrology,* 2nd ed., Wiley, New York, 1988.

H. Randy Sweet,[1] Gerritt Rosenthal,[2] and Dorothy F. Atwood[3]

Water Level Monitoring—Achievable Accuracy and Precision

REFERENCE: Sweet, H. R., Rosenthal, G., and Atwood, D. F., **"Water Level Monitoring— Achievable Accuracy and Precision,"** *Ground Water and Vadose Zone Monitoring, ASTM STP 1053,* D. M. Nielsen and A. I. Johnson, Eds., American Society for Testing and Materials, Philadelphia, 1990, pp. 178–192.

ABSTRACT: Measurement of the depth to ground water is a basic element in all hydrogeologic investigations providing data for gradient, flow direction, seepage velocity, and aquifer constant calculations.

The U.S. Environmental Protection Agency (EPA) Technical Enforcement Guidance Document (TEGD) specifies a measurement accuracy goal of ± 0.01 ft for Resource Conservation and Recovery Act (RCRA) facilities. This accuracy goal may be unrealistic, since measurements are limited by their precision, and both accuracy and precision are affected by random and systematic sources of error and uncertainty.

Random precision uncertainties include instrument sensitivities, the measuring point location, and operator technique. Random accuracy problems include short-term climatic effects (precipitation, temperature, barometric pressure) and instrument calibration. Experience has demonstrated that these accumulated uncertainties range from ± 0.02 to ± 0.20 ft.

Systematic errors are both anthropogenic and site related and include surveying accuracy, well deviation from vertical, instrument deterioration (e.g., cable stretching), and special site problems (multiphasic liquids, high gas pressures, foaming, and other problems).

These errors may increase inaccuracy or make readings highly variable. The cumulative uncertainty from both random and systematic error sources is ± 0.10 to ± 0.30 ft for a "pristine" shallow, unconfined aquifer, while for difficult installations or where anthropogenic factors are not well controlled, the accumulated error may be several feet.

This paper describes sources of error and uncertainty and reports on several practical experiments to quantify the uncertainty in water table measurements. The importance of understanding these sources in setting accuracy goals is stressed.

KEY WORDS: ground water, conductive probes, water level indicators, transducers, data loggers, measurement accuracy, precision, error

Measurement of the depth to ground water is the most basic element in all hydrogeologic investigations. These measurements provide the foundation for essentially all interpretations, calculated constants, and projections with respect to aquifer properties such as transmissivity and storativity. When coupled with a common surveyed datum, the short-term response measurements provide a basis for gradient, flow direction, and seepage velocity determination. Measurements between shallow and deep systems provide a basis for interpretation of vertical potential gradients and the definition of regional, intermediate, and local flow systems. Long-term ground-water measurement data are a key tool in regional or basin management projects.

[1] President and principal hydrogeologist, Sweet-Edwards/EMCON, Kelso, WA 98626.
[2] Manager, Environmental Services, Sweet-Edwards/EMCON, Kelso, WA 98626.
[3] Senior project hydrogeologist, EMCON Associates, San Jose, CA 95131.

Accuracy goals must be set early in the design of an investigation project, and the efficacy of these goals must be considered in light of the project precision capabilities. The U.S. Environmental Protection Agency (EPA) Technical Enforcement Guidance Document (TEGD) published in 1986, specifies a measurement accuracy of ±0.01 ft for Resource Conservation and Recovery Act (RCRA) facilities. Specifying an accuracy of measurement, without considering overall data-base needs, measurement precision, or random and systematic sources of error or uncertainty, fails to meet the first tenet of scientific experimentation. The objectives of this paper are to describe sources of error and uncertainty and to provide specific information on the limits to precision and accuracy in ground-water level measurements.

Error

When a unit is measured with the greatest exactness that the available instruments and measurement method can provide, and with the greatest care and skill an individual can exercise, the results of successive measurements will still differ among themselves. The error of a measurement is the difference between the observed or measured value and the "true value" for that measurement. Not all, and perhaps none, of the values obtained are "correct" within the limits of measurement precision. Therefore, the average value of a series of measurements is generally accepted as the most probable estimate of the "true value."

The related terms of accuracy and precision, which are sometimes used interchangeably and therefore incorrectly, must also be defined with respect to error. Accuracy is defined as the error of a measurement, or the deviation from the true value. Precision relates to the reproducibility and number of significant figures of these measurements.

Error is generally expressed as "relative error" and presented as a percentage value. The accuracy of a measurement is given by stating the relative error. A standard method for estimating the "true value" is to calculate the arithmetic mean of a large number of measurements. The magnitude of the deviations of a series of measurements from the average is a measure of the precision of the measuring instrument.

It is evident then that accuracy expresses the correctness of a measurement, and precision the reproducibility of a measurement. Accuracy without precision is obviously impossible, but precision by no means implies accuracy. Precision, and consequently accuracy, are affected by both random and systematic sources of error and uncertainty.

Sources of Error and Uncertainty

Equipment

Random precision uncertainties result from instrument sensitivities, measuring point location, and operator technique. A wide range of instruments has been used over the past one hundred years in measuring the depth to water. Frank Riley, of the U.S. Geological Survey (USGS) at Menlo Park, California [1], shared his thoughts on water level measurement technique evolution during the 36 years he has been working for the Water Resources Division of the USGS. He has noted that an industry, as well as USGS, standard for water level measurements is the chalked steel tape. Standard accuracy for the steel tape method is considered by the USGS to be ±0.01 ft. The best precision for repeatability is ±0.005 ft. The most serious problems encountered are kinking and pump line or casing wall condensate obscuring the chalk line. Precision of the steel tape is affected by both temperature and tension. Most surveying technique handbooks describe temperature and tension corrections. A 100-ft tape reportedly stretches 0.1 ft over a temperature range of 0 to 100°F. The tapes are typically scaled to be accurate at 70°F.

Electric well probes (also called conductive probes or water level indicators) are perhaps the most common instrument used in measuring the depth to water. Literature provided by several manufacturers of well sounding devices (water level indicators, well probes, electric well probes, dip meters, and M-Scopes)[4] was reviewed. The manufacturers evaluated in this review include the following: Slope Indicator Co. (SINCO), Powers Electric Products Co. (M-Scope), Fisher Division—Underground Detection Instruments, U.O.P. Johnson (Watermarker), and Actat Corp. (Olympic well probe), among others.

None of the product literature reviewed identified a specific level of precision, but they all stated that their devices would allow "accurate" water level measurements. The product literature reviewed made no mention of wire or cable stretch due either to cable weight or temperature change. One manufacturer (Fisher) referred to a tensile strength of "approximately 300 pounds" while others refer to an unspecified "high strength" or "high tensile strength" for the cable used.

Individuals at Actat and SINCO referred to "as manufactured" accuracy and cable stretch. Both manufacturers use coaxial cable with a copper-clad, steel center wire. Both feel that the solid steel wire minimizes stretch. Actat claims an accuracy of ±0.5% when manufactured, but suggests that the actual accuracy is better than ±0.1% error.

SINCO reports an error tolerance of ¼ in. in 50 ft (0.04%). They have found that their cable markings are accurate to approximately 0.1 ft in 500 ft (0.02%). The cable weight, with no estimate of thermal effects or additional weight applied, reportedly results in a stretch of 0.02%.

Operators

In order to factor the random operator error into the electric water level indicator data base, an experiment was developed which allowed 26 operators using four different instruments to make a series of measurements (N) at two monitoring wells with a combined total of $N = 62$ at Site A and $N = 28$ at Site B.

At Location A, two relatively new 100-ft Actat (Nos. 1 and 2) water level indicators were used. All measurements were taken by professional geologists with graduate degrees and experience levels ranging from <1 to >20 years. At Location B, the water level measurements were taken using a 100-ft SINCO sounder (No. 1) and a 700-ft SINCO probe and wire connected to an Olympic instrument (No. 2). The graduate geologists conducting the measurements at Location B had experience levels ranging from <1 to >5 years. A transducer/data logger setup was installed in the borehole at both sites for continuous measurement of relative water levels throughout the experiments. A statistical evaluation of the data from Locations A and B indicated that uncertainty (as measured by the standard deviation from average) varied from ±0.17 to 0.45% for each meter series, and that the site uncertainty, comparing averages for each meter, ranged from 0 to 0.63%. Comparison of manual and transducer data produced a 0.02-ft (0.21%) discrepancy in values at Site A. The transducers measured water table variations of approximately 0.2% in 100 min for Site A (see Fig. 1) and nearly 1% (0.98%) at Site B. In both cases, the calculations screened out obviously erroneous data (>10 standard deviations from the mean).

The transducer measurements were made using a 10-psi transducer (Hermit Model 1000B) with a nominal accuracy level (Hermit Environmental Data Logger Owners' Man-

---

[4] The M-Scope is a trademark of the Powers Electric Products Co.

ual, June 1986) of ±0.11 ft (±0.5% of full scale). The absolute accuracy depends on the accuracy of the setup. The nominal internal precision of these devices is listed as ±0.003 ft, but the data printout (resolution) precision is usually to a level of ±0.01 ft.

In another experiment conducted at a semi-arid site with water levels over 400 ft below the surface, Atwood and Lamb [2] reported a precision of ±0.05 ft when two operators were using the same instrument at the same location within a short time period. A comparison of electric sounder measurements with those taken by a downhole transducer showed an overall uncertainty of ±0.10 ft, which is generally attributed to sounder sensitivity and operator error.

Float-activated, continuous recorders have been in use for some time. According to Frank Riley (USGS) [1], the Stevens Type F recorder is the most accurate and simplest to use for continuous water level measurements. He reports that the overall accuracy achievable with this type of recorder is ±0.001 ft. However, depth, borehole deviation, well diameter, and well access difficulties limit the usefulness of this device. Riley has noted that he has had good results in wells of 4-in. (and greater) diameter to depths of 200 to 300 ft. Special designs have been developed to use these recorders in wells of smaller diameter [3].

Jacob [4] was interested in the precision of a Stevens Type-F recorder while using different gage-to-height ratios. His initially recorded data employed a gage-to-height ratio of 1 to 1. For his more detailed studies, he modified the recorder to expand the time and water level scales and obtained more refined readings at a gage-to-height ratio of 5.1 to 1. This did, however, require a smaller float, which was observed to dampen the readings (i.e., the float lag was inversely proportional to the diameter of the float wheel). This effectively reduced the average recorded fluctuation from 0.03 to 0.016 ft. This information is emphasized in order to point out that when the causes of data fluctuation are identified, it may be possible to correct for the apparent error. Sources of type-F recorder error are discussed in more detail in the Leupold and Stevens data book [5].

Transducers

Currently, transducers are the instrument of choice for most continuous water level measurement. Riley [1] noted that he used transducers 30 years ago when battery life was a significant limitation on the usefulness of transducers. Modern electronic circuitry, better power sources, and the coupling of data loggers to computer systems for easy data reduction have greatly enhanced the value of these instruments. One shortcoming of transducer units is that they measure relative pressure or depth. The absolute or initial depth-to-water measurement is generally taken with a conductive probe or steel tape, thus limiting the absolute accuracy of the transducer/data logger to that of the pre-start calibration in most applications. Transducer types include the diaphragm transducer plus newer designs made up of bonded strain gage, vibrating wire, or vibrating crystal units.

When using transducers/data loggers, the precision of both units must be considered. The *Ground Water Monitoring Review* for summer 1987 [6] lists the types and manufacturers' claims of resolution or precision for a variety of transducers. This information has been corroborated by interviews with manufacturers' representatives. Achievable precision is dependent upon the scale required and the cost of the unit, and one should note that most manufacturers of transducers assume that the transducers are in an isothermal environment. The precision of transducers ranges from 0.25 to 0.10% of the full scale (FS). Transducers are commonly available in the 2 to 50-psi range. At 2.31 ft/psi, the FS range is 4.6 to 116 ft for the aforementioned transducers, with precision ranges from 0.01 to 0.005 ft for the 2-psi transducer and 0.29 to 0.12 ft for the 50-psi transducer.

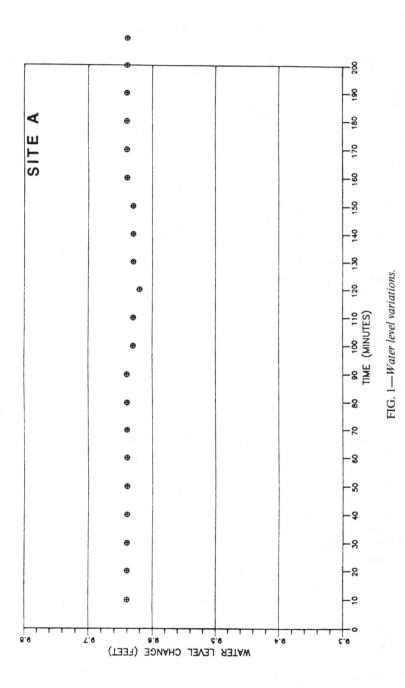

FIG. 1—*Water level variations.*

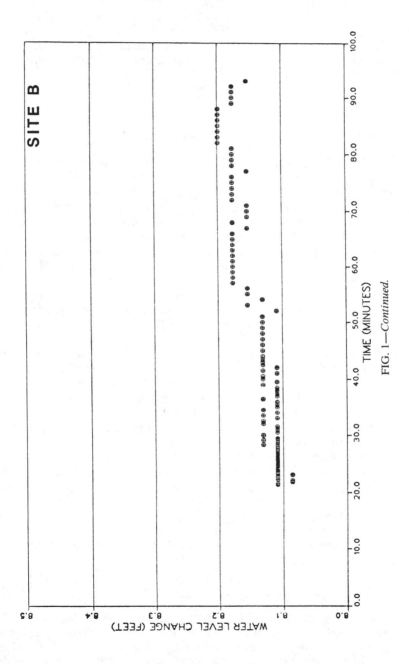

FIG. 1—*Continued.*

Transducers are commonly coupled to data logger units. If not properly paired, the resolution or readability of the data logger can be a limiting factor in the precision of the transducer/data logger system. As an example, the Terra 8 (Northwest Instrumentation) is capable of reading 1 unit/1024. For a 2-psi transducer with a 4.6-ft FS, the data logger precision is 0.004 ft (i.e., 4.6 ft/1024). Since the precision range for the 2-psi transducer is 0.01 to 0.005 ft, the data logger is adequate for this application. The Fig. 1 transducer plot for Site B illustrates the scale steps for a 10-psi transducer with 0.10% precision, or a plotting resolution of 0.02 ft.

Water Level Fluctuations—Natural

Another source of random accuracy problems is short-term water level fluctuation. As early as 1906, Veatch [7] compiled a comprehensive listing of water level fluctuation sources. In his listing he identified the following general categories: fluctuations due to natural causes and those due to "human agencies." Natural causes include rainfall, evaporation, barometric and temperature changes, rivers, ocean level changes (tides), and direct tides from solar or lunar attraction. Human causes included settlement, deforestation, cultivation, irrigation, dams, water supply developments, cities, and freight trains. He noted that many of the causes of fluctuations, such as precipitation, rivers, and tidal changes, are not due to an actual change in ground water storage but to a change induced by plastic deformation caused by the loading and unloading of the aquifer. Ferris et al. [8] expanded the list of causes and summarized more recent work in explaining effects. Freeze and Cherry [9] have also tabulated and discussed mechanisms that lead to fluctuations in ground water levels. As an example of natural fluctuations, in the cases of the two test experiments shown on Fig. 1, the water levels varied by 0.03 ft in 1000 min (0.3%) at Site A and 0.081 ft (0.99%) in 80 min at Site B.

A case can be made for categorization of barometric or diurnal water level fluctuations as random in short-term, or systematic in longer term, experiments. Making corrections for barometric pressure changes during long-term pump tests has been a standard practice since the early 1940s [10]. In order to emphasize the time dependency as well as the magnitude of natural water level fluctuations in relatively deep aquifers in a semi-arid region, Atwood and Lamb [2] placed downhole transducers in two wells, A and B, completed at different water-bearing zones for a two-month period. Two superimposed water level fluctuation patterns were observed in both wells (See Figs. 2 and 3). The first pattern observed is cyclic and diurnal with average fluctuations of ±0.20 ft. The second pattern has an irregular cycle which spans two to seven days with fluctuations averaging ±0.1 ft. However, several larger fluctuations exceeded 0.5 ft, with a maximum of 0.8 ft. These large changes in water levels were interpreted as being caused by barometric response to passing weather fronts. These responses are temporally synchronized (Figs. 2 and 3) in the wells, but the magnitude of the fluctuation is different in each well. Figure 4 illustrates natural and diurnal fluctuation interpretations.

Water Level Fluctuations—Anthropogenic

Random anthropogenic fluctuations of water levels are also well documented. In 1939, Jacob [4] wrote a now-classic paper on the water level changes induced by a passing train. The magnitude of fluctuations that he observed ranged from 0.024 to 0.035 ft and averaged 0.030 ft. The differing response depended primarily on the weight and velocity of the train, with increased weight and velocity increasing the magnitude of the water level change.

During site characterization and testing of a contaminant capture system at a RCRA site in western Oregon, an analogous effect was noted in a shallow semiconfined aquifer. At this site, transducer/data logger sets were installed in seven monitoring wells and piezometers surrounding a contaminant capture well. Figure 5 shows the drawdown response in one observation well over a 15-day period. What initially appeared to be a random fluctuation in the water levels recorded by the transducer/data logger sets was later interpreted to be a systematic change, apparently driven by diurnal parking lot loading and unloading over the shallow but partially confined aquifer. The tightly packed data set, therefore, allowed an interpretation and correction of the long-term measured and projected drawdown.

Special Problems

At highly contaminated facilities, special problems arise when attempting to measure water levels. Multiphasic liquids (e.g., floating oil in irrigation wells from deep turbine pumps) have long posed a problem in measuring true water levels. It is not uncommon to encounter several feet of accumulated oil in such a system. Light nonaqueous-phase liquids, such as those from petroleum tank leaks, have become a common problem in many underground storage tank (UST) investigations. Experience has shown that the use of petroleum paste on a steel tape can provide a measurement accuracy of ± 0.10 ft. Clear acrylic bailers used in the measurement and collection of floating products can measure from a film or sheen of 0.001 ft (in shallow floating plumes) to ± 0.10 ft (in thicker plume layers).

Another special case encountered in landfill monitoring wells is the interference resulting from foaming leachate and methane gas pressures. Water and leachate levels, as well as foam generation in highly contaminated wells, are dependent upon the rate of gas generation, gas pressures built up within the waste, the barometric pressure confining that gas at any given point in time, or a combination of these factors. Our experience has shown that the measurements within these landfill wells can be virtually impossible to obtain with either conductive or transducer devices and may "naturally" vary as much as 10 ft in a matter of hours.

Systematic Errors—Well Problems

Systematic errors are primarily anthropogenic and include surveying accuracy, well deviations, instrument deterioration (e.g., cable stretching), and poor calibration. Wellhead surveying problems have been discussed by Atwood and Lamb [2], with a reported accuracy resolution of 0.01 ft. As discussed earlier, cable stretch data are not generally available but may be on the order of 0.02 to 0.10%. The calibration and care of the instruments are operator dependent and very difficult to quantify.

Well deviation, especially in deeper monitoring systems can be a major concern. At a semi-arid site in California's Central Valley, this was studied in some detail. The site is underlain by dipping water-bearing sandstone units which are confined between low permeability claystones. The depth to ground water is typically greater than 400 ft from the surface. The hydraulic gradient in the water-bearing sandstones is very small, averaging less than 1 ft over 1000 ft. With such a shallow gradient, it was imperative that the highest possible accuracy be maintained in order to define ground water movement. Natural water level fluctuations were evaluated at the site and have been discussed earlier. Well deviation from the vertical was suspected as a major element of inaccuracy. In a well which is not exactly vertical, the actual water depths are shorter than those measured with a conductance probe or other measuring instrument. Minor well deviation occurs in most site wells

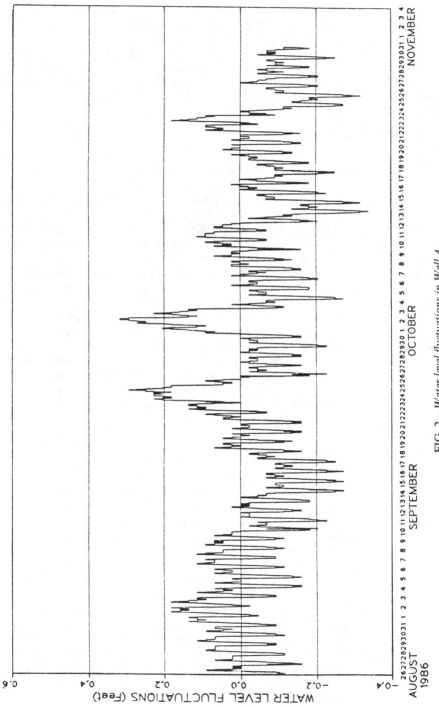

FIG. 2—*Water level fluctuations in Well A.*

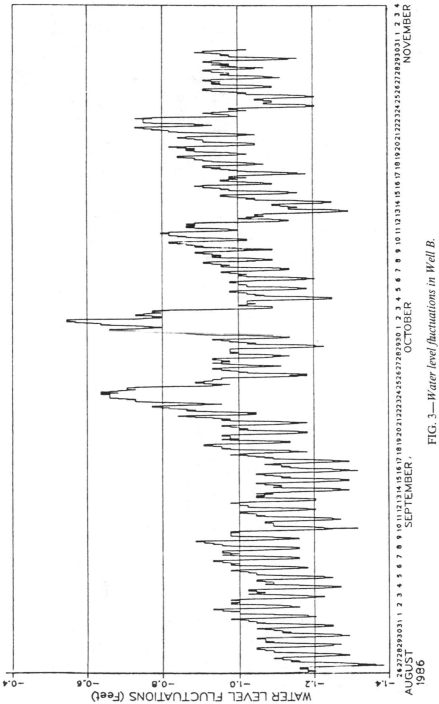

FIG. 3—*Water level fluctuations in Well B.*

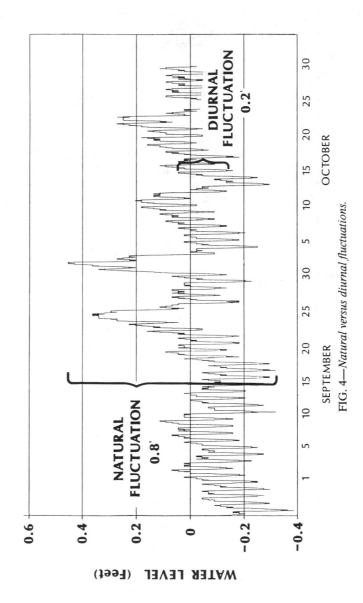

FIG. 4—*Natural versus diurnal fluctuations.*

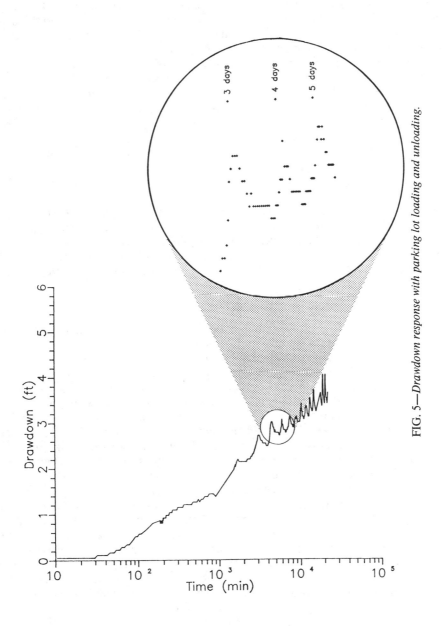

FIG. 5—*Drawdown response with parking lot loading and unloading.*

because of the great depth of the wells and the dipping strata, which cause the boreholes to generally drift into the dipping beds.

Over the length of these wells, the error introduced by deviation can be significant. Fortunately, this type of error can be corrected by measuring actual deviation with one of several available downhole instruments. The borehole drift tool was used most frequently during this site investigation. This downhole probe contains a transducer for sensing degree and direction of inclination. The magnetic-multishot survey is another technique used for deviation determination and consists of a 1.75-in.-diameter probe which records a compass and inclinometer reading on a film as the probe is lowered down the hole. A gyroscopic survey, the third type utilized during the site investigation, was employed in wells with steel casing materials. This 1.75-in.-diameter probe has a spin motor which is aligned to a surface reference of known orientation. Readings are then recorded on film as the probe is lowered down the hole.

Deviation from true vertical increases with depth. In this particular site investigation, the vertical correction at the piezometric surface ranged from −0.01 to −5.97 ft. If uncorrected, this systematic error could completely obscure the true gradients between adjacent wells.

A deviation survey corrects for a well's variation from the vertical. However, the probes and instruments used to measure the well deviation have a limited precision. The survey companies have reported the precision of the equipment and survey technique to be ±0.30 ft at this site. Even when corrected, the precision of these instruments creates a systematic precision error that limits overall water level measurement accuracy.

Cumulative Error

Random errors are generally not cumulative, but can lead to large uncertainties. Systematic errors may or may not accumulate to produce even greater inaccuracies. An example of opposing systematic errors is the combined effects of uncorrected well deviation and measurement device stretching. Cumulative errors may combine anthropogenic factors, weather effects, and slow temporal variations. An important principle is that while inaccuracies may cancel, they are often not predictable, while uncertainties are generally additive. That is, the uncertainty in any set of calculations will be larger than the poorest measurement unless the number of measurements increases to compensate. This is particularly a problem when an aquifer parameter is calculated from a set of variables or parameters, each of which, like a depth to water measurement, is subject to its own characteristics of inaccuracy and imprecision. Calculations using precise values for some factors (e.g., gradients) will be limited by the imprecisions of other factors.

For calculated values using additive or subtractive values, the uncertainties are approximately additive, while for multiplicative factors, the percentage errors are roughly additive. As an example, in situations involving very shallow gradients, additive uncertainties in water levels may lead to percentage uncertainties in the gradient of a similar magnitude (e.g., ±50%) to hydraulic conductivity uncertainty. In such cases, the 0.1 to 0.25% uncertainties or inaccuracies (0.1 to 0.25 ft/100 ft), which are nearly impossible to avoid in water level measurement, may result in ±100% uncertainties in calculated flow velocities.

Conclusions

Table 1 summarizes instrument precision data from various manufacturers' literature. These range from ±0.005 ft/100 ft (0.005%) to ±0.25 ft/100 ft (0.25%). Conductive probe data from shallow well measurements (Sources 3 and 4 in Table 1) represent true mea-

TABLE 1—*Water level measurement—relative precision data.*

Method	Source[a]	Precision, %
Steel tape	(1)	0.005
Conductive probe	(2)	0.04
	(3)	0.21
	(4)	0.34
	(5)	0.01
Transducer	(6)	0.25
	(7)	0.10
	(8)	0.25
	(9)	0.10
	(10)	0.01

[a] Sources:

(1) U.S. Geological Survey (Frank Riley, personal communication, 1987).
(2) Manufacturers' literature and interviews.
(3) Shallow water table experiment, Location A.
(4) Shallow water table experiment, Location B.
(5) Atwood and Lamb (1987), deep wells.
(6) 2 psi at 0.25%.
(7) 2 psi at 0.10%.
(8) 50 psi at 0.25%.
(9) 50 psi at 0.10%.
(10) 1 unit/1024 as per the Terra 8 specification.

surement precision and include operator plus instrument factors, with uncertainties of ± 0.21 to $\pm 0.34\%$. Source 5 in Table 1 is for a deep well experiment and includes operator plus instrument factors, at $\pm 0.01\%$ uncertainty. The percentage error decreases with depth since the operator precision is independent of depth. This implies a maximum precision of ± 0.02 ft attributable largely to operator error.

Accuracy goals must consider the limiting instrument and measurement precision as well as random and systematic sources of inaccuracy. As a rule of thumb, with few data points, accuracy is one order of magnitude (i.e., one decimal point to the left) lower than the limiting precision. As the measurement data population increases, accuracy limits approach precision limits.

This is not to imply that an infinite data base is necessary for all scientific investigations or projects. It does point out that the size of the data base should be designed in consideration of instrument and operation precision as they relate to a desired accuracy goal. As Cedergren [11] once said,"An approximate solution to the correct problem is far better than a refined solution to the wrong problem."

Acknowledgments

Although the mention of trade names and products does not constitute an endorsement, the authors are grateful to the various manufacturers' representatives for their information, cooperation, and candid comments. Staff in the various offices are thanked for assisting the experiments described herein.

References

[1] Frank Riley, U.S. Geological Survey, Menlo Park, CA, personal communication, 1987.
[2] Atwood, D. F. and Lamb, B., "Resolution Problems with Obtaining Accurate Ground Water

Elevation Measurements in a Hydrogeologic Site Investigation," *Proceedings,* First National Outdoor Action Conference on Aquifer Restoration, Ground Water Monitoring, and Geophysical Methods, National Water Well Association, May 1987, pp. 185–193.

[*3*] Shuter, E. and Johnson, A. I., "Evaluation of Equipment for Measurement of Water Levels in Wells of Small Diameter," *U.S. Geological Survey Circular,* No. 453, 1961.

[*4*] Jacob, C. E., "Fluctuations in Artesian Pressure Produced by Passing Railroad Trains, as Shown in a Well on Long Island, New York," *American Geophysical Union Transactions,* Part IV, 1939, pp. 666–674.

[*5*] *Stevens Water Resources Data Book,* 2nd ed., Leupold and Stevens, Portland, OR, 1978.

[*6*] *Ground Water Monitoring Review,* Vol. 6, No. 3, Summer 1987.

[*7*] Veatch, A. C., "Fluctuations of the Water Level in Wells with Special Reference to Long Island, New York," Water Supply Paper 155, U.S. Geological Survey, Washington, DC, 1906.

[*8*] Ferris, J. G., Knowles, D. B., Brown, R. H., and Stallman, R. W., "Theory of Aquifer Tests," Water Supply Paper 1536-E, U.S. Geological Survey, Denver, CO, 1962.

[*9*] Freeze, R. A. and Cherry, J. A., *Groundwater,* Prentice-Hall, Englewood Cliffs, NJ, 1979.

[*10*] Jacob, C. E., *American Geophysical Union,* Vol. 21, 1940, pp. 574–586.

[*11*] Cedergren, H. R., "Seepage, Drainage and Flow Nets," Wiley, New York, 1968.

Roger J. Henning[1]

Presentation of Water Level Data

REFERENCE: Henning, R. J., **"Presentation of Water Level Data,"** *Ground Water and Vadose Zone Monitoring, ASTM STP 1053,* D. M. Nielsen and A. I. Johnson, Eds., American Society for Testing and Materials, Philadelphia, 1990, pp. 193–209.

ABSTRACT: Most common methods used to present water-level data are graphical. These include various kinds of hydrographs, such as those showing water level versus time, drawdown versus time or distance, or the level in one well versus that in another. Contour plots in map view are essential for evaluating ground-water flow. Water levels on cross sections are useful in evaluating the relationship between stratigraphic units and water levels. Flow nets can be used in plan or cross-sectional view to estimate flow lines and possible pathways for seepage.

KEY WORDS: ground water, water level, data presentation, graphical presentation, tabular presentation

It is not enough for an investigator to collect water-level data. Once measurements have been made, they must be organized, evaluated, and interpreted [1]. As with many environmental measurements, those investigating frequently do not have a sampling framework for the populations under study. Some basic assumptions can be made, based on physical measurements of the system, but one is never sure that he understands fully the system he is attempting to quantify. For example, it is often believed that the ground surface is the hypothetical top of the ground water; however, numerous flowing artesian wells have pressure levels above the land surface regime even though the top of the physical water is below the ground at the aquitard. The water that rises in a cased well is the expression of the pressure of the aquifer.

The three broad objectives for gathering engineering data are to discover (*a*) the physical constants and frequency distributions, (*b*) the relationships—both functional and statistical—between two or more variables, and (*c*) the causes of observed phenomena [2]. Most techniques for collecting water-level data depend on three assumptions. First, we assume that water levels at a particular point are continuous in space. Inherent in this first assumption is the assumption that the point monitored represents actual conditions and bears a relationship to the water levels surrounding it in three dimensions. Second, we assume that the water level at a particular time is autocorrelated to previous water-level measurements; that is, some information is repeated or carried over from one observation to the next. Third, we assume that water-level populations have a statistical distribution and that we can sample representative values from a population distribution.

This paper will concentrate on the common presentation forms often used in the general literature. Because English units are still the standard in the United States, all data in the illustrations are presented exactly as originally collected.[2] Numerous statistical and prob-

---

[1] Chief hydrogeologist, Woodward-Clyde Consultants, Wayne, NJ 07470.
[2] Note that 1 ft = 0.3048 m, and 1 in. = 25.4 mm.

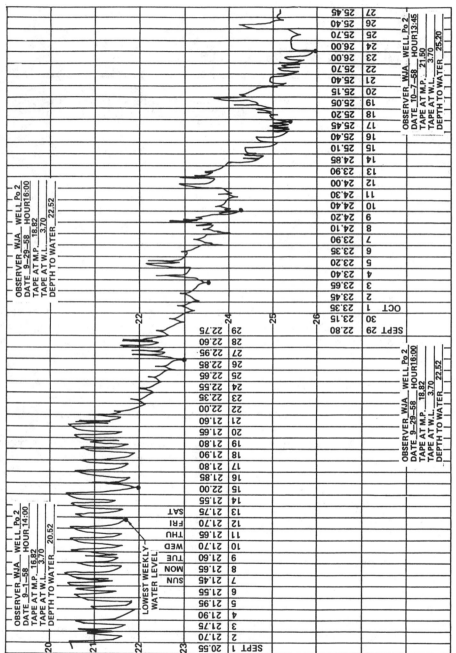

FIG. 1a—Monthly continuous recorder charts [3]. The measurements are in feet (1 ft = 0.3048 m).

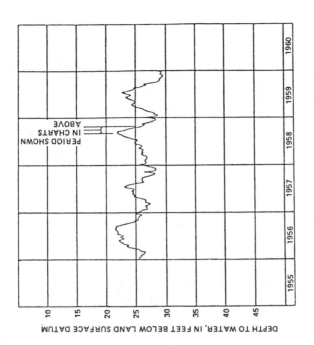

FIG. 1c—Portion of the ten-year-by-month hydrograph (the monthly lines are omitted) [3] (1 ft = 0.3048 m).

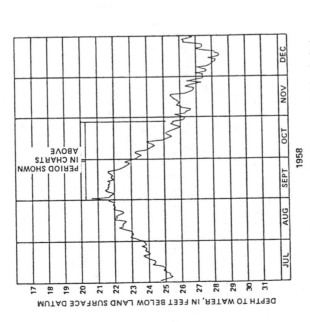

FIG. 1b—Portion of the year-by-day hydrograph (the daily lines are omitted) [3] (1 ft = 0.3048 m).

HUNTERDON COUNTY

402644074563601. Local I.D., Bird Obs. NJ-WRD Well Number, 19-0002.
LOCATION: Lat 40°26'44", long 74°56'36", Hydrologic Unit 02040105, at U.S. Post Office, Sergeantsville.
   Owner: Phillip Fleming,
AQUIFER: Stockton Formation of Triassic age.
WELL CHARACTERISTICS: Dug water-table observation well, diameter 3 ft. (0.9 m), depth 21 ft. (6.4 m), lined with stone.
INSTRUMENTATION: Water level recorder.
DATUM: Land-surface datum is 342.00 ft. (104.242 m) above National Geodetic Vertical Datum of 1929.
   Measuring point: top edge of recorder shelf, 1.50 ft. (0.460 m) above land-surface datum.
PERIOD OF RECORD: June 1965 to July 1970, May 1977 to current year. Periodic manual mesurements, September 1970 to September 1976. Records for 1965 to 1976 are unpublished and are available in files of New Jersey District Office.
EXTREMES FOR PERIOD OF RECORD: Highest water level, 6.91 ft. (2.106 m) below land-surface datum, March 28-29, 1978 and April 2, 1980; lowest, 17.04 ft. (5.194 m) below land-surface datum, January 26-28, 1981.
EXTREMES FOR CURRENT YEAR: Highest water level, 10.11 ft. (3.082 m) below land-surface datum, January 9; lowest, 16.18 ft. (4.932 m) below land-surface datum, October 24.

WATER LEVEL, IN FEET BELOW LAND SURFACE DATUM,
WATER YEAR OCTOBER 1981 TO SEPTEMBER 1982
MEAN VALUES

DAY	OCT	NOV	DEC	JAN	FEB	MAR	APR	MAY	JUN	JUL	AUG	SEP
5	15.65	15.08	14.36	-	-	13.43	10.90	11.80	13.95	14.25	14.69	15.19
10	15.63	15.30	14.22	10.59	-	11.98	10.86	13.07	14.70	14.62	15.24	15.34
15	15.91	15.70	-	12.17	-	12.32	10.55	13.93	12.64	14.43	14.96	15.72
20	16.14	15.51	-	13.28	-	12.58	11.13	14.37	12.47	15.04	15.38	15.90
25	16.01	15.56	-	13.84	12.46	13.17	12.17	14.35	13.72	15.29	15.61	15.48
EOM	14.67	15.89	-	-	12.86	13.85	10.75	13.85	14.49	13.89	15.60	15.03
MEAN	15.72	15.42	-	12.82	-	12.88	11.38	13.41	13.71	14.70	15.15	15.52

WTR YR 1982   MEAN 14.10   HIGH 10.23 JAN 9   LOW 16.17 OCT 23 AND OTHERS

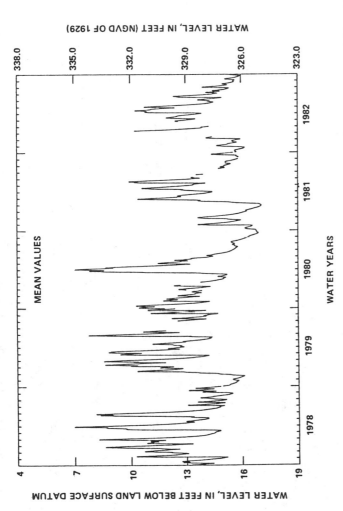

FIG. 2a—*Water level data for WRD Well No. 19—0002, Hunterdon County, NJ [4] (1 ft = 0.3048 m; 1 in. = 25.4 mm).*

OCEAN COUNTY

395714C74223401. Local I.D., Crammer Obs. NJ-WRD Well Number, 29-0486.
LOCATION: Lat 39°57'14", long 74°23'34", Hydrologic Unit 02040301, about 800 ft. (244 m) east of Central Railroad of New Jersey, Whiting.
Owner: Frank Reynolds.
AQUIFER: Kirkwood-Cohansey aquifer system of Miocene age.
WELL CHARACTERISTICS: Water-table observation well, diameter 8 in. (203 mm), depth 69 ft. (21.0 m), slotted steel casing, gravel packed.
INSTRUMENTATION: Water-level recorder.
DATUM: Land-surface datum is 179.00 ft. (54.559 m) above National Geodetic Vertical Datum of 1929.
  Measuring point: top of 8-in. coupling, 090 ft. (0.270 m) above land-surface datum.
REMARKS: Originally a dug well which casing was inserted on March 31, 1966, and the well deepened from 60 to 69 ft. (18.3 to 21.0 m).
PERIOD OF RECORD: May 1952 to current year.
EXTREMES FOR PERIOD OF RECORD: Highest water level, 47.80 ft. (14.569 m) below land-surface datum, June 9-14, 20-29, 1973; lowest, 57.58 ft. (17.550 m) below land surface datum, June 1, 1966; well was dry, November 1957 to February 1958, and December 1965, before deepening.
EXTREMES FOR CURRENT YEAR: Highest water level, 56.40 ft. (17.191 m) below land-surface datum, July 13-17; lowest, 57.40 ft. (17.496 m) below land-surface datum, January 2-9.

WATER LEVEL, IN FEET BELOW LAND SURFACE DATUM,
WATER YEAR OCTOBER 1981 TO SEPTEMBER 1982
MEAN VALUES

DAY	OCT	NOV	DEC	JAN	FEB	MAR	APR	MAY	JUN	JUL	AUG	SEP
5	57.06	57.24	57.31	57.40	57.28	57.04	56.82	56.76	56.70	56.45	56.50	56.70
10	57.09	57.25	57.34	57.39	57.24	57.00	56.81	56.76	56.66	56.41	56.52	56.73
15	57.11	57.26	57.35	57.38	57.19	56.96	56.78	56.77	56.61	56.40	56.54	56.73
20	57.15	57.27	57.37	57.36	57.15	56.91	56.76	56.79	56.56	56.41	56.58	56.79
25	57.19	57.29	57.38	57.34	57.11	56.88	56.76	56.81	56.51	56.46	56.62	56.80
EOM	57.23	57.31	57.39	57.31	57.09	56.85	56.77	56.75	56.48	56.50	56.67	56.87
MEAN	57.13	57.26	57.35	57.37	57.20	56.95	56.79	56.78	56.60	56.44	56.56	56.76

WTR YR 1962   MEAN 56.93   HIGH 56.40 JUL 14 AND OTHERS   LOW 57.40 JAN 2 AND OTHERS

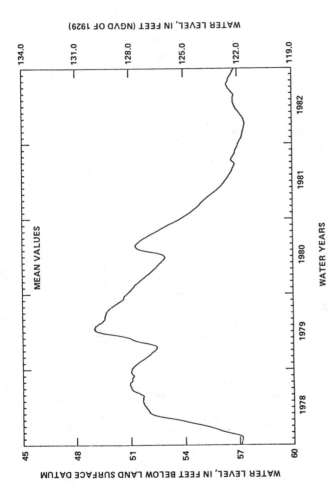

FIG. 2b—*Water level data for WRD Well No. 29—0486, Ocean County, NJ [4] (1 ft = 0.3048 m; 1 in. = 25.4 mm).*

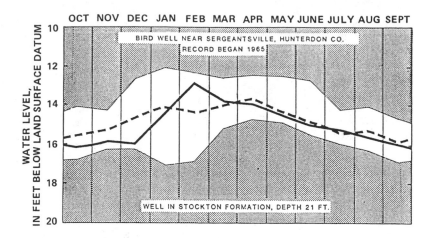

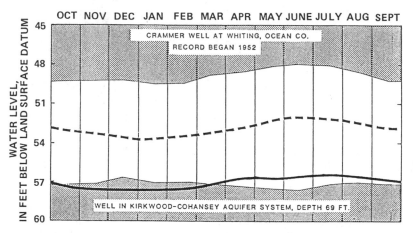

UNSHADED AREA —— INDICATES RANGE BETWEEN HIGHEST AND
LOWEST RECORDED MONTHLY MINIMUM WATER
LEVELS, PRIOR TO THE CURRENT YEAR.

DASHED LINE —— INDICATES AVERAGE OF THE MONTHLY MINIMUM
WATER LEVELS, PRIOR TO CURRENT YEAR.

SOLID LINE —— INDICATES MONTHLY MINIMUM WATER LEVEL FOR
THE CURRENT YEAR.

FIG. 2c—*Monthly ground water levels at the two key observation wells in Figs. 2a and 2b*
[4] *(1 ft = 0.3048 m).*

ability graphical presentations have been used in the literature to interpret and present
ground-water data, but these are beyond the scope of this paper.

We know from continuous water-level recorders that water levels are a continuous function in almost all natural aquifers and are controlled by boundaries or imposed stresses or
both. Numerous studies and thousands of water-level records indicate that ground-water
levels are probably normally distributed, especially in a dynamically stable natural system.

If the aquifer experiences stresses that cause boundary conditions to exert a control, then a bimodal distribution is likely to be produced. This is a result of the combination of the normal distribution and a random distribution reflecting the stress and boundary effects.

Accuracy and precision of measurements are important issues, but these are not discussed in this paper. The data presentation here assumes that the representative field data have been carefully collected and recorded in a consistent manner, converted, if necessary, to a standard engineering unit, and checked for accuracy and precision. The data set is then ready for processing into a working or final presentation form.

Hydrographs

Most of the common methods used to present water-level data are graphical. These include various kinds of hydrographs. Hydrographs are x-y plots of water-related data shown in relation to some other measurable continuous variable, such as time or distance. These graphs commonly depict the water level versus time, drawdown versus time or distance, or some calculated value, such as elevation above mean sea level (or some other convenient datum point). These simple plots allow the analyst to compare information visually. The graph can easily be examined to determine the minimum and maximum values and trends. Often, a trend is complex and may be easier to evaluate visually than to analyze mathematically.

The most common hydrograph is a continuous water-level recorder chart, illustrated in Fig. 1a. This figure illustrates copies of two successive monthly water-level recorder charts [3]. The initial quality assurance checks and manual digitization of the daily average water level, reported as a depth below a known reference point, have been completed. From these digitized data, we can show the average daily water level, illustrated in Fig. 1b. Filtering the data set again to remove monthly perturbations produces a long-term plot covering many years (Fig. 1c).

Combination presentations, such as those found in the U.S. Geological Survey (USGS) Water Resources Division (WRD) reports, present both text, tabular, and graphical information on the stations. Figure 2a illustrates the record of a dug well located in Hunterdon County, New Jersey, which responds quickly to recharge and probably to local discharge to streams [4]. The tabular results presented, although potentially more accurate, do not give the analyst the ability to infer time relationships. It is readily apparent from the graphical presentation that this well exhibits wide variations with time. It would probably not be appropriate to measure the water level monthly and expect the results to represent the true population.

Figure 2b illustrates a water-table well located in Ocean County, New Jersey, where the ground-water levels are much deeper. It is not dominated by short-term recharge and discharge events, but shows a more muted response consistent with time. The record for this well clearly shows that the drought, which began in late 1980, had a severe impact on water levels during 1981 and 1982. The monthly ground-water level data for both of these wells (Figs. 2a and 2b) are summarized in Fig. 2c, which combines period-of-record monthly minimums and maximums, with plots of monthly minimums for the previous year and the current year. Comparison of these two key observation wells shows that the drought caused the Cramer well in Ocean County to drop below historic lows from October through March of the water year 1982 (1982–83). However, the Bird well in Hunterdon County showed monthly minimums for both 1981 and 1982 that apparently did not differ appreciably from the historic record.

Figure 3 illustrates a presentation of data from an artesian aquifer that is instrumented

OCEAN COUNTY

395609074124001. Local I.D., Toms River TW 2 Obs. NJ-WRD Well Number, 29-0534.

LOCATION: Lat 39°56′09″, long 74°12′14″, Hydrologic Unit 02040301, about 200 ft. (61.0 m) east of Double Trouble Road on the north side of Jakes Branch, South Toms River. Owner: U.S. Geological Survey.

AQUIFER: Englishtown aquifer of Cretaceous age.

WELL CHARACTERISTICS: Drilled artesian observation well, diameter 8 in. (203 mm), depth 1146 ft. (349.3 m), screened 1080 to 1146 ft. (329.2 to 349.3 m).

INSTRUMENTATION: Water-level extremes recorder, February 1977 to current year. Water-level recorder, December 1965 to March 1975.

DATUM: Land-surface datum is 18.34 ft. (5.590 m) above National Geodetic Vertical Datum of 1929.

Measuring point: front edge of cutout in recorder housing, 1.70 ft. (0.518 m) above land-surface datum.

PERIOD OF RECORD: December 1965 to March 1975, February 1977 to current year. Records for 1965 to 1975 are unpublished and are available in files of New Jersey District Office.

EXTREMES FOR PERIOD OF RECORD: Highest water level, 48.37 ft. (14.743 m) below land-surface datum, May 28, 1966; lowest, 104.34 ft. (31.803 m) below land-surface datum, between June 8 and September 29, 1982.

EXTREMES FOR CURRENT YEAR: Highest water level, 102.05 ft. (31.105 m) below land-surface datum, between October 1 and December 8; lowest, 104.34 ft. (31.803 m) below land-surface datum, between June 8 and September 29.

WATER LEVEL, IN FEET BELOW LAND SURFACE DATUM,
WATER YEAR OCTOBER 1981 TO SEPTEMBER 1982

WATER-LEVEL EXTREMES

PERIOD	HIGHEST WATER LEVEL	LOWEST WATER LEVEL
1 Oct 1981 to 8 Dec 1981	102.05	102.94
8 Dec 1981 to 9 Mar 1982	102.44	103.37
9 Mar 1982 to 8 Jun 1982	103.10	103.48
8 Jun 1982 to 29 Sept 1982	102.91	104.34

MEASURED WATER LEVEL

DATE	WATER LEVEL
8 Dec 1981	102.53
9 Mar 1982	103.37
8 Jun 1982	103.16
29 Sept 1982	104.34

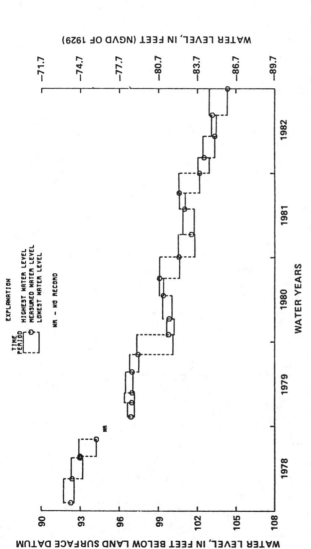

FIG. 3—*Water level data for WRD Well No. 29—0534, Ocean County, NJ [4] (1 ft = 0.3048 m; 1 in. = 25.4 mm).*

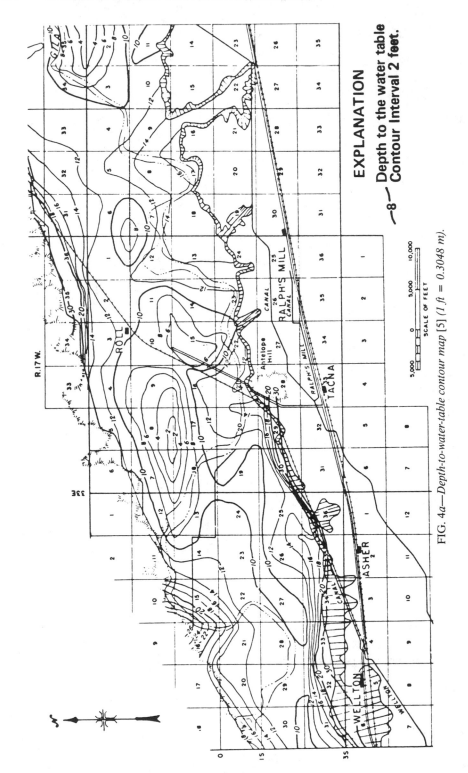

FIG. 4a—Depth-to-water-table contour map [5] (1 ft = 0.3048 m).

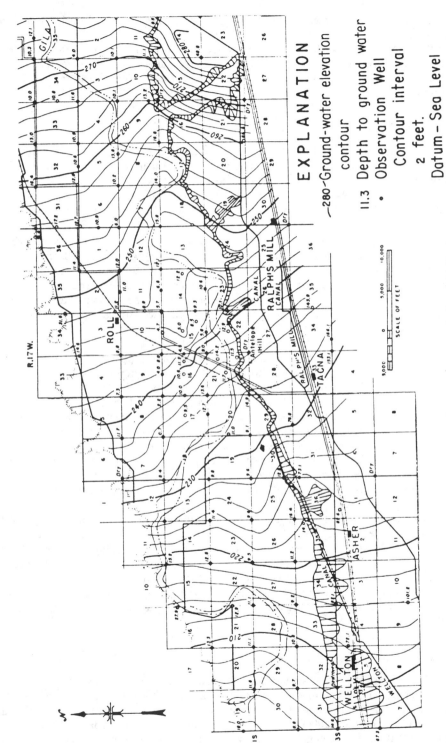

FIG. 4b—Water-table contour map [5] (1 ft = 0.3048 m).

with a water-level extremes recorder. This graph clearly shows that water levels continued to decline between 1978 and 1982. The quarterly measured water levels would depict a very different story if the extreme values were not shown. If one connects the quarterly values, it would appear that a considerable amount of recharge was taking place, occasionally superimposed on a steady decline. The extremes indicate less variation and fewer apparent upward trends.

Contour Plots, Flow Nets, and Sections

Contour plots in map view combine representative water levels from many monitoring points at a specified point in time. Synoptic water-level measurements are essential when an evaluation of ground-water flow is necessary. Figure 4a illustrates a depth-to-water-table map [5]. The same data have been converted to elevations above the sea level datum and presented in Fig. 4b. These two maps can be used for very different purposes. The depth-to-water-table map can be very important for geotechnical engineering and construction. It also gives the analyst an idea of where potential wetlands or other sensitive environments may be located. It can also be useful in planning and construction of deep foundations. The water-table elevation contour map is useful in evaluating ground-water flow. By measuring the spacing of the equipotential (contour) lines, the gradient between two points can be calculated. This information, when combined with the aquifer characteristics, allows the flow volume and velocity to be calculated. Lines constructed perpendicular to the equipotential lines indicate flow direction.

A plot with both equipotential lines and flow lines defines a flow net (Fig. 5). Figure 5 illustrates a ground-water mound caused by deep percolation (upper right corner). Water flows out radially from the mound. The distance between adjacent flow lines, W, increases with distance from the mound and with the flattening gradient. The size of the cells increases as the gradient decreases. The flow, V, through any cell equals the flow through any other cell. A flow net can be constructed in plan (map) view or in sectional (vertical) view. Flow nets are often used in engineering studies, especially when there is a contrast in material characteristics. They are also most useful when the horizontal or vertical gradients are steep and velocity profiles are of concern. Commonly, they are used to define seepage pathways either under or through an engineered structure, such as a dam. Flow nets can provide simplistic presentation of recharge and discharge relationships in a system.

Hydrogeologic profiles are constructed by combining geologic data from borings, test pits, and outcrops with water level data from observation points. Figure 6 shows a representative profile illustrating the relationship between a canal recharging the soils and the seepage discharging to a river. The differences caused by variations in material characteristics can be determined from the calculated slope of the water table at different points along the flow path. Presentations of this type are commonly found in geotechnical engineering studies, particularly because of the importance of water content in the strength of soil materials.

Summary

In summary, there are many ways to present ground-water level data so that the information can be understood and used. Commonly used x-y plots (hydrographs) are invaluable in visually evaluating trends in relation to time or distance. Contour maps give a snapshot view of the water table in terms of space. Flow nets and sectional views illustrate relationships that have wide engineering or hydrogeologic application.

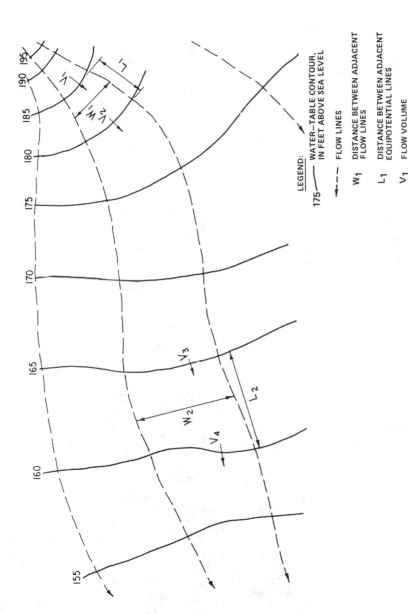

LEGEND:

175—— WATER–TABLE CONTOUR,
       IN FEET ABOVE SEA LEVEL

- - - FLOW LINES

→   FLOW LINES

W_1 DISTANCE BETWEEN ADJACENT
        FLOW LINES

L_1 DISTANCE BETWEEN ADJACENT
        EQUIPOTENTIAL LINES

V_1 FLOW VOLUME

FIG. 5—*Flow net analysis* [5]. *(The measurements are in feet (1 ft = 0.3048 m).*

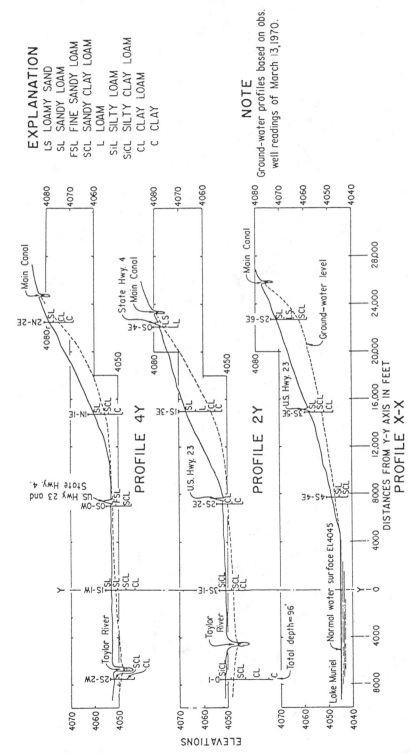

FIG. 6.—*Hydrogeologic profiles [5]. The elevations are the number of feet above sea level (1 ft = 0.3048 m).*

References

[*1*] Barry, B. A., *Engineering Measurements,* Wiley, New York, NY, 1964.
[*2*] *ASTM Manual on the Presentation of Data and Control Chart Analysis, ASTM STP 15D,* American Society for Testing and Materials, Philadelphia, 1976.
[*3*] Klein, M. and Kaser, P., "A Statistical Analysis of Ground-Water Levels in Twenty Selected Wells in Ohio," *Ohio Department of Natural Resources, Division of Water Technical Publication,* No. 5, 1963.
[*4*] "Water Resources Data—New Jersey—Water Year 1982," USGS Water-Data Report NJ-82, U.S. Geological Survey, Denver, CO, 1982.
[*5*] *Ground Water Manual,* Bureau of Reclamation, U.S. Department of the Interior, U.S. Government Printing Office, Washington, D.C.

Monitoring Well Purging and Ground-Water Sampling

W. R. Ridgway¹ and D. Larssen²

A Comparison of Two Multiple-Level Ground-Water Monitoring Systems

REFERENCE: Ridgway, W. R. and Larssen, D., "A Comparison of Two Multiple-Level Ground-Water Monitoring Systems," *Ground Water and Vadose Zone Monitoring, ASTM STP 1053,* D. M. Nielsen and A. I. Johnson, Eds., American Society for Testing and Materials, Philadelphia, 1990, pp. 213–237.

ABSTRACT: In the context of the Canadian Nuclear Fuel Waste Management Program, Atomic Energy of Canada Ltd. (AECL) is conducting research into the underground disposal of nuclear fuel waste deep in stable plutonic rock. As part of this research, AECL has developed and tested many geotechnical instrumentation systems. Two multiple-level ground-water monitoring systems have been used to evaluate hydrogeological conditions in fractured rock and overburden materials: the Waterloo system and the Westbay system. Both instruments are designed to isolate several sections of a single borehole for ground-water measurements. The instruments were tested in adjacent boreholes drilled to about 60-m depth in granitic rock at the AECL's Atikokan research area in Ontario, Canada. Various aspects of these instrument systems, including the design, testing and sampling methods, and ground-water measurement data are compared. The information contained in this paper may assist future users to determine if these systems, or other systems, will be effective for their applications.

KEY WORDS: Waterloo system, Westbay system, ground water, multiple-level monitoring systems, instrumentation design, installation activities, pressure measurements, fluid sampling, hydraulic conductivity measurements

Atomic Energy of Canada Ltd. (AECL) has been conducting hydrogeological research as part of the Canadian Nuclear Fuel Waste Management Program [1]. The objective of the program is to assess the concept of deep underground disposal of nuclear fuel waste in stable plutonic rock. Since instrumentation will be required to monitor ground-water conditions in the rock surrounding the underground disposal vault, AECL geoscientists have developed and evaluated many different instrumentation systems.

In 1985, AECL and Westbay Instruments Ltd. conducted a joint project to compare the Westbay and Waterloo multiple-level ground-water monitoring systems. The project was carried out at a study area at AECL's research area, located on the Eye-Dashwa granitic pluton, about 30 km north of Atikokan, Ontario, Canada (Fig. 1). The study area is situated on the Eye-Dashwa lineament (Figs. 2a and 2b), a northeast-trending regional topographic depression passing through the bottom of the Eye and Dashwa Lakes.

In this paper, the authors compare and evaluate various aspects of these two monitoring systems, including the instrument design, testing and sampling methods, and ground-water measurement data. We also review quality assurance procedures and costs. Although

¹ Engineering geologist, Whiteshell Nuclear Research Establishment, Atomic Energy of Canada Ltd., Pinawa, Manitoba, Canada R0E 1L0.
² Geological engineer, Westbay Instruments Ltd., North Vancouver, British Columbia, Canada V7L 1G4.

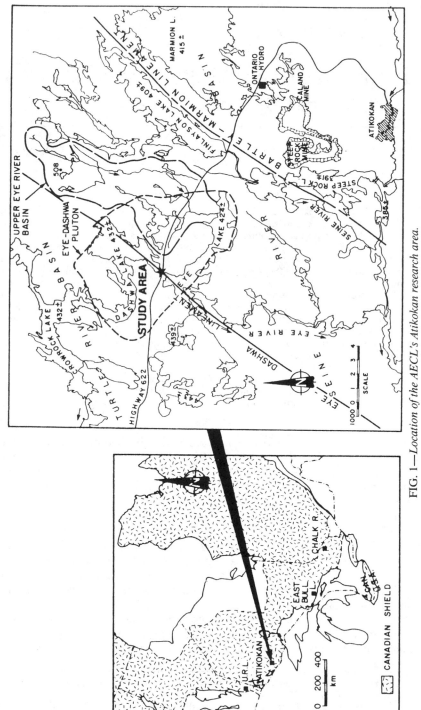

FIG. 1—*Location of the AECL's Atikokan research area.*

design changes and improvements have recently been made to both instrumentation systems, the basic operating principles and components are similar to those described in this paper.

Geology and Drilling

Two adjacent boreholes, ML-11 and ML-16, were placed 8 m apart on the portage separating the Eye and Dashwa Lakes. Figure 2b shows the local topography, the Eye-Dashwa lineament, the lake levels, and the ground-water springs, and Fig. 2a shows a cross section with the location of the two boreholes.

The two comparison boreholes, ML-11 and ML-16, were drilled vertically to a depth of 56 and 59 m, respectively, using a track-mounted Longyear 38 drill rig with NQ-size (76-mm-diameter) diamond coring equipment. Clear surface water (that is, containing no drilling mud or drilling fluid) was used to minimize contamination of the formation ground water. During drilling, the drill water was labelled with rhodamine dye at selected depth intervals, to aid in subsequent purging activities. The drill core was logged, and lithology and discontinuity data were collected [2].

The bedrock in ML-11 and ML-16 consists of fractured, partly weathered to unweathered granite. The fracture discontinuity index ranges from 0 to more than 20 per metre and the rock quality designation (RQD) values [3] range from about 50 to 100%. The fracturing is discussed in the sections on ground-water chemistry and hydraulic conductivity testing, further on in this paper. Overburden materials consist of sandy and silty glacial outwash deposits, 5.9 m deep at ML-11, and 9.0 m deep at ML-16. The Waterloo and Westbay instruments installed in these boreholes had access ports located in both bedrock and overburden materials.

Instrument Components and Configuration

Figure 3 is a schematic diagram showing components of the Waterloo and Westbay instrumentation systems (based on figures presented by Cherry and Johnson [4] and by Black et al. [5]).

Waterloo System

The ground-water monitoring system of the University of Waterloo [4], Waterloo, Ontario, Canada, was installed in borehole ML-11. The installed system was a prototype, different from the current version of the Waterloo system. (Shortly after this installation was carried out, the manufacturing rights for the Waterloo instrument were obtained by Solinst Canada Ltd., and an updated commercial version became available in mid-1985.) The instrument consisted of a bundle of open, flexible polyethylene access tubes. Each tube, 10 mm in inside diameter (ID) and 13 mm in outside diameter (OD), was connected to an individual monitoring zone. The tubes were contained within a 51-mm-ID Schedule 40 polyvinyl chloride (PVC) casing.

The monitoring zones of the Waterloo instrument were isolated by self-inflating chemical packers clamped on the outside of the casing. The packers consisted of an expandable Dowell[3] sealant sleeve covered with a waterproof rubber membrane. When water was

---

[3] Dowell sealant was obtained from the Dow Chemical Co.

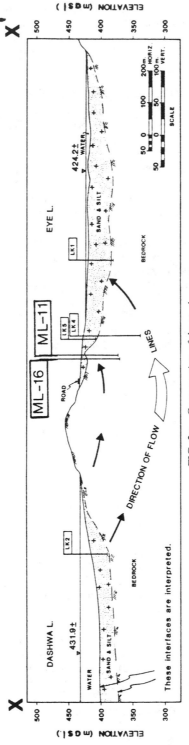

FIG. 2a—Cross section of the study area.

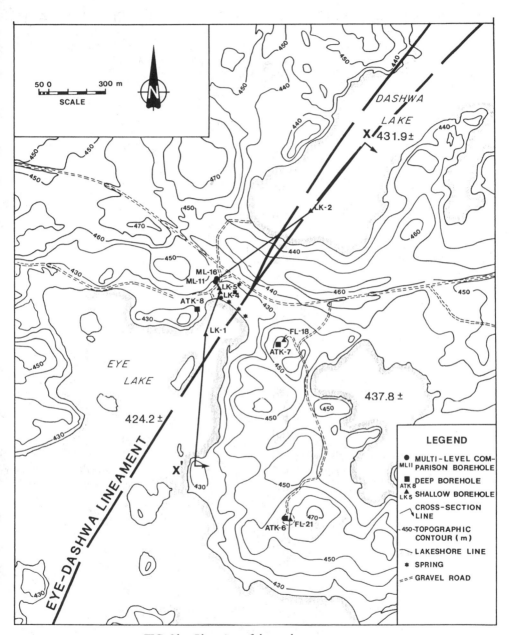

FIG. 2b—Plan view of the study area.

WESTBAY MP MONITORING WELL

WATERLOO MONITORING WELL

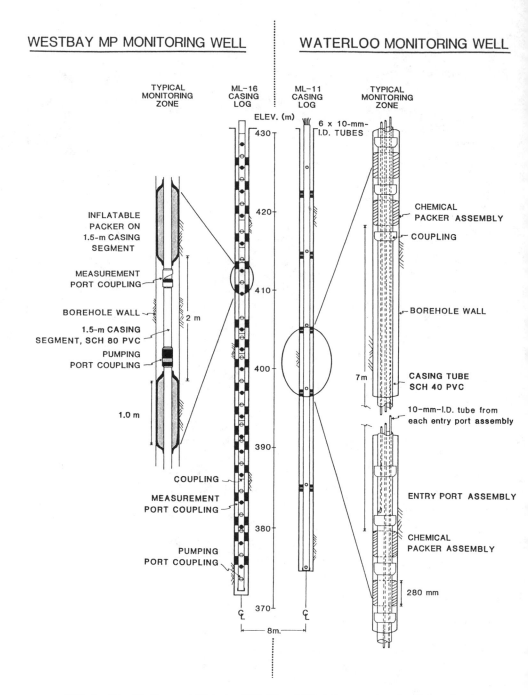

FIG. 3—*The Westbay and Waterloo (ML-16 and ML-11) comparison boreholes, showing the two instrumentation systems.*

poured into the interior of the PVC casing the packers expanded. The water came in contact with the Dowell sealant sleeves through holes in the wall of the casing (not shown in Fig. 3). This water contact caused expansion of the sleeves to more than twice their original volume, sealing off an NQ-size (76-mm) borehole. The packers were fully expanded after about 24 h. The installed instrument consisted of ten packer assemblies (Fig. 3) arranged in five pairs, to isolate six monitoring zones. Each pair of packers had a total length of about 800 mm, and the length of the individual monitoring zones ranged from 7 to 11 m.

The polyethylene tubes provided access for monitoring and sampling ground water from the designated zones. Each access tube was joined to an entry port and extended from the port to the ground surface. The entry port was an open elbow (10-mm OD) of PVC pipe that penetrated through the casing.

The PVC casing segments were threadless and the string was joined together with glued couplings. The installed instrument string had (1) a Schedule 40 plug glued on the bottom end, (2) six entry port/access tube assemblies, (3) five packer assemblies, and (4) several PVC pipe segments. In the updated Solinst version of the Waterloo instrument, double O-ring seals replace glued joints.

Westbay System

Borehole ML-16 was completed with a Westbay ground-water monitoring system (Fig. 3), which was designed and manufactured by Westbay Instruments Ltd. The first commercial version of the Westbay instrument was introduced to the geotechnical-hydrological field in 1978. The current Westbay system components and probes are described by Black et al. [5]. The Westbay installation consisted of a single, closed access tube made of 37-mm-ID Schedule 80 PVC pipe segments with machined ends for coupling attachment. The three different coupling types used in this installation were each formed of acrylonitrile-butadiene-styrene (ABS) plastic with Viton O-ring seals. A shear wire was threaded through the couplings to join the casing segments.

The monitoring zones of the Westbay instrument were isolated by hydraulically inflated packers attached to the outside of the casing segments. The packers consisted of a urethane gland with stainless steel clamps at either end of a PVC casing segment. These were inflated individually with a packer inflation tool, which was lowered into the instrument casing. The tool was seated on a one-way valve, which allowed movement of fluid through the wall of the casing and into the interior of the packer gland. The packer inflation was controlled from the surface by measuring the fluid injection pressure and volume at selected time intervals. The packer sealing pressure was about 690 kPa (100 psi), and each packer had a sealing length of about 1 m.

The instrument was installed (Fig. 3) with two different types of valved couplings for access to each monitoring zone. Fluid samples and fluid pressures were taken through the measurement port couplings using a probe, which was lowered inside the instrument casing. Purging, permeability testing, and large-volume sampling were carried out through the pumping port couplings. A mechanical wire-line tool was used to open and close the valves on the pumping port couplings. When a single pumping port was opened, only one monitoring zone was connected to the casing and the casing behaved like a single conventional standpipe piezometer. The modular design of the couplings and casing segments allowed installation of 19 monitoring zones into borehole ML-16. These zones ranged from about 1.0 to 4.0 m in length and were isolated by 18 packers.

Installation Methods

The monitoring zones and annulus seals for the two boreholes were selected by reviewing the geological drill logs. The instruments were installed during winter, and hence a heated drill shack was required during drilling and instrumentation installation. After completion of the borehole drilling, each hole was flushed for several hours with surface water to remove any drill cuttings. A whole day was set aside for each installation. Considering the winter conditions, which were particularly adverse during the Waterloo system installation, both installations went smoothly and successfully.

Waterloo

For the Waterloo instrument, a casing log was prepared the night before the installation. The instrument casing and tubing were cut to length, and the other parts were assembled and labelled. On the installation day, the sections of PVC casing were laid out in sequence next to the borehole. The couplings, rolls of polyethylene tubes, and packer assemblies were retained in the heated drill shack. The installation was carried out on 27 Feb. 1985, with the outside air temperature at $-30°C$.

The installation began with attaching and clamping the longest polyethylene access tube to the lowest entry port assembly. Then, the tubing was threaded through a 3-m section of casing, and the port assembly and casing were joined with a coupling. The coupling and casing were heated, glued together, and placed into the borehole. Two more segments of PVC casing, a packer assembly, and an entry port assembly were threaded, coupled, and carefully glued together. This was followed by attaching and clamping the next longest polyethylene tube to the next entry port assembly. The threading and coupling of components was continued until the installation was completed.

The instrument string was watertight and buoyant when placed below the water level in the open borehole. During the installation, surface water was poured into the casing to overcome the buoyancy effect. The Waterloo installation was completed in 4 h with four people participating. Next, the casing was filled with water, causing the packers to expand and seal the borehole annulus. The next day, tests were carried out to check the depth of the access ports and the borehole annulus seals.

Westbay

For the Westbay instrument, a casing log describing the position of each of the instrument components was also prepared the night before the installation. The component supply was checked to ensure that all the required parts were available. Each coupling of the instrument string had a serial number, which was recorded. On the installation day, the couplings were attached to the casing segments and these were laid out in correct sequence next to the borehole. The instrument components were coupled together and the string was placed into the borehole under the shelter of a drill shack. During installation of the instrument string, surface water was periodically poured into the casing since the casing was watertight and buoyant.

The Westbay instrument string was installed on 25 March 1985, in about 4 h with three people participating. The outside air temperature was $+2°C$. The following day, an additional 10 h was needed to inflate the 18 hydraulic packers sequentially using a packer inflation tool, beginning at the lowest packer. The volume of water injected into the packers and the packer inflation pressures were recorded for quality control purposes. During the inflation period, the instrumentation couplings were also located with the inflation tool.

When the tool was pulled upward from the bottom of the instrument string, it seated itself on the helical shoulders located on the inside of each coupling.

Fluid-Pressure Measurements

Waterloo

The method for fluid-pressure measurement in the Waterloo instrument was the same as that used in an open standpipe piezometer. A small-diameter (5-mm) electric water-level tape was used to measure the water level in each polyethylene access tube, and these measurements were compared with a known benchmark elevation. The average time required to measure water levels in the six monitoring zones of ML-11 was about 15 min, or 2 min per measurement. More time was needed to test whether the annular seals were performing properly.

During winter, the water level in the interior of the casing was drawn down below the frost line to prevent the instrument from freezing. However, the instrument had to be thawed and a heated enclosure was required for winter monitoring since the water levels in the access tubes were above the ground surface.

Westbay

The method for fluid pressure measurements in the Westbay instrument involved lowering a probe, containing an electric pressure transducer, inside the casing to a measurement port coupling. Initially, the probe was lowered to the bottom of the instrument string and was pulled upward until it was seated on the desired measurement port coupling. Then, it was activated to complete a hydraulic connection to the monitoring zone. During activation, a seal was made isolating the pressure transducer from the fluid inside the instrument casing. Once the transducer had stabilized, the pressure was displayed on the transducer control unit at the ground surface.

The pressure at each monitoring zone was measured in sequence, beginning at the bottom of the casing. The piezometric pressure was calculated based on the difference between the readout measurement and the reference pressure and elevation of the interior fluid column. The average time required to measure the pressures at all 19 monitoring zones was 60 min, or about 3 min per measurement.

During winter, pressure measurements were carried out without the instrument having to be thawed. The water in the access pipe was isolated from the monitoring zones and was maintained below the depth of freezing.

Fluid Pressure Data

The piezometric profile from the Waterloo instrument (Fig. 4) indicates a component of upward flow, from 423 m to the water table at 428 m, stable artesian conditions between 423 m and 406 m, a downward component between 406 and 405 m, stable artesian conditions from 405 m to 385 m, and a slight upward gradient in the lowest interval (all measurements are in metres above mean sea level).

The piezometric profile of the Westbay instrument (Fig. 4) is more detailed since more zones were installed. For the upper part of the hole from about 430 to 412 m, the readings are in agreement with those from the Waterloo instrument. Below 412 m, the readings are slightly different, and from 412 to about 405 m there are stable artesian to weakly downward gradients. This is underlain by a strong downward gradient to about 400 m. From

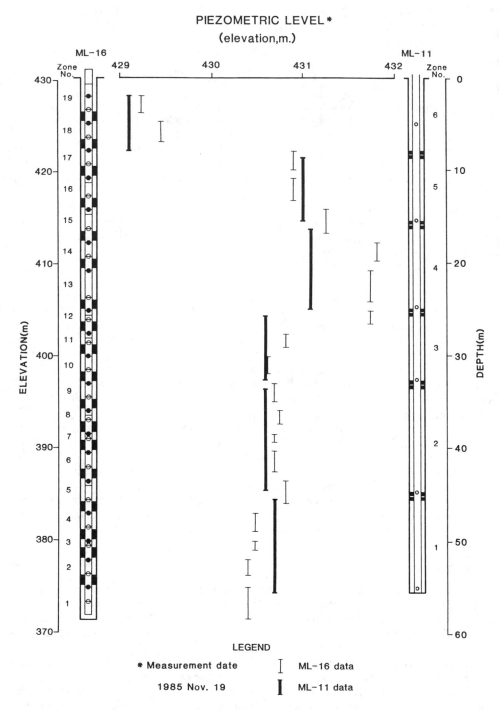

FIG. 4—*Comparison of the piezometric profiles for ML-16 and ML-11.*

about 400 to 384 m, there are stable artesian conditions, and below 384 m the gradients are stable to weakly downward.

The piezometric data from the Westbay instrument showed a wider range of measurements than that from the Waterloo instrument. The range is probably in response to water pressures from discrete fractures, which can be measured with short monitoring intervals. Generally, short intervals minimize interzone flow and therefore reduce any effect of piezometric-level averaging. Averaging may have occurred in ML-11 as a result of the long monitoring zones. There are inflections in the ML-16 profile at depths of 18, 27, 30, and 47 m, and these are consistent with the packer seals in ML-11 at depths of 16, 26, 34, and 46 m. The monitoring zones in ML-11 may have caused short-circuit effects to nearby ML-16. Alternatively, the differences in piezometric readings may have been caused by differences in fracture hydrology between the two boreholes.

Fluid Sampling Methods

When possible, a similar sampling procedure was used for both instrument systems. Development pumping and purging were carried out to remove drilling fluids from each monitoring zone. The absence of rhodamine dye in the purge discharge water from all monitoring zones in both instrument systems indicated that the drilling fluid was no longer present. After development and purging activities had been completed, water samples were taken from most zones of both wells as follows:

ML-11: Large-volume samples were collected from 6 of 6 zones.

ML-16:

    1. Large-volume samples were collected from 6 of 19 zones (Nos. 2, 6, 10, 13, 16, and 17). These zones were selected since they closely correspond to the 6 zones in ML-11.

    2. Small-volume samples were collected from 12 other zones.

    3. Zone No. 9 was not sampled.

Waterloo

The standing-fluid levels of the monitoring zones in the Waterloo instrument were above the ground surface, and thus, purging could be done using a peristaltic suction lift pump. A line from the pump was placed into the polyethylene access tube and used to purge water from the tube and the monitoring zone (Fig. 3).

Purging was carried out during August 1985 and the data from this activity are presented in Table 1. The purge volume[4] for each monitoring zone ranged between 12 and 22 L, and about 0.2 to 1.3 times these volumes was pumped from the zones. The average pumping rate was about 69 mL/min. This rate was limited by the low transmissivity of the monitoring zone and the suction lift capability of the pump. Complete development of this monitoring instrument for removal of drill cuttings was not possible because of the low flow rates and a time constraint. This lack of development, however, was not considered to have hindered the instrument performance since the hole had been extensively flushed prior to installation of the instrument.

---

[4] The purge volume was calculated as the sum of the water in the annulus between the borehole wall and casing, and the riser tube volume.

TABLE 1—*Purging data for the two instruments.*

Well No.	Monitoring Zone Number	Date	Purgeable Volume (L)	Pumping Rate (ml/min)	Pumping Time (min)	Volume Pumped* (L)	Purgeable Volumes Pumped (L)
ML-16 (Nov. 1985)	1 to 12	19,20,21	145	1700	640	1090	7.5
	13 to 19	21	95	2000	360	720	7.6
	13 to 19	22	95	1000	420	420	4.4
Total			240		1470	2230	
ML-11 (Aug. 1985)	1	6	21.5	228	120	27.4	1.27
	2	6,7	22.4	42	150	6.3	0.28
	3	7,8	14.7	33	90	3.0	0.20
	4**	8,9,12	16.6	15	155	2.3	0.14
	5	8,9	14.6	33	95	3.1	0.21
	6	9	12.2	67	60	4.0	0.33
Total			102		670	46.1	

* Prior to sampling.

** Pumped dry, insufficient recharge for sampling.

Westbay

The fluids were purged and the monitoring zones were developed in the Westbay instrument (Table 1) by using an electric centrifugal pump. The pump was attached to a 19-mm PVC intake line placed at the bottom of the instrument casing. The water in the monitoring zones was accessed by opening the valved pumping port couplings (Fig. 3).

The instrument monitoring zones were purged once and in two groups, based on a review of fluid pressure readings. The first group of monitoring zones, Nos. 1 through 12, had similar fluid pressures[5] in each of the monitoring zones. Before these valved couplings were opened, the water level inside the instrument casing had been drawn down to minimize any fluid mixing between the monitoring zones. After completion of purging of the first group of zones, the valves of these pumping port couplings were closed. The second group of monitoring zones, Nos. 13 through 19, were purged using the same method.

The purging data are presented in Table 1. The purge volume of the lower group of zones was 145 L, while that of the upper group was 95 L. During 19–21 Nov. 1985, an amount 7.5 times the purge volume was pumped from each group of zones. The upper group of zones was purged for a second time on 22 November, when an additional 4.4 times the purge volume was removed. With these activities, it is assumed that all monitoring zones, even those with low transmissivity, had had at least one purge volume of fluid removed.

A surge-block technique was used for development and purging of the Westbay instrument, to achieve sediment-free water discharge. This plunging technique caused water to move into and out of a pumping zone and was used to remove particulate matter. Purging of the instrument was carried out just after the installation of the instrument, since the monitoring zones and casing were hydraulically isolated from each other. Successive sampling did not have to be preceded by purging, but further purging could have been done by repeating the described process.

Sample Collection

Water samples were collected from all six monitoring zones in the Waterloo instrument. Water sampling was carried out using a peristaltic pump and with the same method as was described for purging. For each zone, the procedure involved pumping until steady-state water chemistry was achieved. When this occurred, water samples were taken from the pump discharge line. After completion of sampling, the pump lines were rinsed with deionized water.

Water samples were collected from 18 of the 19 monitoring zones in the Westbay instrument. Sampling was done with a stainless steel bottle connected to a wire-line-operated sampling probe. For each zone, the procedure involved obtaining a volume of water (80 to 250 mL) to rinse the collection apparatus. Then, the probe was repeatedly lowered to the same zone until the required volume of water was collected. After collection of a sample from a zone, the sampling apparatus was cleaned with a rinse of dilute nitric acid followed by a rinse of deionized water.

Sampling data for ML-11 and ML-16, based on sampling activities carried out in November 1985, are summarized in Table 2. The collection rate for ML-11 represents the average water discharge of the peristaltic pump calculated when the formation was yielding water. The time required to obtain a sample was greater than the collection rate indicates because of the intermittent pumping time. The collection rate for ML-16 is based on the

---

[5] There was, however, about 1 m of hydraulic head difference between Zones 1 through 11 and Zone 12.

TABLE 2—*Sampling summary for comparison of the instruments.*

	Westbay (ML-16)	Waterloo (ML-11)
Sample collection rate	29 mL/min	37 mL/min
Time required per sample	49 min	167* min
Total volume of sample collected	14865 mL	5700 mL
Total sampling time	875 min	1084 min

* with incomplete purging.

time needed to retrieve samples from the measurement port with the sampling apparatus. The time required for each sample included three to four trips to the port and the preparatory activities.

Ground-water Chemistry

Ground-water samples were analyzed in the field using AECL's mobile laboratory. Field measurements included the pH, temperature, conductivity, and chloride and bicarbonate concentrations. Some of the water samples were acidified and sent to AECL laboratories for detailed analyses.

The chemical results compared are the pH and electrical conductance (Fig. 5) and the total dissolved solids (TDS) and tritium values (Fig. 6). The trends for various chemical parameters are similar for both instruments. These values are typical for shallow ground water in the study area and comparable to a summary of chemical results contained in a report by Underwood McLellan Ltd. [2].

The pH values, which generally increase with hole depths, range from 6.1 to 7.8 and tend to be higher in less fractured rock. At some depths, the values for ML-11 appear to be an average of those values measured in several zones of ML-16. The averaging is probably the result of the longer monitoring zones in ML-11. The TDS and conductivity values also increase with hole depths and the values for both holes correspond closely to each other. However, from about 400 to 380 m in ML-16, there is a decrease in conductivity and TDS. At the same depth interval in ML-11, there is an increase in these values.

The tritium results from both wells are comparable and range from a low of 10 tritium units (TU) to a high of 59 TU. The low values at 400 m may be related to its being a lower hydraulic conductivity zone, since areas of lower hydraulic conductivity appear to have lower tritium content.

Hydraulic Conductivity Testing

The hydraulic conductivity for each monitoring zone of the two instruments was tested after completion of the purging and development activities. Hydraulic conductivity measurements were performed using a variable-head slug test [6].

In the Waterloo installation, either a rising-head or a falling-head hydraulic conductivity test could have been performed. A rising-head test technique was used because the static

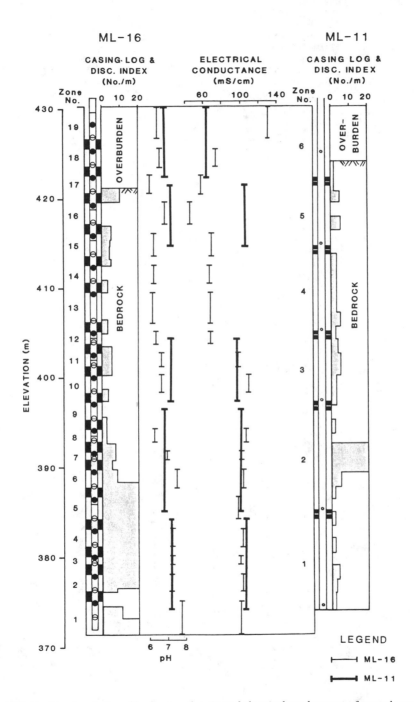

FIG. 5—*Geochemical profile showing the pH and electrical conductance of ground waters from ML-11 and ML-16.*

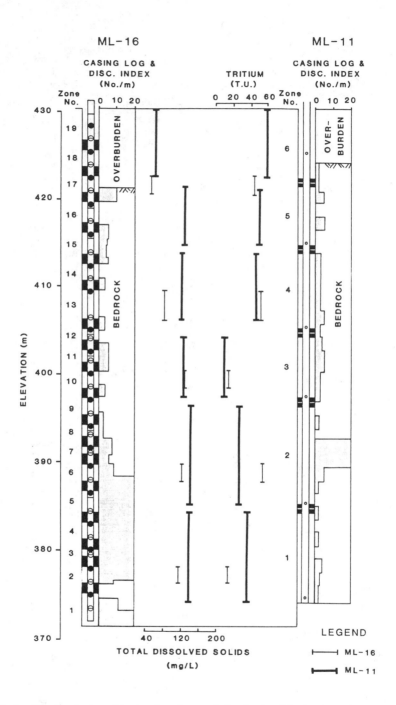

FIG. 6—*Geochemical profile showing the total dissolved solids (TDS) and tritium in ground waters from ML-16 and ML-11.*

water level was above the ground surface. The test was performed using a peristaltic pump and a measured length of suction line. The water level in the access tube was lowered to a depth of about 5 m below the static water level. The test was initiated when the suction tube was removed and substituted with a narrow-diameter water-level indicator. The recovery with time of the water level in the access tube was recorded.

In the Westbay installation, either a rising- or a falling-head hydraulic conductivity test could have been performed, as well, but a rising-head test was chosen so that the test methods in the two installations would be comparable. The test was performed using an electric centrifugal pump and a PVC intake line to lower the water in the instrument casing. Prior to each test, all of the valved couplings were closed, which made the instrument casing watertight. Then, the water level in the casing was pumped to a level of about 5 m below the piezometric level in the zone to be tested. The test was initiated by opening the desired pumping port using a wire-line tool. An electric water-level tape was used to measure the water-level recovery with time in the instrument casing.

Test Data

Hydraulic conductivity test results for ML-11 and ML-16 are shown in Fig. 7. The data from ML-16 correlate better with the degree of fracturing or discontinuity index than the data from ML-11. However, not all zones of intense fracturing in ML-16 have corresponding high hydraulic conductivity values. Hydraulic conductivity values (Fig. 7) from ML-11 are generally lower than those from ML-16. The ML-11 values range from 1×10^{-7} to 7×10^{-10} m/s and average about 3×10^{-8} m/s. The ML-16 values range from 9×10^{-5} to 7×10^{-9} m/s and average about 8×10^{-6} m/s.

Discussion

Table 3 summarizes the comparison between the two instrumentation systems.

Well Design

The downhole components of the Westbay and Waterloo monitoring instruments were mainly plastic (PVC and ABS), with some metal parts. The main differences between the systems were the monitoring zone access methods and the borehole annulus sealing method.

For the Westbay instrument, access to monitoring zones and monitoring was done with wire-line tools and equipment manufactured by Westbay Instruments Ltd. The monitoring equipment was sophisticated and equipment operators required specialized training. The instrument was installed with two different types of access ports for every zone, and the instrument components were arranged in such a way that there was up to one zone every 2 m. These characteristics allowed flexibility in the testing, sampling, and measurement activities taking place in ML-16.

Access to the zones of the Waterloo instrument was gained by placing probes into the polyethylene riser tubes. The equipment for testing and sampling was simple, easy to operate, and available from several suppliers. Two constraints of the Waterloo instrument were (1) the small diameter of the monitoring zone access tubes and (2) the fact that only six monitoring zones could be installed in a single 76-mm-diameter borehole. In formations with high transmissivity, however, the narrow diameter of the access tube would not constrain sampling. Also, six monitoring zones are often more than adequate for a single monitoring borehole.

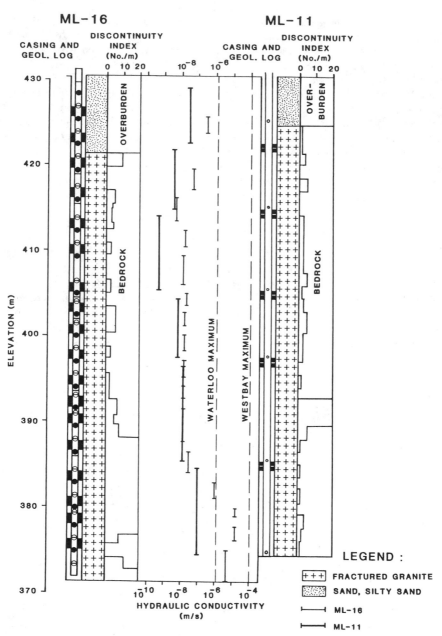

FIG. 7—*Hydraulic conductivity test results for ML-16 and ML-11.*

At the time this paper was prepared, the deepest known Waterloo instrument was installed to about 150 m, while Westbay instruments have been installed to depths greater than 1000 m. The deep Westbay installations contain more stainless steel parts, such as couplings, valves, and packers than were used in the installation for this study.

TABLE 3—*Comparison summary for the Westbay and Waterloo instruments.*

WESTBAY	WATERLOO
1. Well Design	
—isolated many zones in a single well	—isolated 6 zones in a 76-mm well
—modular components allowed design changes during installation	—modular components allowed some design changes during installation
—single access tube, valved couplings for access to each monitoring zone	—multiple-access tubes, each dedicated to a monitoring zone
—PVC and ABS construction with some stainless steel parts	—PVC construction
—one set of monitoring tools was used for many wells	—one set of monitoring tools was used for many wells
—1000-m depth capability	—150-m depth capability
2. Annulus Seals	
—hydraulically inflated packers	—chemically expanding packers
—inflation of individual packers by injection of water through a valve into the packer gland	—inflation of all packers simultaneously by adding water to interior of instrument casing
—deflation of packers is possible	—deflation of packers is not possible
3. Installation	
—simple layout and installation procedures	—simple layout and installation procedures
—short time to install instrument string, inflating the packers took additional time	—short installation time, more labour-intensive
—heated enclosure required in winter	—packers were self-inflating, additional time to confirm packers were inflated
	—heated enclosure required in winter
4. Fluid-Pressure Monitoring	
—transducer probe and transducer control unit were used	—conventional electric water-level sounder was used
—longer time was required for readings	—shorter time was required for readings
—transducer was calibrated during monitoring routine	—sounder was calibrated against a steel tape
—access tube was not hydraulically connected to the monitoring zone, freezing cannot occur	—dedicated access tubes are open and did freeze in winter
—short monitoring zones may have allowed more accurate fluid pressure readings of discrete geological features	—longer monitoring zones may have caused averaging of fluid pressure readings

TABLE 3—*Continued.*

	WESTBAY	WATERLOO
5.	Fluid Sampling	
	—sample collection was complex	—sample collection was simple when water levels were above the suction lift
	—purging was simple, rapid; may not be required after first sampling	—purging was time-consuming since monitoring zones had low transmissivity and yield. Also, purging was required before a new sample could be collected
	—sampling tools were complicated and sampling process was labour-intensive	
	—sampling tool decontaminated by rinsing	—conventional probes and monitoring equipment
	—samples can be collected at formation pressure	—pump lines decontaminated by rinsing
	—winter sampling required a heated enclosure	—a new tool is available to collect samples at formation pressure
	—purged 7.5 monitoring-zone volumes	—purged 0.1 to 1.0 monitoring-zone volumes
6.	Hydraulic Conductivity Tests	
	—test method was simple, rapid	—test method was simple, rapid
	—rising head test by applying rapid head change	—rising head test by removing slug from riser tube
	—measured the recovery of water level in the casing	—measured the recovery of water in the riser tube
	—maximum measurable hydraulic conductivity 1×10^{-4} m/s using liquid level techniques	—maximum measurable hydraulic conductivity 2×10^{-6} m/s using liquid level techniques
7.	Quality Control Checks	
	—couplings had serial numbers	—no serial numbers since prototype system
	—packer inflation pressure was recorded for each packer	—packer inflation was checked tests to verify the annulus seal can be included in the monitoring routine
	—tests for annulus seal verification are part of the monitoring routine	
	—checks possible on depth of downhole components after installation	—checks possible on depths of downhole components after installation
8.	Cost	
	—cost of downhole components several thousands of dollars	—cost of downhole components several thousands of dollars
	—two days for installation, less labour required	—one day for installation more labour required
	—probes more than $20k, but can be rented or leased	—probes less than $2k, but optional probes and equipment are extra
	—operating cost varied with the test activity	—operating costs varied with the test activity

Annulus Seals

The Westbay packers were individually inflated using a hydraulic inflation tool. This was a controlled operation and, hence, packer inflation pressure data were collected. After a few years of service, if needed, the packers can be deflated, and the instrument string can be removed and replaced. However, packer deflation is destructive and all the packers would have to be replaced.

The Waterloo packers were chemically inflated and formed a tight seal against the borehole wall. The inflation was confirmed by monitoring and testing the water levels in the access tubes and the instrument casing to ensure that there were no leaks or interzone connections. The borehole diameter, however, could have an effect on the sealing capability of a packer. For example, if a packer had been located at a point in the borehole where the diameter was enlarged, the sealing pressure might have been reduced. Second, if a packer in the instrument string system had failed, deflation of the other packers would not have been possible since all the packers were inflated simultaneously. Removal of the string for repair of a packer would destroy the remaining packers and might destroy other components of the instrument.

Installation

The Westbay system was installed with the assistance of a casing log. Many parts were preassembled and laid out on the ground next to the borehole. Next, the instrument string was installed and packer inflation was done on the following day. The Waterloo string was installed using similar techniques, but no wire-line tools or extra time were required to inflate the packers. Because of the winter conditions, a heated enclosure was needed for both installations, and the Westbay packer inflation equipment had to be protected from freezing. Care was required when these instrument strings were installed. Errors in the assembly and installation of the components could potentially cause problems when the instruments were to be monitored.

Fluid-Pressure Monitoring

Fluid-pressure monitoring of the Westbay instrument was done by lowering a transducer probe to a valved coupling and reading a transducer control unit, while a water-level sounder was used to monitor the Waterloo instrument. The pressure transducer and the sounder were of similar accuracy. The pressure transducer was calibrated as part of the monitoring routine. Similarly, the sounder was routinely checked and calibrated. Routine tests for zone interconnectivity should be done with both instruments. This test was part of the Westbay monitoring routine and may be instituted as part of the Waterloo routine. Readings were easier and faster with the Waterloo instrument and the equipment operator required less training.

A higher percentage of annular sealing may improve pressure monitoring detail in a complex hydrogeological environment [7]. As installed, about 28% of the Westbay borehole and 7% of the Waterloo borehole were sealed. The discrepancy in pressure readings from the two instruments may be due in part to the differences in sealing lengths. However, during the installation of ML-11, the percentage of annular sealing could have been increased simply by adding more chemical packers.

The length of the monitoring zones can also affect a piezometric pressure profile [8]. The Westbay instrument was installed with much shorter monitoring zones than the Waterloo instrument. The short monitoring zones in ML-16 may have allowed more accurate fluid pressure readings of discrete geological features, while the longer zones in ML-11 may have caused averaging of pressure readings.

The length, density, and flexibility of the monitoring zones in the Waterloo instrument could have been improved by three different methods. First, the monitoring zones could have been shortened by adding more packers. Second, a larger-diameter hole could have been drilled to accommodate a larger casing holding more riser tubes. Finally, two holes, a short one and a longer one, could have been drilled, allowing more flexibility in locating zones.

Fluid Sampling

Fluid sampling using the Westbay instrument required the use and operation of tools and probes. For development and purging, a wire-line tool was used to open the valved pumping port coupling, and the water was removed using a pump attached to a suction line. When development and purging were completed, the pumping port valves were closed and the measurement port couplings were used for sampling the monitoring zones. No further development was required for subsequent sampling of the monitoring zones.

Samples were collected from the Westbay instrument monitoring zones at atmospheric pressure using a probe and sample bottle system. The system was decontaminated with dilute acid and deionized water rinses, followed by a formation water rinse. Depending on the procedure, samples may be collected either at the formation pressure or at the atmospheric pressure. Samples collected at formation pressure can be analyzed for the various dissolved gases.

Sampling at the Waterloo instrument was easily carried out with conventional equipment. Water samples were removed from the access tubes using a suction tube attached to a peristaltic pump. Several monitoring zones can be pumped synchronously when additional units are added to the pump. Sampling of ML-11, however, was time-consuming and the instrument could not be properly purged. This may have been due to the combined effects of the narrow port diameter, low yield, and low transmissivity of the formation.

Samples were collected from the Waterloo instrument at the water discharge line from the peristaltic pump. Prior to sampling, pumping was done at each zone until steady-state ground-water chemistry was achieved. This technique was suitable for most chemical analyses but not for dissolved gas sampling, which required collection of the sample at formation pressure. A downhole canister sampler, described by Johnson et al. [9], can be used to collect a sample at formation pressure. Also, sampling below the suction lift can be accomplished using a small-diameter gas-drive unit, described by Robin et al. [10].

Winter sampling or pressure monitoring in both instruments required a heated enclosure, such as an ice fishing tent. During winter, water in the casing annulus of the Waterloo instrument was drawn down below the frost line. However, the water in the access tubes could not be drawn down and had to be thawed prior to monitoring. The Westbay instrument did not require thawing since the water in the casing annulus was isolated from the monitoring zones and was drawn down below the frost line. Recently, modifications have been made to the commercial version of the Waterloo instrument to allow winter sampling and more flexibility in other sampling and pressure-monitoring activities.

Hydraulic Conductivity Tests

Rising-head hydraulic conductivity tests were simple and rapid with both instruments. Tests in the Westbay instrument required the use of downhole tools, while tests in the Waterloo instrument were done with conventional equipment. In the Westbay instrument, the water level was drawn down, a tool was used to open the monitoring zone port quickly, and the rate of water level recovery was recorded. In the Waterloo instrument, the water level was drawn down with a peristaltic pump, the pump suction line was removed, and the rate of water level recovery was recorded.

The size of the entry port area and the access tube limits the maximum hydraulic conductivity that can be measured using these two instruments. When using a water-level tape, the maximum measurable limit was calculated[6] to be about 2×10^{-6} m/s for the Waterloo instrument and about 1×10^{-4} m/s for the Westbay instrument. These measurement limits are shown in Fig. 7.

Quality Control Checks

Quality control records were generated during installation and operation of Westbay instrument components. First, the serial numbers of all of the couplings were checked and recorded so they could be traced back to the manufacturer. Second, a leak test can be performed on each port and joint prior to placement of these components into a borehole. However, this test was not performed on the installed instrument. Third, during the packer inflation operation, the volume of water injected into each packer and the final inflation pressure were recorded. After the packers were inflated, the depths of all couplings were checked against the instrument log using downhole tools. Finally, routine checks of the annular seals and casing integrity were conducted as part of the monitoring routine.

Quality control records were also generated during the installation and operation of the Waterloo instrument. The depths of the measurement ports were sounded through the access tubes. The packer inflation and casing string integrity were checked by withdrawal of water from a zone and then monitoring any effects on the water levels in the adjacent zones or the casing annulus. A review of the water-level monitoring results also gave an indication of the packer inflation and casing string integrity. There were no numerical data generated on the packer inflation pressure since the Waterloo packers are self-inflating. Finally, there were no serial numbers recorded for access ports or other key instrument components since the installed instrument was a prototype.

Costs

Compared with a conventional single standpipe piezometer, the cost[7] for the downhole components of the two systems used was high. The cost for either system would be several thousands of dollars to equip a 60-m-deep hole with six monitoring zones. The main advantage over a single standpipe is that only one borehole is required for multiple-level information, therefore lowering the total cost, including drilling of the boreholes.

The installation costs for both systems were about the same. The instrument strings were installed in about the same amount of time, although more people were required for the Waterloo installation. An extra day was needed to inflate the Westbay packers individually, while an extra hour was needed to confirm that the packers in the Waterloo instrument were properly inflated.

---

[6] The calculation is based on a variable-head hydraulic test. The following assumptions are made: a recovery of 5 m of water, a maximum recovery of one reading per 15 s, and five readings used to define the recovery response. Thus, a minimum test duration of 75 s was needed.

[7] The cost comparison is based partly on data supplied by Westbay Instruments Ltd. and Solinst Canada Ltd.

The monitoring probes, which were required for both instruments, added to the cost. A complete set of Westbay probes was more than $20 000, but these probes can be rented or leased. The Waterloo instrument was monitored with conventional equipment at a cost of less than $2000; however, optional equipment and probes for the commercial version of the Waterloo instrument could increase the cost of monitoring to several thousands of dollars. The advantage in the acquisition of probes for either instrument is that they are portable and can be used to monitor many instruments in a network or at other sites.

Operating costs differ depending on the activities. Hydraulic conductivity testing took less time per zone with the Westbay system. Pressure readings were more time-consuming in the Westbay instrument. The time needed to develop the monitoring zones of the Waterloo instrument far exceeded the time needed to develop those of the Westbay instrument. However, the extra time needed for the Waterloo instrument was partly due to the low transmissivity of the formation at the test site.

Conclusions

These two multiple-level ground-water systems have completely different designs and operating approaches. The systems differ in cost, design, installation techniques, testing methods, complexity, and flexibility. Both systems were easily installed and used reliably to evaluate the ground-water chemistry, permeability, and fluid pressure at different depths within a single borehole.

These multiple-level systems have many advantages over conventional standpipe piezometers or open-borehole testing techniques. First, the cost of information per monitored zone may be lower than that for several standpipe installations, particularly when deep boreholes are required. Second, a variety of quality control tests can be performed to confirm the annular seals and check the integrity of downhole components. Finally, routine monitoring generates data on the performance of the instrument as well as the required ground-water data.

The capabilities, complexity, and cost of these or other ground-water monitoring systems should be evaluated to select the most appropriate instrument for an application.[8] Table 3 may provide some assistance in the selection process. When evaluating an instrument, the performance aspects should be considered in the context of the overall technical requirements and objectives of a project.

References

[1] Pearson, R., "An Overview of Hydrogeological Field Research in the Canadian Nuclear Fuel Waste Management Program," *Proceedings,* International Groundwater Symposium, International Association of Hydrogeologists, Canadian Chapter, Montreal, Quebec, Canada, May 1984.

[2] Underwood McLellan Ltd., "Hydrogeologic Investigations, Research Area 4, Atikokan, Ontario," unpublished AECL report, available from Applied Geoscience Branch, Whiteshell Nuclear Research Establishment, Pinawa, Manitoba, Canada, 1985.

[3] Deere, D. U., "Geological Considerations," *Rock Mechanics in Engineering Practice,* K. G. Stagg and O. C. Zienkiewcz, Eds., Wiley, New York, 1968, pp. 1–20.

[4] Cherry, J. A. and Johnson, P. E., "A Multilevel Device for Monitoring in Fracture Rock," *Ground Water Monitoring Review,* Vol. 2, No. 3, 1982, pp. 41–44.

[5] Black, W. H., Smith, H. R., and Patton, F. D., "Multiple-Level Ground Water Monitoring with

---

[8] Note that the instruments are available through the following companies. For the Westbay instrument: Westbay Instruments Ltd., 507 East Third Street, North Vancouver, British Columbia, Canada V7L 1G4. For the Solinst instrument: Solinst Canada Ltd., Williams Mill, 515 Main Street, Glen Williams, Ontario, Canada L7G 3S9.

the MP System," *Proceedings,* NWWA-AGU Conference on Surface and Borehole Geophysical Methods and Ground-Water Instrumentation, Denver, CO, 15–17 Oct. 1986, pp. 41–61.

[6] Papadopulos, S. S., Bredehoeft, J. D., and Cooper, H. H., "On the Analyses of "Slug Test" Data," *Water Resources Research,* Vol. 9, No. 4, August 1973, pp. 1087–1089.

[7] Patton, F. D. and Smith, H. R., "Design Considerations and the Quality of Data from Multiple-Level Groundwater Monitoring Wells," *Ground-Water Contamination: Field Methods, ASTM STP 963,* American Society for Testing and Materials, Philadelphia, 1988.

[8] Patton, F. D., "Groundwater Instrumentation for Determining the Effect of Minor Geologic Details on Engineering Projects," *The Art and Science of Geotechnical Engineering of the Dawn of the Twenty-First Century,* a volume honoring Ralph B. Peck, University of Illinois, Urbana, IL, 1989, pp. 73–95.

[9] Johnson, R. L., Pankow, J. F., and Cherry, J. A., "Design of a Ground-Water Sampler for Collecting Volatile Organics and Dissolved Gases in Small-Diameter Wells," *Ground Water,* Vol. 25, No. 4, July–August 1987, pp. 448–454.

[10] Robin, M. J. L., Dytynshyn, D. J., and Sweeney, S. J., "Two Gas-Drive Sampling Devices," *Ground Water Monitoring Review,* Vol. 2, No. 1, 1982, pp. 63–66.

James D. Pennino[1]

Total Versus Dissolved Metals: Implications for Preservation and Filtration

REFERENCE: Pennino, J. D., "**Total Versus Dissolved Metals: Implications for Preservation and Filtration,**" *Ground Water and Vadose Zone Monitoring, ASTM STP 1053,* D. M. Nielsen and A. I. Johnson, Eds., American Society for Testing and Materials, Philadelphia, 1990, pp. 238–246.

ABSTRACT: Review of ground-water quality data obtained from wells in southwestern Ohio revealed the presence of lead, cadmium, and other heavy metals. Often, no pollution source could be discovered to account for the presence of the heavy metals. In order to establish whether sampling methods were a possible source of contamination, three possible sources of spurious heavy metal values were investigated: (1) the sample preservative, (2) the sampling equipment, and (3) suspended matter in the samples.

The results of this study indicate that painted labels on nitric acid preservative ampules, used during the period 1978 to 1980, contributed to false indications of ground-water pollution. Tests of galvanized steel and polyvinyl chloride bailers showed that this equipment was not a source of lead or cadmium. Statistical analysis of a random sample of ground-water analyses from 70 wells revealed a significant relationship between sediment and lead values. Furthermore, lead values generally increased as the amount of sediment in the sample increased. This relationship probably reflects the presence of lead in particulate matter in the water sample.

Particulate matter found in well water samples can be pieces of corroded well casing or fine sediment from the formation in which the well is screened. Since very little information is available on heavy metals in Ohio's glacial deposits, additional studies of these materials should be done to provide background data for the evaluation of heavy metals in ground water.

KEY WORDS: total metals, dissolved metals, ground water, heavy metals, sample filtration, sample preservation, bailer, arsenic, cadmium, lead, iron, manganese, aluminum, chromium, copper, nickel, titanium, zinc

In the past two decades, improvements in analytical techniques have made it possible to detect heavy metals in concentrations of a few parts per billion. During this same period, many states have designated an agency to collect ground-water quality data and maintain records of these data. In most states the data represent the first analyses of ground water for heavy metals, and it is often not known what background levels (natural concentrations) of heavy metals are normal in the ground water or what levels may be due to the activities of man.

Among workers in the ground-water field, there is uncertainty over what constitutes a representative sample. Since background heavy metal concentrations are usually not known, and since these data may be questionable because of undocumented quality control and turbidity in the samples [1], it is important to evaluate existing heavy metal data carefully.

---

[1] Hydrologist, Minnesota Pollution Control Agency, St. Paul, MN 55155.

The Ohio Environmental Protection Agency (OEPA) has maintained a ground-water quality monitoring program for both background water quality and pollution effects since its inception. Arsenic, cadmium, chromium, and lead have been monitored and detected in concentrations ranging from less than 1 ppb (the lowest detection level of the available instruments) to several times the maximum contaminant levels allowed in drinking water by the interim primary drinking water standards [2]. Lead, cadmium, and other heavy metals were detected in background monitor wells, landfill and industrial waste disposal site monitor wells, and residential wells. In many cases no obvious source could be found for the heavy metals detected in the ground water. Over a period of eleven years a considerable number of questionable analyses had been accumulated in the OEPA files.

A study was initiated with the objective of determining the validity of the heavy metal data. This study is especially important since some of the questionable data were obtained from wells used for drinking water. Another objective of the study was to evaluate sampling procedures and other conditions which could cause false heavy metal values to appear in the data. Since lead and cadmium are the heavy metal parameters that are most often unexplained in the data, this study deals mainly with them.

The data used for this study were generated by experiment or were selected from the OEPA files. The data consisted of analyses of samples from wells at facilities or areas where heavy metal pollution was unlikely. Therefore, the occurrence of lead or cadmium found in these wells was probably the result of contamination from the sampling procedure, the sample handling, the well casing, the formation in which the well was screened, or laboratory error. The study described in this paper was designed to investigate the sampling equipment, sample preservation, and formation sediment in the sample.

General Sampling Procedures

Ground-water samples were collected twice a year from all of the regular monitoring stations. Samples were obtained using a bailer (steel or plastic), pitcher pump, or portable submersible pump or from a tap using the facility's or homeowner's turbine, submersible, or jet pump. The well was usually bailed or pumped until at least one casing volume was removed. Household wells were pumped until both the casing and pressure tank volume were removed. When the casing volume was not known, the well was pumped until the water temperature reached the normal ground-water temperature for the locality. If the well water was cloudy, pumping or bailing was continued until the water became as clear as possible before the sample was collected. It was necessary to obtain a clear sample since there was no field filtration equipment available in the ground-water program at that time.

The samples for metal analyses were collected in disposable plastic bottles that had been rinsed with some of the sample water before being filled. Samples for metal analysis were preserved with nitric acid to obtain a pH of 2 or less [3]. The preservatives were furnished in 1.5 or 5-mL glass ampules.[2] The samples were kept on ice or in a refrigerator during storage and transportation to the Ohio Department of Health (ODH) Laboratories. At the laboratory the samples were analyzed according to U.S. Environmental Protection Agency (USEPA) procedures [3].

---

[2] The preservative ampules used by the OEPA during the period of this study were made by Poly Research Corp., New York, NY. Mention of the manufacturer's name does not necessarily constitute endorsement.

Evaluation of Nitric Acid Preservative

Introduction

From late 1978 to early 1980 many wells in the monitoring program exhibited abnormally high lead concentrations. These included ambient monitor wells and wells installed to monitor ground-water pollution. In 1979, chemists at the ODH laboratories and at the U.S. Geological Survey were concerned that the paint used to label the acid ampules and to coat the neck of the ampule might contain lead and cadmium. The ampules break apart at the neck which is etched or scratched to facilitate separation. The preservation technique involved snapping the neck of the ampule and shaking or tapping the acid into the sample bottle. Some of the paint was scraped from the ampule and tested at the ODH laboratory and found to contain lead at several hundred parts per million. However, no tests of the acid itself or of a white film found on the outside of many of the ampules were performed. This white film appeared to be a precipitate which formed on the ampules during transportation and storage. If any of the ampules were broken, acid vapors from the broken ampule seemed to cause precipitation of substances in the atmosphere onto other ampules in the shipping carton. This precipitate appeared as a white film on the ampules. Further investigation was needed to determine whether use of the preservative ampules was resulting in lead contamination of the samples.

Investigation Procedure

Several nitric acid ampules with the painted labels and necks were obtained. These ampules were divided into three groups. The first group was used to test the paint. The second group was used to test the white film on the ampules. The third group was tested for the presence of metals in the acid itself. A sample of distilled water treated with 1.5 mL of nitric acid from an unpainted batch of ampules (furnished by the same manufacturer) was used as a control.

Ampules from the first group were wiped and rinsed clean. Three of these ampules were then placed in 500 mL of distilled water which had been treated with 1.5 mL of nitric acid from an unpainted batch of ampules that had also been used for the acid in the control sample. After 5 to 7 min of immersion in the distilled water, the ampules were removed and the water was poured into a rinsed sample bottle.

The procedure was repeated for the second group, except that the outside of the ampules was not cleaned so that the white film would be exposed to the acidified water. One ampule from the third group was cleaned, opened, and added to a sample bottle filled with distilled water. The four samples, including the control, were sent to ODH laboratories for analysis.

Results of the Investigation

Table 1 shows the results of the experiment. In addition to lead and cadmium, the metals aluminum, arsenic, chromium, and copper were also tested for. The control sample did not show the presence of any of the metals at the detection levels listed in Table 1. The lead concentration in the sample from the cleaned ampule was 130 μg/L, and the cadmium concentration was 8 μg/L. The uncleaned ampules contained 140 μg/L of lead and 16 μg/L of cadmium. In the test of the acid preservative itself, the concentrations of all metals were below the detection limit levels. The experiment indicated that both the paint and the white film were sources of lead and cadmium. The painted ampules with a white film appeared to be a greater source of lead and cadmium. When the ampules were opened and poured into the sample container, droplets of the acid often came in contact with the paint

TABLE 1—*Nitric acid preservative study.*[a]

Metal	Detection Level	Ampule Uncleaned with White Film	Ampule Cleaned, with White Film Removed	Acid Preservative	Control
Aluminum	200	*[b]	*	*	*
Arsenic	10	*	*	*	*
Cadmiun	5	16	8	*	*
Chromium	30	*	*	*	*
Copper	30	*	*	*	*
Lead	5	140	130	*	*

[a] All values are in micrograms per litre.
[b] Asterisks indicate that the values are below detection level.

or white film, or both, on the outside edge of the ampule neck. Apparently, this was the mechanism for some of the spurious results found in previous analytical data.

Based on these results, wells which had shown high lead concentrations in 1978 and 1979 were resampled. Both lead and cadmium were found to be below the detection levels in many of these wells.

Sampling Equipment Study

Investigation Procedure

Another study was initiated to determine whether metal contamination could be introduced into the sample from the sampling equipment.

Two galvanized steel bailers (1.90 and 4.44 cm in diameter), a 10.16-cm-diameter titanium bailer, and a 10.16-cm-diameter polyvinyl chloride (PVC) plastic bailer were tested in this study. Each bailer was examined for any residues, scrubbed if necessary, rinsed once with tap water, and then rinsed with distilled water. This was the cleaning procedure followed prior to use of the bailers in the field. Each bailer was filled with distilled water and allowed to stand for 1 to 3 min to simulate field conditions, since it generally takes 1 to 3 min to retrieve a sample from a well. The sample bottles were rinsed with distilled water from the bailers to simulate field rinsing with sample water and then the sample bottle was filled with the remaining water in the bailer. A control sample of distilled water was also prepared. Each sample was preserved with 5 mL of nitric acid to obtain a pH of 2 or less. The samples were analyzed for aluminum, arsenic, cadmium, chromium, copper, iron, lead, manganese, nickel, and zinc. A sample from the titanium bailer was also tested for titanium.

Results of the Investigation

The results of analyses of the water samples from the bailers are presented in Table 2. Both galvanized steel bailers indicated aluminum concentrations at the detection limit level of 200 μg/L. The 1.90-cm-diameter bailer and the 4.44-cm-diameter bailer showed iron concentrations at 1200 μg/L and 100 μg/L, respectively. The 1.90-cm-bailer also showed manganese at the detection level of 30 μg/L. All the other parameters tested in the steel bailer samples were found to be below the detection level.

The presence of aluminum and manganese in the steel bailers is difficult to explain. These metals may be present as trace contaminants in the galvanized steel. The aluminum

TABLE 2—*Bailer study results.*[a]

Metal	Detection Level	1.90-cm Bailer	4.44-cm Bailer	Titanium Bailer	PVC Bailer	Control
Aluminum	200	200	200	400	*[b]	*
Arsenic	10	*	*	*	*	*
Cadmium	5	*	*	*	*	*
Chromium	30	*	*	*	*	*
Copper	30	*	*	*	*	*
Iron	30	1200	100	*	*	*
Lead	5	*	*	*	*	*
Manganese	30	30	*	*	*	*
Nickel	100	*	*	*	*	*
Titanium	2000	NA[c]	NA	*	NA	NA
Zinc	30	*	*	*	*	*

[a] All values are in micrograms per litre.
[b] Asterisks indicate that the values are below detection level.
[c] NA = not analyzed.

concentration of 400 μg/L in the titanium bailer is even more difficult to explain since that bailer was specifically designed to be free of all metals except titanium and steel in the ball valve. No aluminum or manganese was detected in the sample of water from the PVC bailer or from the control sample. The presence of aluminum and manganese was attributed to traces of these elements in the metals used to manufacture the bailers or to experimental error. No other metals were detected in the sample from the titanium bailer. All of the parameters tested were below the detection levels for the water sample from the PVC bailer. The control sample also showed all concentrations below the detection levels.

Iron appeared to be the only significant contaminant in the water sample from the galvanized steel bailers. No lead, arsenic, cadmium, chromium, copper, nickel, or zinc were detected in this portion of the study. Therefore, it was unlikely that the bailers were a source of these metals.

Sediment Study

Introduction

Another objective of this study was to determine whether suspended matter, noted in many of the samples, contained lead and other heavy metals. As previously discussed, the samples were usually preserved in the field without filtering. Since some wells would not produce a clear sample, cloudy samples were occasionally obtained. However, acid preservation of the sample could leach metals from formation sediments or other suspended material in the sample, thereby contributing dissolved metals to the water sample.

Most of the wells used for background water-quality data were municipal wells equipped with turbine or submersible pumps. Wells installed for monitoring pollution typically consisted of galvanized steel casings with brass or steel drive points, plastic screens and casings, or black iron casings with preforated sections as screens. Samples were obtained from these wells using steel or plastic bailers, a hand pump (pitcher pump), a portable 30.3 to 37.9-L/min submersible pump, or existing pumps installed by the property owner or facility operator (usually a submersible pump).

Investigation Procedure and Results of the Investigation

Seventy wells were selected for this part of the study. The analyses from these wells were selected from data for the period 1980 to 1982. Data from 1978 through 1980 were not used because of the preservative problem described above. The wells were selected from monitoring stations where there were no known sources of lead other than the well itself or the formation material. The wells included both shallow and deep wells with various diameters and casing materials (plastic, iron, and galvanized steel). All of the wells were screened in glacial deposits.

One to five analyses were selected from OEPA files for each of the 70 wells (a total of 150 analyses). The analytical reports included the lead concentrations, a pollution code, the type of well, the condition of the sample, and the identification of the well. The pollution code is a subjective term indicating whether or not the well is considered to be contaminated by some pollution source. The sample condition is a coded value judgement signifying that the sample was observed to be either clear or cloudy when it was collected.

The Statistical Analysis System (SAS) computer package [4,5] was chosen to analyze the data. The data were entered into the IBM 3031 computer at Wright State University, in Dayton, Ohio. The first step in the analysis was to select randomly one analysis for each of the 70 wells. This was done with a random selection program [4]. The data were then simplified as follows: (*a*) Concentrations of lead above an arbitrary value of 15 μg/L were considered significant. Values below the 15 μg/L level were not used because of a greater chance of sampling and laboratory error at these lower concentrations. If the concentration was 15 μg/L or greater, the value for lead was entered as "yes"; if not, it was entered as "no." (*b*) The types of wells were categorized as "metal" or "plastic." (*c*) The pollution code was entered as a "U" for unpolluted and a "P" for polluted. (*d*) The sample condition was listed as "clear" or "cloudy."

The new data set was then subjected to a log-linear model analysis performed on a contingency table [6]. This analysis procedure was selected because of the necessity of using descriptive terms to represent the data rather than numerical values. The model contained four variables: (1) LEAD (no, yes); (2) SITE (unpolluted, polluted); (3) TYPE (metal, plastic,); and (4) CONDITION (clear, cloudy). The resulting best fitting log-linear model produced by the computer was

$$\log m_{ijkl} = u + u_{1(i)} + u_{2(j)} + u_{3(k)} + u_{4(l)} + u_{14(il)} + u_{23(jk)}$$

where m_{ijkl} represents the expected value of lead (absent or present) listed in a cell of the contingency table; $u_{1(i)}$, $u_{2(j)}$, $u_{3(k)}$, and $u_{4(l)}$ are computer-generated numerical values for the four variables listed above; u_{14} and u_{23} are the interactions between Variables 1 and 4 and between Variables 2 and 3, respectively. The model was operated at the 95% confidence level. The interpretation of this model indicates that Variables 1 and 4 are related and Variables 2 and 3 are weakly related, and there are no other significant associations. Hence, the presence of LEAD is related to the CONDITION of the water sample. The relatively weak relationship between SITE and TYPE is probably due to the fact that some of the wells were selected from polluted sites (although lead was not suspected as a contaminant from activities at the chosen sites) and because most of the wells at contamination sites are plastic monitor wells. Therefore, there would be an artificial correlation between sampling SITES and well TYPE.

TABLE 3—*Contingency table.*

Condition	Presence of Lead		Probability of Lead Presence, %
	No	Yes	
Clear	31	1	0.03
Slightly cloudy	7	3	0.30
Cloudy	7	8	0.53
Very cloudy	1	12	0.92

In order to confirm the relationships found with the log-linear model, a logit analysis was performed. The logit analysis indicated that the only factor that was significant in affecting the logit LEAD content is Variable 4, CONDITION [6,7]

$$\text{logit}_{jkl} = \log \left(\frac{m_{1jkl}}{m_{2jkl}} \right) = w + w_{4(l)}$$

Since only LEAD and CONDITION were related, the parameters SITE and TYPE were dropped from the log-linear model and a more sensitive analysis was obtained by subdividing the CONDITION parameter into four levels. These levels were clear, slightly cloudy, cloudy, and very cloudy. These parameters were based on qualitative observations made at the time the samples were collected. The contingency table resulting from this log-linear model is shown as Table 3. The numbers in the right-hand column of Table 3 represent the probability of finding lead in a sample of water under the given CONDITION.

The value of the Pearson chi-square statistic for this table is 35.713 with 3 degrees of freedom. This is a highly significant value indicating that the parameters CONDITION and LEAD are strongly associated. This means that, as cloudiness increased, the probability that the well water contained lead increased. In fact, a test of the significance for the slope of a simple linear regression on these percentages (Table 3) for the CONDITION variable rendered a highly significant result: chi-square = 35.347 with 1 degree of freedom.

The association between LEAD and CONDITION was strongly positive. For relatively clear water samples there was a small likelihood of lead in the sample; for relatively cloudy water samples there was a high likelihood that lead would be found in the water sample. Furthermore, this relationship was linear in nature. This linear relationship is illustrated in Fig. 1.

Conclusions

The results of this study indicate that several batches of nitric acid preservative used during 1978, 1979, and 1980 could have contributed to erroneous lead values in ground-water quality data. It was shown that part of this contamination could be attributed to paint and, to a lesser extent, to a white film on the outside of the preservative ampule. To minimize the possibility of such contamination, unpainted ampules should be used,[3] and the ampules should be thoroughly rinsed with distilled water or some of the sample water to remove any soil or precipitates before breaking them open.

---

[3] The previously referenced manufacturer now provides an unpainted preservative ampule.

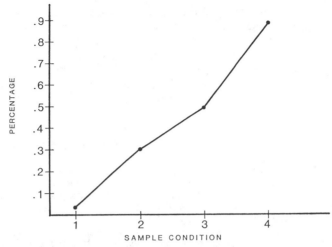

FIG. 1—*Probability that the water contains lead* (P) *versus the sample condition.*

Tests of the bailers used to collect samples showed that this equipment was not a source of lead or cadmium. Iron may have contaminated some samples collected with galvanized steel bailers, and previous iron concentration data obtained by this sampling method should be evaluated accordingly. Furthermore, these bailers should not be used where iron analysis is required.

Statistical analysis of a random sample of analyses from 70 wells revealed a significant relationship between sediment and lead values. Furthermore, lead values increased as the amount of sediment in the sample increased. This relationship probably reflects the presence of lead in the glacial deposits of southwestern Ohio. Since very little information is available on heavy metals in Ohio's glacial deposits, additional studies should be done to provide background data for the evaluation of heavy metals in ground water.

Field filtration may be necessary for samples from wells which yield cloudy water if it can be shown that any heavy metals present in the sample are contained only in the sediments which contributed to the sample's cloudiness. This may be determined by collecting both a filtered and an unfiltered sample for the metals of concern and comparing the results of analyses. Both samples must be acidified in the field, with the filtered sample acidified immediately after filtering.

It may also be advisable to analyze for metals of concern in samples of soils representative of the zone where the ground water is being sampled. This is especially valuable at a ground-water contamination investigation site. An analysis of the soils taken from a nearby area where ground-water contamination is not present will provide direct evidence of the presence of natural heavy metals in the soils. Soil samples from the vicinity of the contaminated wells can also be analyzed for total metals to obtain an indication of the metals that may have been released by the contaminated ground water and subsequently adsorbed onto the aquifer materials.

There are many sources for heavy metals in ground-water samples. This study reveals that some of the heavy metals found in ground-water samples are unrelated to pollution sources. The findings of this study substantiate the need for careful quality control measures in any ground-water monitoring program. For instance, if field blanks had been used in the late 1970s, the lead and cadmium contamination from the preservation technique would have been caught and rectified much sooner. This study has also demonstrated the need for careful field documentation during sampling. Documentation of the sample condition provided data for a statistical analysis which led to conclusions about lead in the formation sediments.

Historical ground-water quality data is essential in the evaluation of man's impact on ground water and in developing ground-water protection programs, but such data must be carefully screened for spurious values which are the result of the sampling technique or well design rather than the actual ground-water quality.

References

[1] Strausberg, S., "Turbidity Interferences with Accuracy in Heavy Metals Concentration," *Industrial Wastes,* March/April, 1983, pp. 20–21.
[2] *National Interim Primary Drinking Water Regulations,* Publication No. EPA-570/9-76-003, U.S. Environmental Protection Agency, Office of Water Supply, Washington, DC, 1976.
[3] *Methods for Chemical Analysis of Water and Wastes,* Publication No. EPA-600/4-79-020, U.S. Environmental Protection Agency, Office of Research and Development, Environmental Monitoring and Support Laboratory, Washington, DC, 1979.
[4] Council, K. A., Ed., *SAS Applications Guide,* SAS Institute, Inc., Cary, NC, 1980.
[5] Helwig, J. T. and Council, K. A., Eds., *SAS User's Guide,* SAS Institute, Inc., Cary, NC, 1979.
[6] Fienberg, S. E., *The Analysis of Cross-Classified Categorical Data,* 2nd ed., MIT Press, Cambridge, MA, 1977, pp. 95–119.
[7] Bishop, Y. M. M., "Full Contingency Tables, Logits, and Split Contingency Tables," *Biometrics,* Vol. 25, 1969, pp. 383–400.

G. M. Spreizer,[1] D. Maxim,[2] N. Valkenburg,[1] and M. Hauptmann[1]

How Flat Is Flat?—Termination of Remedial Ground-Water Pumping

REFERENCE: Spreizer, G. M., Maxim, D., Valkenburg, N., and Hauptmann, M., "**How Flat Is Flat?—Termination of Remedial Ground-Water Pumping,**" *Ground Water and Vadose Zone Monitoring, ASTM STP 1053*, D. M. Nielsen and A. I. Johnson, Eds., American Society for Testing and Materials, Philadelphia, 1990, pp. 247–255.

ABSTRACT: Remedial actions are being initiated at many Superfund sites, and the analyses of water-quality data from these sites are being performed. The start of remediation is obvious—the system used for cleaning up is turned on; however, the determination of when this remediation is "complete" may not be clear.

With the initiation of a remedial measure, it is expected that the detected concentrations of characteristic contaminants will decrease rapidly and then level off to some asymptotic concentration. This is expected since the efficiency of remediation is generally considerably greater at higher concentrations than at lower concentrations. It is expected that some form of an exponential model, $y = C_{exp}(-kx)$, will describe the system's behavior. This is especially true when ground-water pumping is the remedial measure used, as is considered in this paper.

With Superfund remediation schemes, cleanup or performance goals are set for key compounds. The cleanup is complete when these goals are attained. However, attainment of cleanup goals may be technically infeasible in certain cases and alternative criteria may be appropriate.

This paper presents a statistical methodology, based on an anticipated exponential decrease in the contaminant concentration resulting from ground-water pumping remediation. The methodology will initially predict when remedial measures can be terminated. As the remediation program continues and additional monitoring data become available, the methodology will determine whether and when preestablished concentration limits will be attained and when the equilibrium concentration can be estimated.

KEY WORDS: ground water, statistics, termination pumping, exponential model, asymptote, empirical model, technical infeasibility, compliance waiver, SARA

SARA (the Superfund Amendments and Reauthorization Act of 1986, Public Law 99-499) requires the U.S. Environmental Protection Agency (USEPA) to use cleanup methods preferentially that reduce the volume, toxicity, and mobility of hazardous wastes at Superfund sites rather than nonpermanent remedial alternatives which only involve "containment" of contamination. Currently, bioreclamation, and pumping and treating systems (with or without caps and slurry walls) are the only two ground-water treatment methods which meet the SARA requirements for reducing the volume, toxicity, and mobility of contaminants. Bioreclamation is beginning to be used more often, but ground-water pumping and treating is still the most common method of remediating ground-water contamination.

[1] Senior scientist, senior associate, and associate, respectively, Geraghty & Miller, Inc., Andover, MA 01810.

[2] President, Everest Consulting Associates, Inc.,

SARA also requires that ARARs (Applicable or Relevant or Appropriate Requirements) from federal and state laws be applied as cleanup standards at Superfund sites. In practice, this has meant that the USEPA has applied the excess cancer risk levels (for carcinogenic compounds) for the cleanup of these compounds in ground water. The 1×10^{-6} (cumulative) excess cancer risk level is a commonly used standard. Table 1 shows the 1×10^{-6} and 1×10^{-7} excess cancer risk concentrations for various carcinogens. The cumulative risk approach, which totals the risks of all compounds present, reduces the 1×10^{-6} excess cancer risk cleanup levels of individual compounds significantly. These cumulative levels are determined from the number of chemicals present and their concentrations. For instance, assuming that benzene and chloroform are present at 0.67 and 0.432 ppb, respectively (an individual 1×10^{-6} excess cancer risk level for each compound), the associated cumulative ground-water risk is 2×10^{-6}. If the cleanup standard is set at the 1×10^{-6} excess cancer risk level, then the cleanup standards will be 0.33 and 0.216 ppb for benzene and chloroform, respectively (50% of the individual cancer risk levels for each).

In addition to the analytical problems connected with detecting concentrations in the sub-parts-per-billion range, these cleanup standards are so low that they may not be achievable (i.e., with available technologies, it may not be possible to reach these standards). SARA contains provisions to address those potential situations in which compliance may be "technically impractical from an engineering perspective," and thus, SARA permits a compliance waiver from the ARARs if infeasibility can be shown. A method for demonstrating the technical infeasibility of achieving a very low cleanup standard is presented in this paper. Emphasis will be placed on those situations in which contaminant concentrations in a ground-water pumping-and-treating system approach an equilibrium level (asymptote) that is higher than the designated ARAR concentration. When concentrations approach an asymptotic value, further real reductions in contaminant concentrations cannot be achieved in a reasonable period of time.

Background

Before remedial pumping begins, chemicals moving through aquifer materials with the natural flow of ground water are expected to be in dynamic equilibrium between the liquid and solid phases of the system. Depending on the chemical properties of the contaminants and on the chemical and physical properties of the water and the aquifer materials, the partitioning equilibrium will favor adsorption to the solid phase or solution in the liquid

TABLE 1—*Excess cancer risk concentrations.*[a]

Compound	1×10^{-6} Excess Cancer Risk Level, μg/L	1×10^{-7} Excess Cancer Risk Level, μg/L
Benzene	0.67	0.067
Chloroform	0.432	0.0432
1,2-Dichloroethane	0.385	0.0385
1,1-Dichloroethene	0.07	0.007
Methylene chloride	5.15	0.515
Tetrachloroethene	0.77	0.077
1,1,2-Trichloroethane	0.68	0.068
Trichloroethene	3.57	0.357
Vinyl chloride	0.017	0.0017

[a] Based on information and procedures in the *Superfund Public Health Evaluation Manual.*

phase. The greater a chemical's affinity for water, the longer it will spend in the liquid phase and the faster it will move.

Similarly, the greater the amount of organic material (natural and anthropogenic) in the aquifer, the slower the organic compounds will move. This is because an equilibrium condition results from the cumulative effect of the chemicals' affinities for the aquifer materials and the water. An example of this type of equilibrium and its cumulative effect is the separation of chemicals during chromatography, in which the substances with greater affinity for the solid phase are eluted more slowly (i.e., their movement through the chromatography column is retarded). Compounds have different retention (retardation) rates; these rates allow the identification of multiple compounds contained in a single sample.

When remedial pumping begins, the rate of movement of chemicals in the system increases because ground water moves faster towards the pumping well (by the influence of a greater hydraulic gradient) than it does under natural flow conditions. As the capture zone of the pumping well expands outward, larger portions of the plume or plumes and increased amounts of less contaminated water are drawn toward the pumping well. The concentrations of chemicals detected at the well will fluctuate, depending on the relative amounts of clean and contaminated water arriving at a particular time. If the pumping well is located at the leading edge of the plume, the concentration is expected to decline immediately because there will be more clean water than contaminated water in the capture zone of the pumping well. However, if the well is located in the center of the plume, the concentrations detected may increase at the start of pumping, because of the initial high concentration of the contaminants, and will decline as pumping continues.

Eventually, the capture zone will no longer expand, because water from recharge or leakage from other aquifers will balance the amount of water removed from the system through the remedial (pumping) well. At this point, a hydraulic equilibrium will have been reached. Under these new conditions, the chemical equilibrium will also have been re-established. In these situations, a portion of the total contaminant mass present in the liquid phase will be gradually removed by the remedial well. A certain proportion of the remaining mass will be associated with the solid phase and, as a result, will replenish the concentration in the ground water as contaminants are removed. The concentration of contaminants detected in ground-water samples will remain relatively constant and will approach an asymptotic level until the contaminant mass associated with the aquifer materials has been removed. The length of time that the concentration remains unchanged will depend on the initial areal distribution of the contaminant mass, the amount of the mass, and the partition coefficients of the various chemicals; this length of time can be decades for large plumes that contain high concentrations of contaminants. If the concentration approaches an asymptote, it would require an unreasonably long pumping period to achieve further real reduction in the contaminant concentration.

The exponential decrease in contaminant concentration to an asymptotic value in the ground water is similar to the time profile of a contaminant concentration in a chamber. The contaminant emissions from the surfaces in the chamber are assumed to occur at a rate E (measured in grams per second). For illustrative purposes, E is assumed to be constant; however, E can be estimated from the product of the unit emissions rate (measured in grams per square centimetre per second) and the contamination chamber surface area, A (measured in square centimetres). If G is the grams of contaminant in dissolved or entrained form in a chamber of a volume, V (in cubic metres), the average contaminant concentration, C, is then G/V (in grams per cubic metre). Water is assumed to be pumped into this chamber with a constant flow rate, Q (in cubic metres per second). It is also assumed that the incoming water contains contaminant at a concentration of C_i.

Given these assumptions, the mass balance equation is

$$\frac{dG}{dt} = E - \frac{GQ}{V} + C_iQ = E - CQ + C_iQ$$

This differential equation can be integrated and evaluated with respect to these boundary conditions:

1. C_o is the contaminant concentration at time O, then integration of the above equation yields an expression for the concentration at time t, (C_t), and is (with intermediate integration steps given)

$$\frac{dG}{dt} = \frac{VdC}{dt} = C_iQ - CQ + E$$

$$\int_{C_o}^{C_t} \frac{dC}{C_i - C + \dfrac{E}{Q}} = \int_o^t \frac{Qdt}{V}$$

$$\ln\left(C_i - C_t + \frac{E}{Q}\right) - \ln\left(C_i - C_o + \frac{E}{Q}\right) = \frac{-Qt}{V}$$

$$C_t = \frac{E}{Q} + C_i + \left[C_o - \left(\frac{E}{Q} + C_i\right)\right]e^{-(Qt)/V}$$

2. As the time t is allowed to increase, the C_t expression has the limit $E/Q + C_i$. If C_A equals $E/Q + C_i$, and Q/V is k, the equation is simplified to a form of an exponential model

$$C_t = C_A + (C_o - C_A)e^{-kt}$$

Therefore, as long as the emissions continue with rate E, the exponential model should describe the concentration in the chamber and, by comparison, the asymptotic decrease in contaminant concentrations in a ground-water pump-and-treat system.

An example of an exponential decline in the 1,1,1-trichloroethane concentration detected in samples collected from a remedial pumping well is shown in Fig. 1. After a large initial decline (about 1500 ppb) and the probable detection of a new slug of contamination (at about 400 days), the concentration detected fluctuates near the 200 ppb level, an asymptotic concentration described by graphical techniques.

Methodology

The following statistical method is used to demonstrate the potential technical infeasibility of achieving ground-water cleanup. The method is based on an anticipated exponential decrease in contaminant concentration with time. This method predicts when the remedial pumping can be terminated. As the pumping program continues and additional monitoring data become available, the method continuously revises the value of the asymptote and the time when a specified percentage of the asymptotic value is expected to be reached (an asymptote is only reached at infinity). The method is not dependent on a specific monitoring frequency; however, spatial, temporal, sampling, and analytical sources of variability must be considered before the statistical analyses are performed

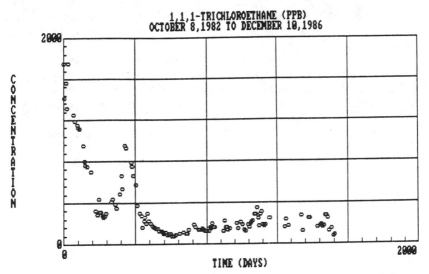

FIG. 1—*1,1,1-Trichloroethane concentrations detected in samples from a remedial pumping well.*

because the degree of uncertainty in conclusions drawn from the monitoring data must be assessed.

The concentrations of the indicator constituents are represented by C_{ijt}, which is used to denote the concentration of constituent i, from well j, at time t. To simplify the notation utilized in the following discussion, the constituent and location subscripts have been omitted and the data are assumed to be a sequential time series of concentrations for one compound at one location. The statistical procedures described below should be followed for each indicator constituent at each well.

Given this simplification of the notation, the data are in the form C_t, where C_t is the concentration (in appropriate units) of the constituent at time t and t equals 0, 1, 2, ... This technical infeasibility statistical plan includes the following two steps to demonstrate that the cleanup standard may not have been achieved:

1. For an initial period, data are collected; these are summarized to monitor the progress of the ground-water remediation.

2. Upon completion of this initial collection period, the statistical procedures described below are employed to determine estimates of the time, t_r (when at least a defined percentage of the asymptotic concentration has been achieved, or when the cleanup standard has been met), for each contaminant at each well.

The data are plotted (with the concentration as a function of time) and an exponential model is fitted. This exponential model is given as

$$C_t = C_A + (C_o - C_A)e^{-kt}$$

where

C_t = concentration at time t,
C_A = asymptotic concentration,
C_o = initial concentration at time O, and
k = decay constant.

The exponential model is expected to be a useful empirical descriptor (see the example discussed in the section on background) and should provide an adequate fit of the concentration data.

One difficulty in using a simple exponential model is that it has the property that the asymptotic value is reached only as time grows infinite. Therefore, it is necessary to define a limiting endpoint other than infinity. In this case, the endpoint is defined to be the time when the difference between the constituent concentration, C_t, and the initial concentration, C_o, is equal to a fraction, θ, of its limiting value $(C_o - C_A)$. For this model, θ is defined to be 0.9, i.e., at least 90% of the asymptotic concentration. The time when this occurs, denoted t_r, is given by

$$t_r = -\ln \frac{1 - \theta}{k}$$

$$t_r = -\ln \frac{1 - 0.9}{k}$$

$$t_r = \frac{2.303}{k}$$

where

$$\theta = \frac{C_t - C_A}{C_o - C_A}$$

The value for θ must lie between 0 and 1.0. Both of these endpoints are unrealistic. The proposed value of 0.9 attempts to provide a balance between the competing, yet interrelated, objectives of constituent removal and realistic cleanup (pumping) times.

The value of k is estimated from the data. A lower confidence limit estimate of k can be used so that the upper confidence limit for t_r can be calculated using nonlinear regression methods. This provides a conservative estimate of t_r so that the remediation effort is not terminated prematurely. The following values are also calculated from the data:

(a) estimates of the asymptotic concentration, C_A, the pumping time required, t_r, and their confidence intervals; and
(b) statistics to measure the "goodness of fit" of the model.

A simulation of this model is given in Table 2; it assumes that the standard error of replicate observations is 10%. The first set of estimates for t_r and the upper confidence limit (UCL) for the time are made after nine years. At this time t_r is estimated as 9.5 years with the UCL indicating that it may take at least 23 years to reach 90% of the asymptotic ratio, θ; hence, remedial pumping would continue. Note that as additional data are collected, the UCL of t_r decreases, reflecting the increased precision of this data analysis effort. After 14 years, the UCL of t_r is reduced to 12.3 years and pumping can be terminated (provided that the concentrations of other parameters have also reached 90% of the asymptotic ratio and that the UCL of t_r equals or exceeds the number of years the remedial measure has been operational).

Examples of other exponential contaminant decline models are given in Table 3. These additional forms of the exponential model may be substituted for the model discussed

TABLE 2—*Simulation of the exponential model and fitting procedure.*

INPUTS:

QUANTITY	UNITS	VALUE	REMARKS
INITIAL CONCENTRATION	PPB	1000	ASSUMED FOR ILLUSTRATIVE PURPOSES
EQILIBRIUM CONCENTRATION	PPB	150	ASSUMED FOR ILLUSTRATIVE PURPOSES
"HALF-LIFE"	YEARS	2.5	ASSUMED FOR ILLUSTRATIVE PURPOSES
CALCULATED RATE CONSTANT	1/YEAR	0.27726	CALCULATED FROM EXPONENTIAL MODEL
THETA	FRACTION	0.9	POLICY INPUT--FRACTIONAL APPROACH TO EQUILIBRIUM
REQUIRED PUMPING TIME	YEARS	8.3	TIME TO REACH THETA--TO BE DETERMINED FROM ANALYSIS
COEFFICIENT OF VARIATION	NA	0.1	RATIO OF STD. ERROR TO MEAN-USED IN SIMULATION

TIME PROFILE   SIMULATION**   NONLINEAR FITTING FITTED ESTIMATES*

TIME YEARS	CONCENTRATION (PPB)	FRACTION REMOVABLE CONCENTRATION REMAINING	RANDOM NORMAL DEVIATE	"SAMPLE" RESULT	CONCENTRATIONS LIMIT PPB	INITIAL ESTIMATE PPB	KINETIC CONSTANT INITIAL ESTIMATE 1/YRS	LCL 1/YRS	TIME TO THETA ESTIMATE YEARS	UCL YEARS
0	1000.0	1.00	0.048	1004.8	*					
1	794.2	0.76	-0.521	752.8	*					
2	638.2	0.57	-1.407	548.4	*		INITIAL COLLECTION PERIOD			
3	520.0	0.44	1.822	614.7	*					
4	430.4	0.33	1.346	488.3	*	NO ESTIMATES MADE IN THIS PERIOD				
5	362.5	0.25	0.420	377.7	*					
6	311.0	0.19	-1.760	256.3	*					
7	272.0	0.14	-0.959	246.0	*					
8	242.5	0.11	0.561	256.1	*					
9	220.1	0.08	-0.717	204.3	107.8	972.2	0.24128	0.09960	9.54	23.12
10	203.1	0.06	-2.097	160.5	86.3	969.2	0.22949	0.11570	10.03	19.90
11	190.3	0.05	0.443	198.2	114.2	973.9	0.24580	0.14503	9.37	15.88
12	180.5	0.04	0.481	189.2	128.9	976.9	0.25573	0.16583	9.00	13.89
13	173.1	0.03	-2.219	134.7	119.0	974.6	0.24868	0.17047	9.26	13.51
14	167.5	0.02	1.723	196.4	135.7	979.0	0.26144	0.18663	8.81	12.34

* ESTIMATES FROM MARQUARDT'S METHOD AS IMPLEMENTED IN PLOTIT (COPYRIGHT) SOFTWARE.
** RANDOM NORMAL DEVIATES TAKEN FROM NATRELLA, M., EXPERIMENTAL STATISTICS, NBS HANDBOOK 91, NATIONAL BUREAU OF STANDARDS, WASHINGTON, D.C. 1963, TABLE A-37.

TABLE 3—*Candidate contaminant decline models.*

Model	Initial Value	Limit Value	Time to Reach A Fraction, Θ, of The Limit Value, t_r
$C_t = C_a + (C_o - C_a)e^{-kt}$	C_o	C_a	$t_r = \dfrac{\ln(1-\Theta)}{-k}$
$C_t = C_a + (C_o - C_a)t^{-k}$	C_o	C_a	$t_r = \exp\left(\dfrac{\ln(1-\Theta)}{-k}\right)$
$C_t = a/(1 + be^{-kt})$	$\dfrac{a}{(1+b)}$	a	$t_r = \dfrac{-1}{k}\ln\left(\dfrac{1-\Theta}{1+b\Theta}\right)$
$C_t = C_a + (C_o - C_a)k^t$	C_o	C_a	$t_r = \dfrac{\ln(1-\Theta)}{\ln k}$
$C_t = kab^t$	ka	k	NA

NOTE: Additional models may be employed in the analysis. Potentially useful models can be found in Frost and Pearson, or Draper and Smith (op. cit.).

above, with the associated equations revised accordingly. All of the exponential model-fitting exercises can be done with standard curve-fitting techniques (e.g., least squares, robust regression).

Limitations of the Method

As with all analysis techniques and models, there are limitations associated with the method. The methodology discussed in this paper is simple but not naive. The only inputs required are concentration values collected over time. However, these concentrations all have uncertainties associated with the analytical results.

As discussed previously, it is necessary to determine the degree of uncertainty associated with the estimates calculated. A replicate sampling program should be undertaken to assess the sampling and analytical variability. This program should include the use of both replicate and duplicate samples and analyses. (In this usage, replication means the collection and analysis of multiple observations at a given time from one or more wells, whereas duplication is the multiple laboratory analysis of a single sample.) It may be advantageous to replicate at least selected measurements for the following reasons:

(*a*) to provide a basis for selecting a useful transformation to enhance the appropriateness of the exponential model;
(*b*) to permit the use of statistical tests for lack of fit of the time-series models employed; and
(*c*) to optimize subsequent data collection efforts.

Estimates of the error variance components (e.g., sampling error, analytical error) are then determined. The data are analyzed using the Box and Tidwell variant of the usual analysis of variance (ANOVA) methods. These estimates are then used to characterize the reliability of the data and ultimately of the conclusions (i.e., C_A and t_r).

Conclusions

Many cleanup standards are being set at very low levels; these low levels may not be technically achievable. The method presented here, which is empirical, can be used to demonstrate that remediation has been completed even though cleanup standards have not been reached.

Edward A. McBean[1] and Frank A. Rovers[2]

Flexible Selection of Statistical Discrimination Tests for Field-Monitored Data

REFERENCE: McBean, E. A. and Rovers, F. A., **"Flexible Selection of Statistical Discrimination Tests for Field-Monitored Data,"** *Ground Water and Vadose Zone Monitoring, ASTM STP 1053,* D. M. Nielsen and A. I. Johnson, Eds., American Society for Testing and Materials, Philadelphia, 1990, pp. 256–265.

ABSTRACT: The capabilities and limitations of a number of statistical methods for statistical discrimination testing are examined. Arguments are developed for flexibility in selecting alternative statistical models. Several alternatives for including detection limit data are developed.

KEY WORDS: ground water, significance tests, t-test, detection limits

In ground-water studies, there is a frequent need to determine whether a significant difference exists between two or more sets of data. This requirement arises, for example, when examining whether a landfill site has impacted the ground-water quality. However, the numerous sources of variability implicit in ground-water quality measurements (e.g., sampling errors, inadequate sample storage and preservation techniques, analytical errors, and the vagaries of nature) require that statistical discrimination tests be employed.

Compounding the problems of uncertainty/variability are features of ground-water quality records including the following:

1. A number of constituents may be of concern, and laboratory costs are high, which tends to act against development of sizable monitoring records.
2. Many phenomena take years to evolve, and therefore the available time frame for sampling programs represents only a "window" of a temporally varying process.

To consider the problem of assessing the significance of the impact of such variables, a number of different statistical tests might be appropriate. Each test involves different assumptions, and there are no absolute rules stipulating which test to apply, only guidelines. Note that one must always temper statistics with a clear understanding of the problem, so that spurious information is not introduced, nor valid information omitted. Statistical analyses of data are not interpretations of the facts; the analyses are just another way of making the facts easier to see, and therefore to interpret. The various statistical methods are tools for data analysis, which, like any tools, have proper and improper applications. The intent of this paper is to argue for the flexibility that will allow the alternative statistical models to be used for a problem.

---

[1] Professor of civil engineering, University of Waterloo, Waterloo, Ontario, Canada N2L 3G1.
[2] President, Conestoga-Rovers and Associates Ltd., Waterloo, Ontario, Canada N2V 1C2.

Alternative Tests

When a finding that is very unlikely to have arisen by chance is obtained, the result is referred to as being statistically significant. In other words, the finding is difficult to ascribe to chance, and the difference must, in all common sense, be accepted as a real difference.

In response to the need to identify statistically significant differences, a number of discrimination testing procedures have been proposed. No absolute rules are available for selecting the discrimination procedures to use in a specific application, only guidelines. To a large extent, the selection of the best procedure involves careful scrutiny of the characteristics of the problem at hand and the assumptions implicit in the particular discrimination technique being considered.

t-Tests

The first of these statistical significance tests to be developed was the student's t-distribution. Mathematically, the t-test procedure, as presented by Fisher [1], allows the testing of whether the means from two sets of measurements, say X (where the elements of X are x_i, where $i = 1, 2, \ldots n_x$) and Y (where the elements of Y are y_j, where $j = 1, \ldots n_y$) are the same. Assuming that X and Y are normally distributed, with the same variance, but that their population means, μ_x and μ_y, may be different, then the difference between the sample means, $\bar{X} - \bar{Y}$, will be normally distributed with the mean $(\mu_x - \mu_y)$. Then

$$t = \frac{|\bar{X} - \bar{Y}|}{\sqrt{\dfrac{(n_x - 1)S_x^2 + (n_y - 1)S_y^2}{n_x + n_y - 2}} \sqrt{\dfrac{1}{n_x} + \dfrac{1}{n_y}}} \tag{1}$$

where $||$ denotes the absolute value sign, S_x^2 represents the variance of sample set X_i ($S_x^2 = \Sigma(x_i - \bar{x})^2/(n_x - 1)$, and S_y^2 represents the variance of sample set Y_i. The t follows a t-distribution with $m + n - 2$ degrees of freedom.

Noteworthy points regarding the above include the following:

(a) the assumption that distributions of X and Y have the same variance is essential to the argument, and
(b) the t-test is based on the assumption that this underlying distribution is normal or Gaussian.

Unfortunately, these assumptions are frequently violated in monitoring records.

Parametric Versus Nonparametric Tests

Out of the fundamental t-test derivations, a number of alternative tests for statistical discrimination have been developed. One group of tests is referred to as parametric (meaning that they utilize the magnitudes themselves), while the second class of tests is referred to as nonparametric (meaning that they use the relative rankings of the magnitudes).

The parametric tests are very good if the following obtain:

1. The assumptions of randomness, independence, and normal distribution are true.
2. An additional test, which compares the distributions, is completed and ensures that there is no significant difference in the variances at a specified level of significance.

TABLE 1—*Summary of test statistics,*

Test	Two Sample t-Test	Satterthwaite Approximation to the Two Sample t-Test
t statistic	$t = \dfrac{\|\bar{x} - \bar{y}\|}{\sqrt{\left(\dfrac{1}{n_x} + \dfrac{1}{n_y}\right) \cdot \sqrt{\dfrac{(n_x-1)S_x^{\,2} + (n_y-1)S_y^{\,2}}{n_x + n_y - 2}}}}$	$t_1 = \dfrac{\|\bar{x} - \bar{y}\|}{\sqrt{\dfrac{S_x^{\,2}}{n_x} + \dfrac{S_y^{\,2}}{n_y}}}$
Degrees of Freedom	$df = n_x + n_y - 2$	$df = \dfrac{\left(\dfrac{S_x^{\,2}}{n_x} + \dfrac{S_y^{\,2}}{n_y}\right)^2}{\left[\dfrac{\left(\dfrac{S_x^{\,2}}{n_x}\right)^2}{n_x - 1} + \dfrac{\left(\dfrac{S_y^{\,2}}{n_y}\right)^2}{n_y - 1}\right]}$
Comments:	Since σ is unknown it is replaced by S, the sample standard deviation.	Note: round 'df' down to the next nearest integer.

Nonparametric tests do not require a test on their parameters (i.e., equal variances) for these reasons:

1. The data do not need to be normally distributed, since these tests use ranks in place of the original data.

2. The data are ranked, which also tends to suppress the outliers, and thus the tests function reasonably well for highly skewed data. This paper does not provide examples of nonparametric tests. (For details, see, for example, Ref 2.)

Insofar as the parametric tests are concerned, and out of the fundamental *t*-test derivations, a number of alternative tests for statistical discrimination have been developed. The different tests include the following:

1. Modified *t*-tests have been developed (e.g., Satterthwaite's approximation (see Imam and Conover [3]) and Cochran's approximation to the Behrens-Fisher *t*-test [4]. These tests accommodate the situation of unequal variances. One should note, however, that the tests are indeed approximate, as the names of the tests indicate, and the nature of these approximations must be reflected during their utilization (see Ref 5 for further details on the nature of the approximations). The difference between Cochran's test and Satterthwaite's

degrees of freedom, and assumptions.

Cochran's Approximation
to the Behrens-Fisher Test

$$t_2 = \frac{|\bar{X} - \bar{Y}|}{\sqrt{\dfrac{S_x^2}{n_x} + \dfrac{S_y^2}{n_y}}}$$

Paired t-Test

$$t = \frac{\bar{D}}{\dfrac{S_D}{\sqrt{n_x}}}$$

where $D_i = x_i - y_i$ for $i = 1 \ldots n_x$

and $S_D = \sqrt{\dfrac{\displaystyle\sum_{i=1}^{n_x} (D_i - \bar{D})^2}{n_x - 1}}$

$df_x = t$ - tables with $n_x - 1$
        degrees of freedom
$df_y = t$ - tables with $n_y - 1$
        degrees of freedom

$df = n_x - 1$

$$W_x = \frac{S_x^2}{n_x} \text{ and } W_y = \frac{S_y^2}{n_y}$$
with the result the
comparison t-statistic is

$$t_c = \frac{W_x t_x + W_y t_y}{W_x + W_y}$$

$$S_x = \sqrt{\frac{\sum (x_i - \bar{X})^2}{n_x - 1}}$$

$$S_y = \sqrt{\frac{\sum (y_i - \bar{Y})^2}{n_y - 1}}$$

test is strictly in calculating t_c, the critical t-statistic. Cochran's test weighs the individual t values for each sample, while Satterthwaite's test calculates new degrees of freedom.

2. Paired sample t-tests may be used when the sample populations, X and Y, are not independent. For some kinds of correlated data, it is possible to define the population being sampled so as to remove or minimize the effect on the test outcome (e.g., cyclical variations in both background and monitored data are not included in the calculation of the standard error). This is done by pairing the individual observations from the correlated populations and then testing the differences between the observations. Once the differences in the pairs are calculated, the differences are treated as a single, random, independent sample. With the pairwise test, there is one half of the degrees of freedom of the two-sample test, but the standard error calculation is smaller.

A summary of the mathematical equations implied in the above-named tests is included in Table 1.

Comparison of t-Test Versus Cochran's Approximation

One should note that both Cochran's and Satterthwaite's tests are approximate formulas for the t-test that allow relaxation of the equality of variances requirement and should be considered the most appropriate test in these situations. However, there are situations in which the assumptions implicit within the basic t-test are *not* violated, and the utilization of the approximate tests in such situations will imply different findings from those intended. This concern may be quantified, as an example, for the case where $n_x = 3$ (the downgradient location has three samples) and $n_y = 4$ (the upgradient has four samples), with the variance ratio $S_x^2/S_y^2 = 4$. The t-test allows a deviation of $|\overline{X} - \overline{Y}|$ of 2.28 before a significant difference is identified (with a level of significance of 5%, by one-tailed t-test). Translation of the allowable deviation by Cochran's test into the equivalent deviation for the t-test corresponds to $|\overline{X} - \overline{Y}| = 3.56$, which, in turn, gives a significance level of 1.3%. In other words, if Cochran's test is used instead of the t-test, the level of significance is not 0.05 but 0.013. (Concerns such as these are further described by McBean et al. [5].) Therefore, the data analyst should not blindly follow stringent guidelines without matching his or her data concerns with the various test assumptions. A summary of statistical discrimination tests in frequent use in the assessment of ground-water quality is provided in Table 2.

Consideration of Detection Limit Data

The need for flexibility in the approach to analysis is further exemplified by situations involving detection limit data. All laboratory analytical techniques have detection limits below which only "less-than" values may be recorded. The reporting of less-than values provides a degree of quantification, but, even at or near their detection limits, the concentration levels of particular contaminants may be of considerable importance because of their potential health hazards. Also, if the subsequent statistical analyses that utilize the monitoring results indicate a significant difference between two monitoring locations, a greatly increased intensity of monitoring may result. The manner in which the "less-than" information is dealt with can have a substantial impact on monitoring programs.

To examine the nature of such difficulties, consider the utilization of the parametric statistical significance testing procedures. These methods require replacement of less-than values by quantified values. In practice, a frequent approach to allow quantification involves assignment of less-than values as "equal to" either (1) the detection limit, (2) one half the detection limit, or (3) zero. These assignments provide a degree of quantification but may seriously affect subsequent utilization of the parameters used in the parametric tests. As an indication of the consequences, consider the example where the less-than values are assumed to be equal to their detection limit (Case 1, above). In this case, the estimate of the mean is high and the estimate of the standard deviation is low. Alternatively, where the less-than values are assumed to be equal to zero (Case 3, above), the estimate of the mean is relatively low and that of the standard deviation relatively high. Cases (1) and (3) indicate the bounds—if the results of the statistical significance tests are the same for Cases (1) and (3), then the assignment of the values for the less-thans does not matter; however, if the tests result in a different finding, an alternative procedure is necessary.

Results contained in the study by McBean and Rovers [6] indicate several additional alternatives. However, another approach may be of use if more than a single constituent has been monitored and there is a correlative behavior between several of the constituents. The procedure will be demonstrated by application to recent monitoring data, as listed in the second and third columns of Table 3. As indicated in the tabulated data for 1,1,2-trichloroethylene (TCE) and orthonitrianiline (ONA), the TCE data contain 6 less-than

TABLE 2—*Summary of statistical discrimination tests in frequent use.*

Type	Test	Major Assumptions	Example Applications
Parametric	Two sample t-test	(i) the distribution of the two samples have the same variance(**), (ii) the distributions are normal or gaussian	Rovers and McBean (1981)[8] and McBean and Rovers (1984)[6]
	Approximate t-tests (i) Satterthwaite's Approximation (ii) Cochran's Approximation to the Behrens-Fisher t-test (Cochran, 1964)	(i) relaxation of the assumption regarding variances, (ii) the distributions are approximately normal or gaussian. Note: This has been explicitly characterized by the coefficient of variation which must be less than unity, in RCRA (Federal Register, 1982).	McBean and Rovers (1988)[5]
	Data-transformations(*) prior to use of the two sample t-test	(i) the distributions of transformed variables have the same variance(**); (ii) the distributions are normal or gaussian	Rovers and McBean (1981)[8]
	Pairwise t-test – when the sample populations are differenced	(i) the differences between the two samples are distributed as normal or gaussian	
Non-parametric	Mann-Whitney test		

Notes:

(*) data transformation approaches have a degree of arbitrariness with respect to the transformation to be employed for a specific application.

(**) a prior F-test indicates whether the sample variances are sufficiently alike to warrant the assumption that they are independent estimates of the same population variance.

TABLE 3—*Historical data record for ONA and TCE, in micrograms per litre.[a]*

Sample No.	ONA	1,1,2-TCE	1,1,2-TCE (estimated)
1	1.1	<5	6.5
2	1.8	10	10
3	0.53	<5	4.7
4	1.5	7	7
5	1.4	7	7
6	3.7	15	15
7	1.1	6	6
8	1.9	7	7
9	1.2	<5	6.8
10	0.29	<5	3.9
11	0.42	<5	4.3
12	0.9	<5	5.8
13	3.4	14	14
14	0.22	5	5

[a] The data were obtained from monthly samplings for ONA and 1,1,2-TCE concentrations.

detection values from a record length of 14. Utilizing the coincident data record in which both TCE and ONA have levels greater than the detection limit, the degree of correlation between the two chemicals was quantified ($r^2 = 0.91$) and the following linear regression equation was derived.

$$1,1,2\text{-TCE} = 3.04 + 3.11 \cdot \text{ONA} \tag{2}$$

The resulting data are plotted in Fig. 1.

Equation 2 was then used to predict the values for the less-than data for 1,1,2-TCE. The resulting data set for 1,1,2-TCE is listed in the fourth column of Table 3.

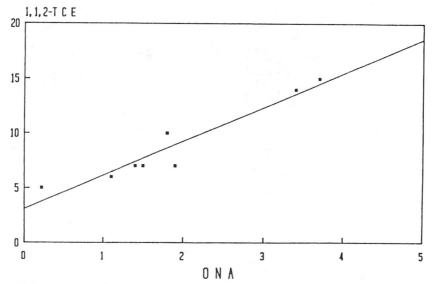

FIG. 1—*ONA versus 1,1,2-TCE data with detection limit data removed. The concentrations are in micrograms per litre.*

The 95% confidence interval (CI) was established (for details on the procedure, see Zar [7]) for the 1,1,2-TCE values and has been plotted, along with the estimated values for the detection limit data, in Fig. 2.

As can be seen from Fig. 2, three of the data points presented as less-thans were estimated, using the regression equation, to be greater than the detection limit. Several comments are appropriate:

The best estimate of a 1,1,2-TCE measurement, given the ONA level, is on the regression line. However, as apparent from Fig. 1, the data pairs with no less-thans reported do not lie along the regression line—there are deviations from the line. It is the purpose of the 95% confidence limits to demonstrate one measure of the degree of variability that exists.

Therefore, the best estimate of individual values is on the line, but the range of possible values (as exemplified by the confidence limits) is considerable. In particular, one should note that many of the potential values for the extrapolated values are less than 5 μg/L for 1,1,2-TCE.

Correlation analyses can be used to extrapolate or estimate values of the less-than data. The viability of the extrapolation requires the following:

(a) high correlation between the constituents, i.e., ONA and 1,1,2-TCE data for the example at hand;

(b) an expectation of similar behavior for the constituents (this assumption is particularly defensible if the chemicals are in the same family but is also potentially useful in the case of the same forcing conditions, e.g., emanating from the same contaminant source); and

(c) if the correlation analysis is performed over temporal data, the data records of each constituent should be examined for persistence of the less-than detection limit data. A persistent trend of less-than data may indicate a disappearance of the compound as a result

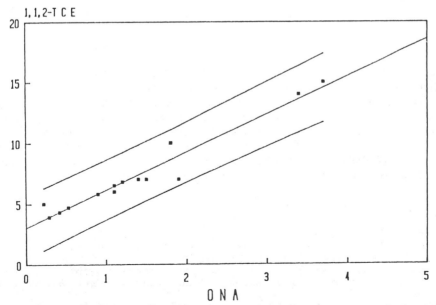

FIG. 2—*ONA versus 1,1,2-TCE with 95% confidence intervals of* Y *estimated. The concentrations are in micrograms per litre.*

TABLE 4—*Estimates of means and variances as determined by alternative methodologies.*[a]

| Statistical Parameter | Correlation Method | Equality Assignments | | | Plotting Method | |
		Equal to Detection Limit	One Half Detection Limit	Zero	Normal Distribution	Lognormal Distribution
Mean	7.4	7.2	6.1	5.1	6.0	5.6
Standard deviation	3.4	3.4	4.3	5.3	6.6	6.4

[a] All concentrations are in micrograms per litre.

of either natural or man-induced causes. Correlation analyses inherently assume the comparable existence or behavior of both compounds.

As a comparison to the correlation method, the mean and standard deviation of the 1,1,2-TCE data were calculated using two of the methods presented by McBean and Rovers [6]. First, the detection limit data were replaced by zero, then, by one half the detection limit, and, finally, by the detection limit. In all three cases, the mean was estimated to be lower and the standard deviation was found to be higher than the values calculated using the correlation analysis technique, as is apparent in Table 4.

The second method involved plotting data on both normal and lognormal distribution paper as per McBean and Rovers [6]. The estimated mean using this procedure equals 6.0 and 5.6 μg/L for the normal and lognormal values, respectively, and the standard deviations were estimated to be 6.6 and 6.4 μg/L, respectively. Note that, based on the limited data set, both the normal and the lognormal plot produce a relatively straight line.

The findings on estimates of the mean and standard deviation are summarized in Table 4. A comparison of several possible tests will provide the reviewer with alternative, but justifiable, means of characterizing ground-water data.

Conclusions

The quantification by statistical analysis, in general, and hypothesis testing, in particular, is intended to reduce and summarize observed data, to present information precisely and meaningfully, and to determine some of the underlying characteristics of the observed phenomena. It should not be presumed that current knowledge of statistical analysis is always sufficient to provide answers to many of the complex problems of ground-water contamination. Nevertheless, the ability to provide answers is enhanced by a flexible approach, using a variety of alternative models, while understanding the limitations of the various models and using a considerable degree of engineering judgment.

References

[1] Fisher, R., *Metron*, Vol. 5, 1926, p. 90.
[2] Unwin, J., Miner, R. A., Srevers, G., and McBean, E., "Groundwater Quality Data Analysis," *NCASI Technical Bulletin*, No. 462, 1985, pp. 148–160.
[3] Imam, R. A. and Conover, W. J., *A Modern Approach to Statistics*, Wiley, New York, 1983.
[4] Cochran, W., "Approximate Significance Levels of the Behrens-Fisher Test," *Biometrics*, March 1964, p. 191.
[5] McBean, E., Kompter, M., and Rovers, F., "A Critical Examination of Approximations Implicit in Cochran's Procedure," *Ground Water Monitoring Review*, Winter 1988, pp. 83–87.

[6] McBean, E. A. and Rovers, F. A., "Alternatives for Handling Detection Limit Data in Impact Assessments," *Ground Water Monitoring Review,* Spring 1984, pp. 42–44.

[7] Zar, J. H., *Biostatistical Analysis,* 2nd Ed., Prentice-Hall Book Co., Toronto, 1984.

[8] Rovers, F. A. and McBean, E. A., "Significance Testing for Impact Evaluation," *Ground Water Monitoring Review,* Vol. 1, No. 2, 1981, pp. 39–43.

Wayne Chudyk,[1] *Kenneth Pohlig,*[1] *Kosta Exarhoulakos,*[1]
Jean Holsinger,[1] *and Nicola Rico*[1]

In Situ Analysis of Benzene, Ethylbenzene, Toluene, and Xylenes (BTEX) Using Fiber Optics

REFERENCE: Chudyk, W., Pohlig, K., Exarhoulakos, K., Holsinger, J., and Rico, N., *"In Situ Analysis of Benzene, Ethylbenzene, Toluene, and Xylenes (BTEX) Using Fiber Optics,"* *Ground Water and Vadose Zone Monitoring, ASTM STP 1053,* D. M. Nielsen and A. I. Johnson, Eds., American Society for Testing and Materials, Philadelphia, 1990, pp. 266–271.

ABSTRACT: *In situ* analysis of ground water is an attractive alternative to conventional sampling and analysis methods, since it eliminates sample handling and chain-of-custody concerns. The prototype instrument recently developed by our research team uses remote laser-induced fluorescence (RLIF) to measure aromatic organics *in situ*. The work discussed here concerns the aromatic organics fraction of gasoline (benzene, toluene, and xylenes). RLIF can theoretically detect over half of the organics on the U.S. Environmental Protection Agency's Priority Pollutants List.

Typical analysis for these contaminants involves routine repetitive analysis, such as quarterly gas chromatography/mass spectroscopy (GC/MS) on samples taken from wells. Such monitoring is labor intensive, requiring a sampling team as well as a laboratory. Samples for GC/MS often take many weeks for analysis, so that remedial action information is late in arriving. An analysis is needed that is quick, simple, and less costly, providing a look at the situation as it currently exists.

The authors propose that routine GC/MS monitoring at characterized sites be replaced with RLIF measurements, supplemented with annual GC/MS analysis as a means of validation. Since RLIF analysis is nonspecific, this approach would be sound only at sites containing known aromatics as the contaminants of interest, or at sites where the aromatics concentration correlates with other problem compounds. Comparisons of RLIF and GC sensitivity for aromatic organics are discussed, with emphasis on the relative advantages and limitations of each method.

KEY WORDS: ground-water monitoring, *in situ* measurements, laser-induced fluorescence, aromatic contaminants, gas chromatography

An interdisciplinary research team at Tufts University, in Medford, Massachusetts, has recently developed a remote laser-induced fluorescence (RLIF) technique which is capable of measuring aromatic organic ground-water contaminants *in situ*. This method uses a laser as a light source, and fiber optics to carry laser excitation light down a well and resulting fluorescent light back up to the surface to a detector [1,2]. Fluorescence methods have been used previously in the laboratory to identify and quantitate contamination of water by petroleum products [3]. Our application uses a similar approach with a portable unit appropriate for pilot or field operation.

Traditional sampling and laboratory analysis can cause changes in sample composition, because of contamination by samplers, and also changes in the sample's pH, redox potential, or loss of volatiles content. *In situ* measurements allow rapid acquisition of informa-

---

[1] Department of Civil Engineering, Tufts University, Medford, MA 02155.

tion at relatively low cost, making them attractive as an alternative method for ground-water monitoring [4].

Theoretical Background

Aromatic organics include the benzene, ethylbenzene, toluene, and xylenes fraction of petroleum fuels, as well as many other ground-water contaminants. Most aromatic organic compounds contain a benzene ring as part of their molecular structure. This part of the molecule can adsorb light of the right wavelength, typically ultraviolet, and later emit the adsorbed energy as fluorescence. The intensity and wavelength of fluorescence are functions of molecular concentration and structure, respectively. Measurement of fluorescence of aromatics dissolved in water has been used to measure their concentration and determine the type of molecule present, leading to identification of sources of petroleum product spills [3]. Such an approach has typically been performed in the laboratory with relatively large equipment. Recent developments in high-energy light sources and fiber optics have allowed us to construct and test a field-portable fluorescence-measuring instrument [1,2].

Procedure

The first prototype RLIF instrument used for measuring aromatic ground-water contaminants has been described elsewhere [2]. The latest prototype RLIF instrument has the light source, detector, and signal-processing electronics mounted in a series of portable modules connected to a fiber-optic sensor placed in a well. In pilot laboratory studies, power is provided by a conventional wall plug, while, in the field, the unit is usually powered by a 3-kW portable generator. The excitation module, containing the light source, is shown in Fig. 1. The components in the excitation module are used to generate ultraviolet light at 266 nm and to focus this excitation light into a fused-silica optical fiber. Pulsed light at 532 nm is emitted by the laser, along with some 1064-nm light. The 1064-nm filter removes the 1064-nm light, and the remaining 532-nm light passes through a frequency-doubling BBO optical crystal. Some of the 532-nm light is frequency-doubled to 266 nm, in the ultraviolet (UV) range. The dichroic mirror selectively reflects only 266-nm light, which is further steered and focused by a prism and lens into the excitation fiber. Two fibers, one excitation and one emission, are joined in the sensor, which is placed into the well to be sampled.

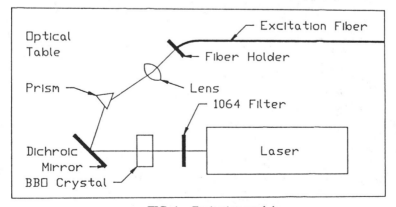

FIG. 1—*Excitation module.*

The sensor is constructed of fused silica glass, stainless steel, and polytetrafluoroethylene (PTFE) so as to minimize contamination of well water by the sensor. The maximum diameter of the sensor is 19 mm (¾ in.), so that it will fit virtually any monitoring well. It may be placed before the well is bailed, and the establishment of formation water can be monitored as purging occurs. For long-term or continuous monitoring, sensors can be left in place in a well. On-site calibration is typically performed with a 1-ppm phenol solution.

Light from the excitation fiber excites aromatic molecules dissolved in the water, causing some of them to fluoresce. The fluorescence is emitted from the molecules randomly in all directions, so some of it is collected by the emission fiber and led back up to the detection module. Scattered emission light is also collected by the detection fiber. The detection module is shown in Fig. 2. About 10% of the collected light is directed onto the power-normalization photomultiplier tube (PMT) to provide an internal laser-power reference signal. The remaining light is spectrally filtered to block interferences and directed into the detection PMT.

The detector module output is processed in the electronics module, which is shown in Fig. 3. Each PMT in the detector produces a current proportional to the amount of light it receives. Current from the power-normalization PMT is proportional to the amount of excitation light sent into the well, which is a function of the laser power during a given pulse. Current from the detection PMT is proportional to the fluorescence of contaminants in the water generated by the same pulse of excitation energy, measured by the power-normalization PMT. Each of the current signals from the PMTs is separately integrated electronically to yield a voltage. The ratio of these voltages is used to determine the intensity of fluorescence from contaminants in the well. The fluorescence intensity is compared with standard curves of fluorescence versus concentration for the contaminants of interest, so that the concentration can be determined. A personal computer (PC) is used for instrument control and data logging. The other parts of the electronics module are power supplies for the laser and detector module, as well as interface electronics for the PC.

Gas chromatography (GC) analysis is used as a check on instrument performance. Since all aromatic molecules are expected to fluoresce, they are all theoretically detectable using RLIF. A list of aromatics detected by U.S. Environmental Protection Agency (EPA) Method 502.2 is presented in Table 1. Since these compounds are all aromatics, they are also theoretically detectable using RLIF. Compounds already tested with RLIF (benzene, toluene, and xylenes) are denoted by an asterisk.

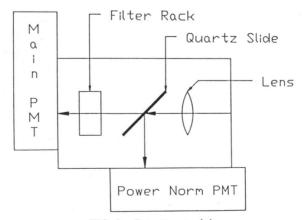

FIG. 2—*Detection module.*

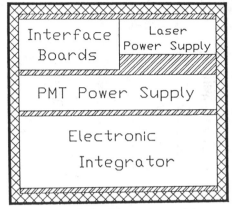

FIG. 3—*Electronics module.*

Results

Previous work has tested model contaminants including phenol, ortho-cresol, toluene, xylenes, and humic acid, as well as both landfill and bark-pile leachates [1,2]. The limits of detection found in the laboratory for these compounds are presented elsewhere [1] and are an indication of the instrument response. Based on those earlier data, single-ring aromatics seem to be easily detected using RLIF. Gasoline is known to contain the single-ring aromatics benzene, ethylbenzene, toluene, and xylenes (BTEX), so that analysis for BTEX

TABLE 1—*Aromatics theoretically detected by both RLIF and EPA Method 502.2.*

Benzene[a]
Bromobenzene
n-Butylbenzene
sec-Butylbenzene
tert-Butylbenzene
Chlorobenzene
2-Chlorotoluene
4-Chlorotoluene
1,2-Dichlorobenzene
1,3-Dichlorobenzene
1,4-Dichlorobenzene
Ethylbenzene
Isopropylbenzene
p-Isopropyltoluene
Naphthalene
n-Propylbenzene
Styrene
Toluene[a]
1,2,3-Trichlorobenzene
1,2,4-Trichlorobenzene
1,2,4-Trimethylbenzene
1,2,5-Trimethylbenzene
o-Xylene[a]
m-Xylene[a]
p-Xylene[a]

[a] Compounds already examined via RLIF.

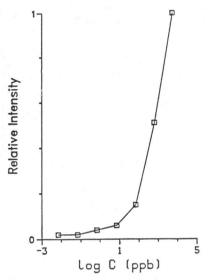

FIG. 4—*RLIF response versus the gasoline concentration.*

gasoline components is a natural extension of earlier work. This study focused on unleaded gasoline. The laboratory instrument response versus the unleaded gasoline concentration is shown in Fig. 4. To obtain the data plotted in Fig. 4, serial dilutions were made of stock gasoline solutions in phosphate-buffered water (pH 7.2). A 25-m sensor was used to record the fluorescence intensity for each solution concentration. The highest intensity was assigned a value of 1.0, while the other intensities were normalized to this number. Beer's law predicts an exponential relationship between the intensity and the log of concentration. The line connecting the data points appears to follow an exponential curve. These data show useful laboratory measurements in the parts-per-trillion range, while field results typically show more scatter, with a useful concentration threshold of detection of about 1 ppb. The RLIF response is to the benzene, ethylbenzene, toluene, and xylenes (BTEX) component of gasoline. More complex aromatic molecules, such as humic acid, tend to fluoresce at longer wavelengths than do the single-ring aromatics. Possible interferences from humic acid can be eliminated by use of glass cutoff filters in the detector module [1]. In this approach, such cutoff filters are placed in the light path before the detection PMT. They screen out the fluorescence from humics, while passing the light from BTEX fluorescence into the detection PMT.

Discussion

Since it measures fluorescence, RLIF is a technique that offers speed, sensitivity, and low cost. Its main defect at this time is its lack of specificity. If a mixture of aromatics is present, the RLIF result will be the total signal from *all* fluorescent species present. Keeping this in mind, there are still many potential uses for this technique. If used in conjunction with other, more sophisticated, more time-consuming, and more expensive methods, useful information about changes in an aquifer system can be obtained in a timely and cost-effective manner. For example, RLIF screening could determine if aromatics are present in a well. The exact makeup of a mixture of aromatics could be found using GC or GC/mass spectroscopy (GC/MS). If, then, a site is characterized as having aromatic organics as the contaminants of concern, the speed and ease of RLIF measurement makes it an attractive alternative to long-term monitoring for changes in the contaminants present. Currently, the limits of detection for RLIF are at or below those for GC analysis, so that the sensitivity of analysis will not suffer.

Limited data from field tests show that the field performance of RLIF is poorer than the laboratory results. Temperature fluctuations in the field, as well as increased vibration, affecting the optics, may be contributing to the variability in field measurements. This problem needs further work.

If a mixture of aromatics is present, the RLIF signal will be a combination of those from each component. If information about each component is needed, separation of the contribution from each component in the mixture would require mathematical deconvolution of the total signal. In baseline monitoring, however, a simple change in the total signal would indicate a change in aromatics concentration in the ground water. Such a simple, inexpensive, and rapid test for change in aromatics concentration in an aquifer would not require extensive deconvolution of the data obtained. Merely following the trend in RLIF results for a given site over time should indicate the presence or absence of a problem involving changing concentrations of aromatic organics, which still may be confirmed by annual GC measurements. The authors hope that the advantages of *in situ* RLIF monitoring will make it an attractive alternative to current methods.

Conclusions

The use of laser-induced fluorescence and fiber optics to measure aromatics contamination of ground water is providing a new tool to the monitoring community. Future pilot and field testing of the current prototype should show the relative advantages of RLIF over traditional methods.

Acknowledgments

The authors wish to thank the National Science Foundation, the U.S. Environmental Protection Agency/Tufts Center for Environmental Management, the U.S. Geological Survey, and the Alexander Host Foundation for support. The dedication and hard work of Jonathan Kenny and George Jarvis of the Tufts University Chemistry Department are deeply appreciated.

Parts of this report were developed under a grant from the U.S. Geological Survey, U.S. Department of the Interior. Those contents, however, do not necessarily represent the policy of that agency, and one should not assume endorsement by the Federal Government.

References

[*1*] Chudyk, W. A., Kenny, J. E., Jarvis, G. B., and Pohlig, K. O., *InTech,* Vol. 34, No. 5, 1987, p. 53.
[*2*] Kenny, J. E., Jarvis, G. B., Chudyk, W. A., and Pohlig, K. O., "Remote Laser-Induced Fluorescence Monitoring of Groundwater Contaminants: Prototype Field Instrument," *Analytical Instrumentation,* Vol. 16, No. 4, 1987, pp. 423–446.
[*3*] Eastwood, D., in *Modern Fluorescence Spectroscopy,* Vol. 4, E. L. Wehrey, Ed., Plenum Publishing, New York, 1981, pp. 251–275.
[*4*] Eccles, L. A., Simon, S. J., and Klainer, S. M., "In Situ Monitoring at Superfund Sites with Fiber Optics (Draft)," U.S. Environmental Protection Agency Environmental Systems Monitoring Laboratory, Las Vegas, NV, March 1987.

Monitoring in Karst

James F. Quinlan[1]

Special Problems of Ground-Water Monitoring in Karst Terranes

REFERENCE: Quinlan, J. F., **"Special Problems of Ground-Water Monitoring in Karst Terranes,"** *Ground Water and Vadose Zone Monitoring, ASTM STP 1053,* D. M. Nielsen and A. I. Johnson, Eds., American Society for Testing and Materials, Philadelphia, 1990, pp. 275–304.

ABSTRACT: Reliable monitoring of ground-water quality in any terrane is difficult. There are many ways in which violation of sound principles of monitoring-network design and good sampling protocol makes it easy to acquire data that are not representative of the water or pollutants within an aquifer. In karst terranes it is especially easy for irrelevant data, which inadvertently misrepresent conditions within the aquifer, to be obtained.

The special problems of monitoring ground water in most karst terranes can be grouped into four major categories of problems that are rarely as significant in other terranes. These categories are the following:

1. *Where to monitor for pollutants:* The only relevant locations are at springs, cave streams, and wells that have been shown by tracing tests to include drainage from the facility to be monitored—rather than at wells to which traces have not been run but which were selected because of their convenient downgradient locations. Wells located on fracture traces and fracture-trace intersections and wells located randomly can be successfully used for monitoring, but only if there is a positive trace from the facility to them. Often successful monitoring can only be done several kilometres away from the facility.

2. *Where to monitor for background:* The only relevant locations are at springs, cave streams, and wells in fractured rock—in which the waters are geochemically similar to those to be monitored for pollutants but which have been shown by tracing tests *not* to drain from the facility—rather than at wells selected because of their convenient locations upgradient from the facility site. This, too, may have to be done several kilometres away from the facility.

3. *When to monitor:* Before, during, and after storms or meltwater events—rather than regularly with weekly, monthly, quarterly, semiannual, or annual frequency.

4. *How to determine reliably and economically the answers to Problems 1, 2, and 3:* Reliable monitoring of ground water in karst terranes can be done, but it is not cheap or easy.

These problems exist because many of the assumptions made for monitoring flow in granular media are not valid for karst terranes. Implicit assumptions made for monitoring in karst terranes with the strategy recommended herein can be stated axiomatically, but they are valid only about 95% of the time.

The monitoring strategy recommended herein is not applicable universally, but it is applicable in most karst aquifers, especially, all those that drain to springs. It is not applicable in terranes that are merely recharge areas of regional aquifers.

KEY WORDS: ground water, ground-water monitoring, karst, water quality, springs, hazardous wastes, limestone, carbonate rocks

---

[1] Senior hydrogeologist, ATEC Environmental Consultants, Nashville, TN 37211.

Approximately 20% of the United States consists of areas underlain by limestone or dolomite; most of these areas are karst terranes. Sinkholes, sinking streams, losing streams, gaining streams, springs, and caves commonly occur, but they may be absent or not obvious. A karst terrane generally includes a karst aquifer, an aquifer occurring chiefly in limestone, dolomite, gypsum, salt, carbonate-cemented clastic rock, or some combination of these rocks. Shale, siltstone, or sandstone may be interbedded in a karst aquifer. The most characteristic feature of a karst aquifer is flow of water through conduits (caves) and along bedding planes and fractures enlarged by solution.

Waste-disposal facilities should not be located within a karst terrane unless those involved are willing to risk sacrificing the use of at least part of the included karst aquifer as a source of potable water. This is a high risk, almost a certainty. Nevertheless, many facilities already exist within karst environments, and hazardous materials are disposed of without the supervision of a hydrologist knowledgeable about karst. Also, hydrologists are not consulted on the selection of the site when there is an accidental spill of hazardous material.

Two major types of flow occur in karst aquifers—conduit flow and diffuse flow, each of which is an end member of a continuum. Springs and cave streams in conduit-flow systems are "flashy"; that is, they have a high ratio between the maximum discharge and the base-flow discharge, typically 10:1 to 1000:1. The discharge responds rapidly to rainfall, and the flow is generally turbulent. The waters possess low but highly variable hardness, and the turbidity, discharge, and temperature of these waters also vary widely [1]. Where a karst aquifer is less developed and is characterized primarily by diffuse flow, its behavior is less flashy; the ratio between the maximum discharge and base-flow discharge of major springs is low (4:1 or less), and the response of their discharge to rainfall is slow. Flow is generally laminar. The hardness in such springs is higher than that in conduit-flow springs, but the hardness, turbidity, discharge, and temperature have low variability [1]. The variations in and relationships among these properties and their variability as a function of aquifer flow, storage, and recharge have been described in an important paper by Smart and Hobbs. [2].

The relationships between conduit-flow aquifers, fractured aquifers, granular aquifers, and diffuse-flow aquifers are shown in Fig. 1. Ground-water flow in conduits and fractures of karst aquifers differs radically from flow in other aquifers. Most commonly, flow in karst is to springs, by way of caves. Such flow is generally faster than in other aquifers; extreme velocities of 2300 m/h (7500 ft/h) have been observed, while a range of 10 to 500 m/h (30 to 1500 ft/h) is common between the same two points (Ref 1, page 202). [The latter two flow velocities are equivalent to approximately 90 and 4400 km/year (55 and 2700 mile/year).] Accordingly, the effects of leakage or a spill of hazardous material on the water quality in a karst aquifer can often be sensed long distances away within less than a day.

To many hydrologists, geologists, and engineers, the flow of ground water in karst is mysterious, capricious, and unpredictable. Few publications adequately discuss the predictive aspects of environmental hydrologic problems of karst terranes or offer practical, experience-based insights for solving them. There are, however, notable exceptions [1–13]. The problems of flow prediction are real, yet they are ignored in nearly all of the environmental monitoring literature and by most current U.S. Environmental Protection Agency (EPA) and state ground-water monitoring regulations and technical enforcement guidance documents. These problems, nevertheless, can be solved. Education and awareness of their existence are the keys for doing so.

This paper discusses where and when to take relevant samples and how to get the data essential for making the where and when decisions; it also discusses implicit assumptions made when monitoring with the strategy recommended herein. No attempt is made to discuss the design or construction of monitoring wells, protocol for sample custody, or

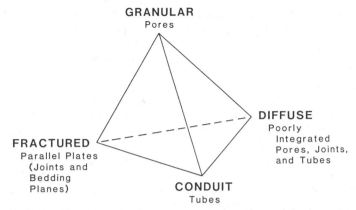

FIG. 1—*Relationship between the four major types of aquifers and the dominant porosity geometry in each. This is a tetrahedral continuum.*

quality assurance/quality control (QA/QC). These topics have been adequately addressed by others, and many are being reviewed and codified by members of ASTM Subcommittee D18.21 on Ground Water Monitoring and Vadose Zone Investigations, a subcommittee of ASTM Committee D-18 on Soil and Rock. This paper is prescriptive and practical rather than philosophical. The methodologies for its recommendations and the rationale for each are discussed in the references cited. Although the various answers to where, when, and how to monitor are strongly influenced by the answers to *why,* this paper does not address the complex questions of why to monitor.

The recommended strategy for traced-spring, traced-cave-stream, and traced-well monitoring, described herein and by Quinlan and Ewers [*1*], is not applicable universally. But it is applicable in all diffuse-flow and conduit-flow aquifers that drain to springs which discharge on land or along the shores of streams, rivers, lakes, or seas. Accordingly, the strategy is applicable in most karsts in the following 16 states, which have significant amounts of karst: New York, Pennsylvania, Maryland, West Virginia, Virginia, Tennessee, Georgia (Appalachians), Alabama, Kentucky, Indiana, Arkansas, Missouri, Iowa, Minnesota, Texas, and Oklahoma. Their karst terranes are characterized by local recharge *and* discharge. Many of these states include some of the more densely industrialized and populated areas of the United States. Karst comprises approximately 25 to 30% of the total area of these states.

The recommended monitoring strategy is applicable only locally in parts of the Floridan aquifer of Florida and in Puerto Rico. In both areas there is significant discharge at springs. The strategy is not applicable in karst terranes that are merely recharge areas of regional aquifers, such as the Upper Floridan aquifer of Florida, Georgia, and South Carolina. It is also minimally applicable in karsts mantled by glacial sediments and in which the discharge is diffuse, flowing into sediment or bodies of water rather than to discrete springs. These karsts comprise an estimated 30% of the total area of these states.

The strategy would be applicable in most of the Edwards aquifer of Texas, much of the Upper Floridan aquifer, and part of the North Coast limestone aquifer of Puerto Rico, where the flow is to springs at the surface and where most springs are diffuse-flow rather than conduit-flow springs. Although much of each of these aquifers is characterized by sponge-like permeability, many of their springs are fed by conduits that are commonly braided (anastomosed). Accordingly, one might use geophysical techniques for trying to find the main conduit and drill to intercept it, but probably would miss it. Nevertheless, monitoring could be successfully accomplished at springs. Wells in these sponge-like aquifers could be used as monitoring sites only if there were a positive trace to them from the vicinity of a facility or from the facility itself during low-, moderate-, and high-flow conditions.

There are numerous small areas of karst in the western United States, but nearly all of them are in isolated, nonindustrialized, unpopulated terranes.

Research is needed on the distribution of and criteria for applicability of the recommended strategy in karsts in the states cited in the four preceding paragraphs.

Ignored here is the fact that most current regulations concerning ground-water monitoring disregard the manifold problems of doing it reliably in karst terranes. The true goal of ground-water monitoring should be to detect the nature and magnitude of changes, if any, in ground-water quality as a result of natural processes or human activity—rather than just to comply with the letter of the law and local regulations. One assumes that federal and state regulations will eventually catch up with reality and will ultimately reflect an emphasis on the *spirit* and *intent* of the law, rather than mere compliance with its letter.

Design of Monitoring Systems in Karst Terranes

Where to Monitor for Pollutants

Conventional minimum monitoring protocol, as required by most regulatory agencies, begins with drilling of a "sufficient" number of wells. Usually this number is a minimum of four—one well upgradient from a facility and three wells downgradient from it, commonly near the property boundary and almost always on the property. Sampling is performed annually, semiannually, and perhaps quarterly, but rarely as often as monthly. Such monitoring wells in a karst terrane may generate great amounts of carefully collected, expensive data, most of which may be of little value because the wells fail to detect the contaminants they are intended to intercept. They fail to do so because the wells do not encounter the flow lines (cave streams) draining from the site, or—if they do intercept them—an insufficient number of samples is taken or sampling is done at inappropriate intervals and at the wrong time [1,3].

The easiest reliable sites at which to monitor ground-water quality in a karst terrane are springs and directly accessible cave streams, shown by dye tracing to drain from the facility being evaluated [1]. Tracing in karsts is usually done with fluorescent dyes because they are the most cost-efficient of many possible tracers. However, use of the term *dye tracing* in this paper is not intended to exclude the use of other tracers.

The preferable alternative for monitoring sites in karst terranes is a suite of wells that intercept cave streams known (by tracing) to flow from the facility to the spring (or springs) that drain the ground-water basin in which it lies. Cave streams may be difficult or impossible to find with traditional geophysical techniques. New geophysical techniques include the use of streaming potential (measurement of the electrical potential gradient caused by displacement of ions from fissures and rock grains as water moves from a recharge area to a discharge area) [14] and the use of acoustic detection (measurement of sound waves caused by the knocking of pebbles against one another during saltation, by cavitating water, or by cascading, riffling, and dripping water in partly air-filled cavities). These methods have been tested with encouraging results [14]. In one situation, however, the results were not supported by the data presented [15]. Other nontraditional geophysical techniques also have promise [14].

A second alternative for monitoring sites is a suite of wells located in fractured rocks, preferably on fracture traces or at fracture-trace intersections. These wells are usable *only* if tracer tests show a connection with the facility under base-flow as well as flood-flow conditions. Randomly located wells could also be used if, again, tracing has first proven a connection from the facility to each of the wells under various flow conditions.

Although some cave passages are coincident with various types of fracture traces and lineaments, not all coincident fracture-related features are vertical and therefore directly above cave passages [16]. More important, many cave streams are developed along bedding planes and are unaffected by vertical fractures. This fact lessens the probability that a well drilled on a fracture trace, a lineament, or the intersection of such linear features will intercept a cave stream. This fact does not challenge the well-known correlation of such linear features with increased water yields [17].

Success in the use of tracing to determine those randomly located wells (domestic, agricultural, or industrial) that intercept flow from a site can be maximized if they are pumped continually to discharge at a low rate, say 4 L/min (1 gal/min), through a passive dye-detector (explained here) which is regularly changed once or twice a week for weeks or months. In most settings this small amount of water can be wasted onto the ground at a reasonable distance from any building or structure with no adverse effect. Pumpage at high rates, say 400 L/min (100 gal/min) or more, may distort the flow field near a well, which is acceptable if the effects of such distortion are recognized. A high pumpage rate for weeks to months is expensive and wasteful; furthermore, disposal of the pumped water can be a problem, especially if it is contaminated by pollutants.

If a well is drilled specifically for monitoring of pollutants, one might argue that it would be more useful if it were to be pumped intermittently on a regular schedule, even when it is not actively being used for a dye trace. This intermittent, nonrandom pumping would allow the monitoring well to simulate more truly the flow conditions around a regularly used well that might be contaminated.

At most locations in karst terranes, proper and reliable monitoring (using springs, cave streams, and wells traced positively) can only be done off the facility. Sometimes this proper monitoring must be, and can only be, done at sites up to several kilometres away from the facility.

Where to Monitor for Background

If randomly located, nontraced wells are irrelevant for monitoring possible contaminants in karst ground water, they are equally irrelevant for monitoring background for contaminants. Therefore, springs, cave streams, and wells in settings geochemically similar to the traced monitoring sites are the only suitable places to monitor background meaningfully. This is true, however, only when these places have been shown by carefully designed, repeated dye traces, done under conditions ranging from base flow and flood flow, *not* to drain from the facility. One must be exceedingly cautious in interpreting negative results of a tracer test [18]. Some randomly located wells that are negative in a judiciously designed, properly executed dye trace are in settings geochemically similar to those of the monitoring wells that intercept cave streams; most are not.

Background data at a new facility can also be obtained from wells that have been monitored before, during, and after storms for at least a year before operations begin. Analysis of continuous records of the water-level stage in these wells can be used for selection of storm-related sampling frequencies and for possible differentiation of wells into several categories. Traces should be run during that year-long period in order to demonstrate the presence or absence of connections between the various wells and the facility to be monitored. If, after allowance is made for flow velocity, a well does not test positive for dye, the well does not comprise an effective monitoring system. If none of the wells, or an insufficient number of wells, to which dye traces have been attempted test positive, an *effective* monitoring system does not exist; probably, the hydrology of the facility was not adequately understood. More wells may have to be drilled and tested by tracing.

It is tempting and all too easy to take false comfort in the interpretation of negative tracing results for a well to mean that a waste disposal facility and its liner (if present) are functioning as designed. More specifically, it is easy to say that a facility is either not leaking or is adequately attenuating everything put into it, but it would be naive to do so. Alternatively, the negative tracing results could just as logically (indeed, more probably) be a consequence of monitoring for dye in wells that do not sense the part of the aquifer in which ground water actively circulates. In most karst terranes, the latter explanation is more likely to be correct. If water is not standing on the ground in pools or ponds, rainfall must be flowing someplace. If it is flowing, an uncapped facility is probably leaking to somewhere.

There are statistical problems associated with the validity of comparing data from background wells with purportedly relevant data from monitoring wells. Some regional offices of the EPA, in recognition of these problems, are looking for ways to make objective, legitimate comparisons. One way they are using is employment of the concept of "trigger levels" for determining whether or not a facility is in compliance with regulations. Nationwide trigger levels for a compound are being set after critical review of data on toxic effects of that compound on health. If the trigger level of that compound is exceeded in a monitoring well for a facility, an investigation of possible sources for it must be made. No matter what concept of background is used, and no matter what the concept of background evolves into, there is still the problem, addressed in this paper, of how to compare legitimately relevant water-quality data.

Where to Monitor for Pollutants and Background: Other Aspects

All six types of monitoring sites and background sites—springs, directly accessible cave streams, wells drilled to cave streams, wells drilled on fracture traces, wells drilled on fracture-trace intersections, and wells drilled randomly—must be tested by tracing, not only during moderate flow but also during flood flow and base flow, in order to prove the usefulness of these sites for monitoring. This must be done during the extremes of expected flow conditions; flow routing in karst terranes commonly varies as the flow stage changes. During flood conditions the water in some conduits may rise, temporarily move through conduits that are dry during low-flow conditions, and be "decanted" (switched) into adjacent ground-water basins, thus being temporarily pirated by them. An example of such hydraulic switching is depicted in Fig. 2, which shows the hydrology of Turnhole Spring in Mammoth Cave National Park, Kentucky. During moderate flow and flood flow, water draining from Park City and from the west boundary of Cave City in the Turnhole Spring ground-water basin of Kentucky may flow via intermediate-level and high-level crossover routes, shown by the north-trending dashed lines with arrows, to as many as three other basins: No. 4, Sand Cave; No. 6, Echo River; and No. 7, Pike Spring (indicated at the south end of Roppel Cave). These three subsurface diversion routes are also shown in Fig. 3, but more schematically. Other, more complex examples of hydraulic switching are known [19–22].

Another peculiarity of water movement in many karst aquifers is distributary flow. An *underground distributary* is a dispersal route analogous to the distributary at the mouth of a major river, but its origin is different (Ref *1*, pages 205 and 207). Figure 3 shows numerous distributaries. The positions and geometry of the underground bifurcations shown are

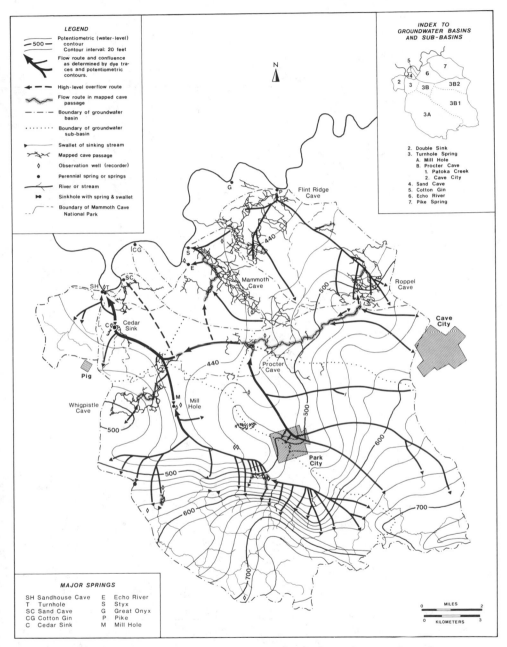

FIG. 2—*Hydrology of the Turnhole Spring ground-water basin, the major basin draining into Mammoth Cave National Park, Kentucky. The beds dip north at about 11 to 40 m/km (60 to 210 ft/mile)* [1].

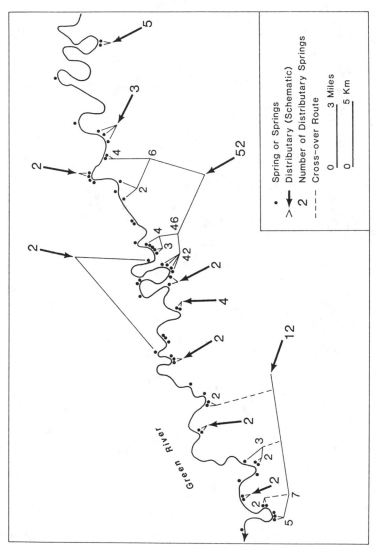

FIG. 3—*Distributary springs along the Green River in the Mammoth Cave area in Kentucky. The numbers indicate the total number of springs in a given distributary system or subsystem. The westernmost large distributary, with twelve springs, is a schematic representation of part of the Turnhole Spring ground-water basin, shown in Fig. 2 [1].*

schematic, but their existence in the map area is a certainty. Some of these underground flow paths have been mapped. Knowledge of the occurrence and functioning of distributaries is important in the design of a monitoring system in a karst terrane because pollutants from a point source in the headwaters or midreaches of a ground-water basin or sub-basin will flow at higher stages to all springs in its distributary system or subsystem. For example, pollutants reaching ground water at a point source in the east-central area of Fig. 2, south of Green River, would—depending upon the flow conditions—disperse to a total of as many as 52 springs in 19 isolated segments along a 19-km (12-mile) reach of the Green River. Not all springs between the extremities of a distributary system are necessarily a part of the system.

The extent to which ground-water basins and their subsurface flow routings can be deciphered by tracing, mapping of the potentiometric surface (water table), and mapping of caves is shown in Fig. 4. The east half of this map summarizes some of the results of a study of the dispersal of heavy metals from a metal-plating plant; the metals had been discharged into the ground (in concentrations of more than 10 mg/L) via a municipal sewage treatment plant at the town of Horse Cave. (The outline of this town is just east of the center of the map.) A study of the chemistry of springs, wells, cave streams, and the west-flowing Green River showed that effluent from the sewage treatment plant flows underground 1.6 km (1 mile) northeast to the unmapped cave beneath the town and then, depending upon the flow stage, to a total of as many as 46 springs in 16 segments along an 8-km (5-mile) reach of the river, about 7 km (4 miles) to the north [23]. None of the 23 wells monitored for heavy metals during base flow showed concentrations higher than background levels, but they should have also been sampled during flood-flow conditions, when water movement might have been reversed from the trunk conduit to some of the wells.

Figure 5 shows complex, radial flow in faulted, flat-lying rocks in the Ozarks of Arkansas. The divergent results of six dye tests, three of which were performed within a mile of a proposed landfill, are summarized. Although the flow routing in Fig. 5 is more complex than that shown in Figs. 2 and 4, it is totally in agreement with the advocated maxim, "Monitor the springs!" Monitoring wells that could be drilled on fracture traces at the proposed landfill site might detect seepage of leachate from it, but there is no means—other than by tracing—to identify correctly and conclusively the places to which leachate would (and probably would not) flow. Stated another way, no matter how superbly efficiently the hypothetical monitoring wells on the landfill property might be able to detect leakage from the landfill, there is no way—other than by tracing or by monitoring of numerous springs and wells off the property—that one could discover the many consequences of leakage from the proposed landfill. This statement would be true even if the monitoring wells were properly constructed and if they functioned efficiently.

All four wells at the town of Pindall, Arkansas (Fig. 5), which were pumped continuously during the dye trace from the east boundary of the landfill site (the first trace run), tested positive for dye. By inference, if other wells in the town had been pumped continuously, many of them, perhaps all, would have tested positive. Note that none of the four wells immediately east and southeast of the first dye-injection site and neither of the two wells immediately west of the western dye-injection site tested positive for dye during the first test. (None of these six wells was sampled for dye during the five subsequent tests.) During the second and third dye tests, in which dye was injected 1.6 km (1 mile) east and west of the proposed landfill site, respectively, only two wells at Pindall were pumped continuously. The southernmost well there tested positive for dye in both tests, as shown; the easternmost well was ambiguous for dye from the east and negative for dye from the west. The high-yield well on the fracture trace tested positive for dye from both the second and

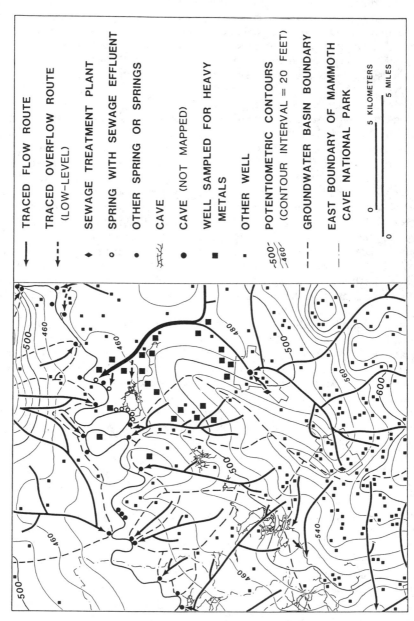

FIG. 4—Ground-water flow routes and pollutant dispersal in the vicinity of the towns of Horse Cave and Cave City, Kentucky. Part of the east half of Fig. 2 overlaps with the west half of this map (modified from Ref 12).

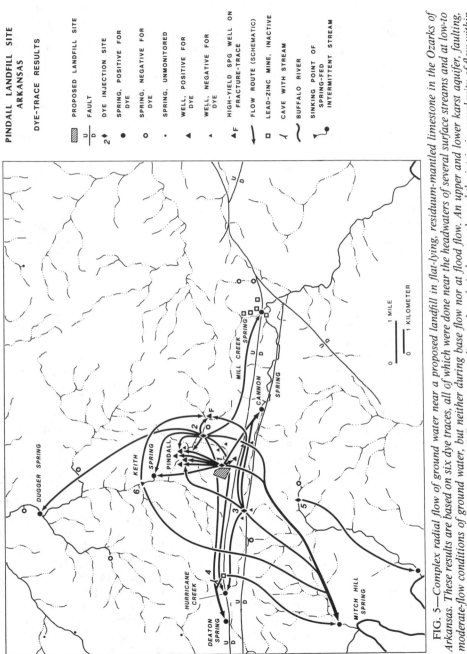

**PINDALL LANDFILL SITE
ARKANSAS**

DYE-TRACE RESULTS

- PROPOSED LANDFILL SITE
- FAULT
- DYE INJECTION SITE
- SPRING, POSITIVE FOR DYE
- SPRING, NEGATIVE FOR DYE
- SPRING, UNMONITORED
- WELL, POSITIVE FOR DYE
- WELL, NEGATIVE FOR DYE
- HIGH-YIELD SPG WELL ON FRACTURE-TRACE
- FLOW ROUTE (SCHEMATIC)
- LEAD-ZINC MINE, INACTIVE
- CAVE WITH STREAM
- BUFFALO RIVER
- SINKING POINT OF SPRING-FED INTERMITTENT STREAM

FIG. 5—*Complex radial flow of ground water near a proposed landfill in flat-lying, residuum-mantled limestone in the Ozarks of Arkansas. These results are based on six dye traces, all of which were done near the headwaters of several surface streams and at low-to moderate-flow conditions of ground water, but neither during base flow nor at flood flow. An upper and lower karst aquifer, faulting, fracture traces, lineaments, spillover levels, differences in sampling and analytical protocols, and the intrinsic complexity of flow within the karst all influence the divergent results of the traces. The results might be different if the dye tests had been run during base-flow or flood-flow conditions [24].*

third dye tests, but negative for dye from the first. Dye recovery from this well during the first test was probably hampered by chlorine added to the well water, which tends to react with and destroy low concentrations of any dye present. The difference in tracer results at this high-yield well may also have been a consequence of greater efficiency in sampling and analysis during these latter tests and of the use of the author's pumped-well dye-sampling device [24]. (Dye trace results from the first test were instrumental in the 1987 defeat of a proposal to use this site as a landfill.)

Radial flow occurs in many karst terranes and has also been documented at waste disposal facilities (Ref 1, pages 214–219); it tends to be associated wth locations on topographic highs. When evaluating a facility by dye tracing, one must keep an open mind and place detectors for dye at not only the logical, obvious places, but also at the illogical, the unlikely, and the "no, it could never go there" places. The cliché "expect the unexpected" applies, no matter how experienced one is in tracing.

The applicability of the spring and cave-stream monitoring strategy and the conventional (randomly located well) monitoring strategy in various types of aquifers is shown in Fig. 6.

The monitoring strategy advocated here and discussed in more detail elsewhere [1,25] works. It is a significant advance over traditional monitoring strategies and has been recognized as such [26], but it is far from the last word on the subject. Much remains to be learned and documented.

When to Monitor for Pollutants and Background

Current conventional monitoring protocol generally requires sampling wells annually, semiannually, occasionally quarterly, and rarely as often as monthly. Such practices are prudent in most nonkarst terranes. In karst terranes, however, even at springs and cave streams judiciously and correctly selected as monitoring sites by the dye-tracing procedures recommended herein and elsewhere [1,25,27], the analytical results of data collected at such regular intervals can be inadvertently misleading, and the net result is a waste of time and money. Why is this? Because, if a karst aquifer is characterized by conduit flow, the chemical quality of the water at a spring to which it drains can be greatly affected by the effects of storms and meltwater events [3,28,29]. Diffuse-flow systems are generally only slightly affected by precipitation events [1].

In order to characterize the natural, storm-related water-quality variability of a spring in a conduit-flow system reliably, sampling must be done much more frequently than has been customary in the past. An attempt to characterize water quality when using typical semiannual sampling for study of a conduit-flow system is analogous to estimating annual rainfall of an area solely on the basis of rainfall data collected on the same two days of each year.

The effect of sampling frequency on the accuracy of characterizing and depicting this storm-related variability in water quality is seen in Fig. 7. Figure 7 is a composite of three published figures [28]. Discharge was recorded continuously. Pesticides were sampled up to 6 times per day, nitrate was sampled up to 20 times per day, and suspended sediment was sampled up to 17 times per day, depending upon the flow stage [30]. For the sake of discussion, however, let us *assume* that these parameters were monitored continuously for the 11-day interval shown. (Assume also that the apparent variations in water quality as a result of its natural variability, the statistics of sampling and analysis, and the analytical error are trivial—even though they possibly are not.) The most important thing shown is that pesticides and suspended sediment have a maximum level that is approximately coincident with that of the discharge. Nitrate, however, has its minimum when the others are

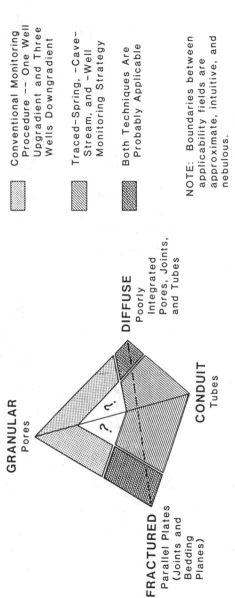

FIG. 6—*Diagram showing a tetrahedral continuum between the four major types of aquifers, the dominant porosity geometry in each type of aquifer, the applicability of the spring and cave-stream monitoring strategy in each type of aquifer, and the applicability of the traditional monitoring technique (using randomly located wells) in each type of aquifer* [51].

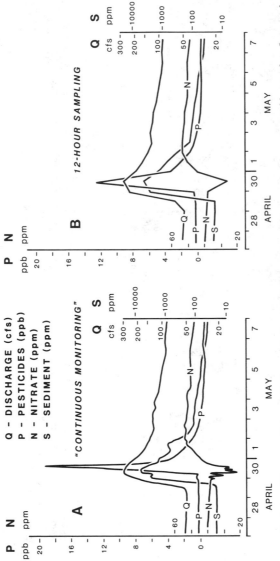

FIG. 7—*Water quality and discharge during a 1984 storm in a karst aquifer characterized by conduit flow, in the Big Spring ground-water basin, Iowa (Q = discharge, in cfs; P = pesticides, in ppb; N = nitrates, in ppm; and S = sediment, in ppm) (modified from Ref 28): (a) continuous monitoring data; (b) 12-h sampling data.*

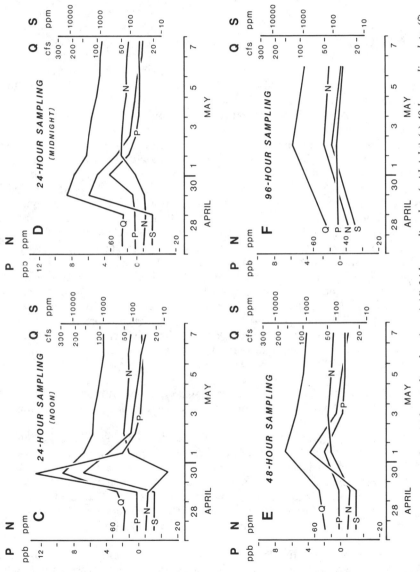

FIG. 7—Continued: (c) 24-h sampling data (noon); (d) 24-h sampling data (midnight); (e) 48-h sampling data; (f) 96-h sampling data.

at a maximum, and it reaches its own maximum several days later when the other pollutants are approaching their low prestorm values. The reasons for this peculiar lack of synchrony are discussed elsewhere [3,28,29]. Note also that pesticide concentrations increased by more than an order of magnitude during and after the storm. Such storm-related variation is common, but longer term, precipitation-related variation in what can falsely be called "background" for pesticide concentration at this site may range over more than three orders of magnitude in just a few months [3]. Such variation in background probably occurs elsewhere.

Figure 7, b through f, is based on the data represented in Fig. 7a, but this figure assumes sampling at intervals of 12, 24, 48, and 96 h. For the storm event that occurred, sampling at 48 and 96-h intervals (Fig. 7, e and f) is totally incapable of suggesting any significant change in the water quality. The 24-h sampling frequency (Fig. 7, c and d) is better, but the midnight samples happen to miss the decrease in nitrate and most of the increase in pesticides. Even a 12-h sampling interval (Fig. 7b) is only a crude approximation of the continuous sampling represented by Fig. 7a.

The author stresses that the "proper sampling frequency," however defined and however determined, will vary with the event to be monitored and with the flow dynamics and flow type (conduit flow versus diffuse flow) of the karst aquifer studied. For example, the sampling frequency necessary to characterize changes accurately in the chemical composition of spring discharge in the Big Spring ground-water basin (as in many other conduit-flow basins in the Mammoth Cave area and elsewhere) will, for similar storms and similar antecedent moisture conditions, be far more than is necessary to provide the same accuracy of characterization at the diffuse-flow springs draining the Edwards aquifer in Texas, the Floridan aquifer, and much of the Ozarks of Missouri and Arkansas.

In September 1987 the EPA announced proposed standards for the concentration of various pesticides in ground water. For atrazine, a herbicide used to control weeds in corn, sorghum, sugarcane, pineapple, and citrus groves, the maximum allowable concentration is 3 ppb [31]. About 90% of the pesticides found in Big Spring basin are atrazine. Therefore, the peak pesticide concentrations shown in Fig. 7, a through c, are about four to six times higher than this 3 ppb limit; midnight sampling barely detected this violation of the 3 ppb maximum (Fig. 7d), and the 48-h and 96-h sampling (Fig. 7, e and f) totally failed to do so.

Statistical procedures for designing and evaluating sampling strategies are available [32–39] but they are complex. A reliable, cookbook approach to this problem for karst areas would be desirable.

Until an acceptable, economically realistic, reliable procedure for sampling ground water in karsts is developed and tested, probably the best protocol is that proposed and discussed in Ref 3 (page 281). In brief, sampling should start at base flow, before the beginning of a storm or meltwater event, and continue until 4 to 30 times the time to the hydrograph peak has elapsed, depending upon the extent to which an aquifer is characterized by conduit flow as opposed to diffuse flow. Sampling may have to be done as often as at 1 to 6-h intervals in the early part of a precipitation event and at 4 to 24-h intervals in the waning part of its hydrograph.

After a precipitation event, the decision about which samples to analyze, if any, should be based on a careful evaluation of the significance of the event, interpretation of the flow stage and conductivity data from it and previous events, and an estimation of the data needed to characterize the monitoring site. Many samples, sometimes all, can be rightfully discarded. Difficult, judicious decisions must be made. These analytical data must be compared with those for samples taken several times per year, during base flow, storms, and meltwater events. Only then can one possibly make a reliable assessment or characterization of the true quality of water draining from a facility. After several years of data have been accumulated and the aquifer behavior is understood, sampling frequency may be decreased.

On the basis of conceptual models, experience, and limited data, the author believes that monitoring in karsts characterized by diffuse flow is easier and cheaper because far fewer samples are required. Also, the more a karst aquifer is characterized by diffuse flow, the higher the probability that fracture-trace-sited wells and randomly located monitoring wells can be used reliably. All wells proposed for monitoring use, however, must still test positive by dye tracing.

A quick, inexpensive way to distinguish between a conduit-flow spring and a diffuse-flow spring is to observe its turbidity (or lack of turbidity) and to measure its specific conductivity before, during, and after several large storms. If the spring is characterized by conduit flow, it will be turbid; the coefficient of variation (standard deviation $\div$ mean $\times$ 100) of its specific conductivity will be from 10% to more than 25%. If the spring is characterized by diffuse flow, its water will always be clear to slightly turbid; the coefficient of variation of its specific conductivity will be less than 5% [1,25]. Remember, however, that conduit flow and diffuse flow are end members of a continuum.

Much remains to be learned about when to sample ground water in karst terranes. The likely possibility of either deliberate or inadvertent acquisition of falsely negative data from these terranes makes it imperative that people in charge of sampling and officials in charge of evaluation of sample data have an understanding of karst problems and carefully scrutinize all analytical results from such terranes.

How to Determine Where and When to Monitor Ground Water in Karst Terranes Reliably and Economically

Where to monitor is typically best determined with the aid of two types of field investigation: dye tracing and mapping of the potentiometric surface. Geophysical studies, such as mapping of the streaming potential of descending waters and mapping of acoustic emissions from cave streams [14], may follow mapping of troughs on the potentiometric surface and precede drilling of wells planned to intercept cave streams on these troughs, which are to be tested by tracing for a connection to a facility.

Under "ideal" circumstances, one can run the dye tests necessary for the design of a monitoring system from a sinking stream on the facility. Often, however, no stream is available for dye injection. How then can introduction of the dye be done? One can use tank trucks of water and inject dye at (in decreasing order of desirability) the following locations:

(*a*) a sinkhole with a hole at its bottom;
(*b*) a sinkhole without a hole at its bottom (excavation may reveal a hole that can be used);
(*c*) a losing-stream reach with intermittent flow;
(*d*) a Class V stormwater drainage well;
(*e*) a well drilled on a fracture trace or a fracture-trace intersection;
(*f*) an abandoned domestic, agricultural, or industrial well; or
(*g*) a well randomly drilled for dye injection.

Wells drilled for dye injection should probably extend about 7 m (25 ft) below the potentiometric surface, if there is one, but this is a site-specific determination. Before going to the trouble and expense of conducting a dye test from injection-site Types *b* and *e* through *g,* however, these sites should first be given a percolation test with a tank truck of potable water to see how rapidly they drain. Alternatively, a pumping test can be run. An electric tape or a pressure transducer can be used to determine the rate of water-level decline in a well; plots of such data can be used to select what is probably the most direct and open connection to the conduits that drain an aquifer. If the percolation test of an injection site shows little or no drainage, dye should not be injected into it. The procedures for dye tracing with trucked water are discussed elsewhere (Ref *1,* page 222) [*27*].

It is economically almost impossible to design a dye trace that simulates the conditions beneath a landfill in, for example, a terrane characterized by 20 m (60 ft) of residual soil. Why is this? Because, the residuum is anisotropic. The distribution of macropores, which may make its permeability several orders of magnitude greater than that of the bulk of the residuum, is unpredictable [*40*]. These larger macropores, through which fluids move most rapidly, may be several centimetres to several metres apart. The problem of intercepting them with a drill hole to be used for dye injection is analogous to the problem of searching for ore bodies with a drill hole; only the size of the targets, and therefore the spacing between holes necessary to achieve the same probability of target interception, is different.

The relative suitability of different drill holes in residuum or bedrock can be evaluated by percolation tests, as discussed above. If the holes are very close to one another it would be prudent to have an electric tape or a pressure transducer in the adjacent holes—just to be sure that the rapid fall of the water level in the tested hole is not a consequence of leakage into one or more of the adjacent holes.

If, instead of using drill holes for dye injection, one were to excavate a 10 by 10-m (30 by 30-ft) pit to the depth of the bottom of a proposed landfill in the example terrane, say 5 m (15 ft), and carefully construct either a simulated compacted or lined bottom of a cell—or even try to make an "undisturbed" bottom—one could not be sure of having simulated or tested the *long-term* permeability conditions at the bottom of the landfill.

Therefore, whether using a sinkhole, a drilled hole, or a pit for dye injection, when designing a dye trace to evaluate a proposed landfill site, one must do the following:

1. Assume that the soil or liner has or will have differential permeability (leakage) that cannot be remedied by economically justifiable construction methods. This assumption is supported by an extensive literature (see Ref *1,* page 199). The long-term permeability of any kind of liner may be affected by chemical changes induced by the leachate. The urgent questions about leakage are *when, to where,* and *how fast.* (This is consistent with the EPA policy of assuming a worst possible case scenario.)

2. Test for the consequences of the leakage that is certain to occur.

Since leakage is reasonably assumed to be a certainty, the dye test should be designed to maximize the probability of getting the dye through the soil or residuum as rapidly as possible. Few consulting firms or their clients can afford to wait a year or more for test results that, until the dye is recovered, are negative. The argument that a dye trace (assuming it is well designed and properly run) is irrelevant to evaluation of a particular landfill because the dye was not injected at the bottom of the actual landfill and precisely at its location is specious.

It is naive, erroneous, and dangerous to assume that all or most of the springs necessary for tracer tests and ground-water monitoring are shown on U.S. Geological Survey 7.5-min topographic quadrangles. The author's experience in numerous karst areas has shown that only about 5% of the springs discharging non-isolated *local* flow are shown on maps. Field-work for finding springs is mandatory; there is no substitute!

All springs within a radius of perhaps 8 to 25 or more kilometres (5 to 15 or more miles) from a facility, especially those within $\pm 90°$ of the likely vector of the hydraulic gradient from it, should be found and probably monitored during dye tracing. A prudent designer of a dye trace will, however, at the beginning of an investigation, generally assume the possibility of radial flow and will have 360° of coverage with dye detectors—if only to defend the test design from criticism of inadequacy. The radius of the spring search is determined by evaluating the stratigraphy, structure, and physiography of the terrane and by proposing various tentative hypotheses about possible flow routes and resurgences. These working hypotheses must be tested.

Dye traces to springs at the bottom of sinkholes are especially important for recognizing hitherto-isolated segments of the plumbing system of a karst aquifer. These segments between a facility and the spring to which it drains can be used for monitoring; they offer two advantages over monitoring of springs: less dilution of pollutants or surrogate compounds, and earlier detection of them.

Imagine how extremely different the tracing results in Fig. 5 would appear if the designer of the dye tests shown had followed a "hunch" and monitored only the springs in one particular direction from the landfill rather than in all directions! At the administrative hearing on whether or not to grant a permit for construction of the proposed landfill—held after the dye test results from an area adjacent to the site were available, but before the other five dye tests had been run—the state's expert witnesses alleged that it was impossible for dye (or pollutants) to flow in opposite directions. They vigorously but erroneously impugned the validity of the dye test. The five subsequent tests resolved all questions about the alleged impossibility of radial flow.

Prediction of flow within a ground-water basin characterized by local flow is usually very much like prediction of flow within a surface-water basin. It moves "downhill" (downgradient) to the trunk drain at the local base level. If the boundaries of either basin are known, it can be confidently stated that although small-scale local flow from an area (or facility) may be in almost any direction, larger-scale local flow will be to tributaries and ultimately to the trunk that drains the basin. For example, in Fig. 2, any dye input (or pollutant) injected south of Mill Hole and west of Park City can confidently be predicted to flow to Mill Hole. Similar predictions can be reliably made for other potential facilities elsewhere in the Turnhole Spring ground-water basin and for anywhere in the area shown in Fig. 4.

The monitoring system for a facility and the consequences of leakage from it should be tested by tracing dye movement from the facility itself. This may not be possible. If it is not, traces should then be run from sites adjacent to the facility, preferably from opposite sides of it, and at points lying on a line approximately perpendicular to the suspected flow direction. This increases the probability of discovering whether a facility is near the boundary of a ground-water basin and whether it consistently flows to the same spring (or springs). Such tracer results are relevant to objective evaluation of the facility. This principle for selecting sites for injection of tracers is called hydrologic juxtaposition. Its validity is strained by the tracer results shown in Fig. 5 but not contradicted by them.

To summarize the principle of hydrologic juxtaposition, if the geology of a dye injection site is similar to that of a site immediately adjacent, it is highly probable that tracer results from the two sites will show discharge to the same spring. Obviously, this may not be so in the immediate vicinity of the boundary between two ground-water basins, but it is the reason a second test is recommended for the opposite side of a facility. In fact, predictions based on the results of a tracer test run from each side of a facility (on or off it) should provide greater confidence than those based on only a single tracer test run from the middle of the area.

Ground-water tracing can be done with many different tracers, but in general the cheapest, most efficacious ones are fluorescent dyes such as fluorescein (CI Acid Yellow 73),

Rhodamine WT (CI Acid Red 388), CI Direct Yellow 96, and optical brighteners such as CI Fluorescent Whitening Agents 22 and 28. Brief summaries of practical techniques for dye tracing have been published [1,41,42], but a comprehensive, plain-English guide to the use of dyes as tracers is scheduled for publication during 1990 [27]. This guide will include many practical hints and suggestions that will enhance the rigor of test designs with any kind of tracer and will increase the reliability of test results. A sampling device for efficiently recovering various types of dyes from continuously pumped wells has been invented and tested by the author. A review summarizing the nontoxicity of dyes commonly used for tracing has been made by Smart [43].

Three types of dye tracing can be used for evaluating the suitability of springs, cave streams, and wells for ground-water monitoring; they are as follows:

1. Qualitative tracing, using either of the following:
    (a) Visual observation of the dye plume. Generally this is wasteful of dye and may cause aesthetic and public-relations problems. Also, there is a great risk of missing the dye pulse when it arrives at the monitoring site, especially at night.
    (b) Passive detection (with passive detectors consisting of activated charcoal or cotton, depending upon the tracer used) plus either visual observation of dye eluted from the charcoal or ultraviolet observation of the cotton [27,41,42].

Qualitative tracing is sufficient for most dye tests; when it is done with passive detectors, it is generally the most cost-efficient tracing technique.

2. Semiquantitative tracing, using passive detectors and instrumental analysis of the elutant or cotton with a fluorometer or a scanning spectrofluorophotometer [27,44,45]. The many variables associated with changes in spring or stream discharge, with the reaction kinetics of sorption of dye onto passive detectors and with elution of dye from the charcoal, all make it impossible to quantify precisely the varying concentrations of dye that passed any specific monitoring site during a given period of time. Instrumental analysis can identify dye concentrations several orders of magnitude smaller than those detectable visually.

3. Quantitative tracing, using instrumental analysis of water samples (either grab samples or those taken with an automatic sampler) or of water continuously flowing through an instrument, preferably (for either option) with continuous measurement of the discharge [19–21,27,46]. Instrumental analysis makes possible more precise determination of flow velocity, the breakthrough curve characteristic of a tracer's arrival and retardation, and the aquifer dispersivity. It also allows calculation of dye recovery. Many interwell traces are done with this type of quantification; some evaluations of wells as potential sites for monitoring can only be done with interwell traces. For a given trace, quantitative tracing is the most expensive procedure, but it can give answers not available by any other technique [19–21,46].

If quantitative tracing results are needed in the design of a monitoring system—and generally they are not—it is commonly far more cost-efficient to first do a qualitative or semiquantitative study. This eliminates substantial costs of sampling and analysis of numerous sites to which no dye travels.

Each of these three types of dye tracing is sufficient and satisfactory for establishing a hydrologic connection between two points. Semiquantitative and quantitative tracing techniques are more sensitive to detection of small concentrations of dye; for litigation, they are more convincing. Quantitative tracing techniques are the most sensitive for detection of small, temporary changes in dye concentration.

No matter which of the three types of dye tracing is used for an investigation, it is important to avoid two of the more common mistakes of people inexperienced in tracing: not sampling at enough sites at which dye could possibly be recovered (generally using passive detectors) and not sampling long enough. If not enough sites are sampled or if sampling is stopped too soon after the first positive results, one would fail to detect dye at the other places to which it also goes (if, indeed, it goes elsewhere) at either the same velocity or a different velocity. Also, one would fail to discover if some of the dye had become stored in the *epikarst* (as discussed subsequently under *Exception 3*) and was being released over a long period of time. Any of these consequences of inadequate sampling for dye would prevent discovery of aquifer properties that adversely influence the adequacy of the design of a monitoring system.

Ideal tracers are conservative. They do not react with soil, bedrock, or ground water and they do not undergo microbial decay. Most real tracers, including dyes, are slightly reactive and may undergo adsorption-desorption and cation-exchange reactions. Organic pollutants may undergo similar reactions, which affect their rate of migration. This rate is predictable and generally correlated with their octanol-water partition coefficients (Ref *47*, pages 397–405, and Ref *48*); these pollutants may also undergo microbial decay. Depending upon their mobility, pollutants may travel faster or slower than a dye. Although tracing velocities can be used as a reliable guide for prediction of flow velocities of pollutants under similar antecedent moisture conditions, especially in conduit-flow aquifers, velocities of conservative tracers in diffuse-flow aquifers (as well as in granular aquifers) will be significantly higher than those of most pollutants.

In dye tracing, wisdom is knowing what essential questions need to be asked, and asking them; experience is knowing the most expedient, most prudent way to get the answers to these essential questions.

Although a dye test is like the birth of a baby—no matter how many men and women are put on the job, it will take however long is necessary to complete the task—it is also quite different. The birth of a baby can confidently be predicted to occur probably about nine months after conception. A dye test, however may be completed within a few hours or days after injection, but it could as easily take weeks, months, and even years. One must be patient while waiting for tracing results—or risk malpractice litigation and loss of valuable data. It might take just a few weeks to do the dye traces necessary to design and test the monitoring system for a facility, but it is likely to require six to nine months of intermittent, careful tracing. Warning: it could take even longer.

The potentiometric surface of a karst aquifer should be mapped with as much data as possible. For basin analysis, a minimum of 1 well per square kilometre (2.5 wells per square mile) is recommended for most aquifers. A facility analysis could require more than 40 times this well density. A carefully contoured potentiometric map, if based on valid measurements of an adequate number of wells, can be used to accomplish the following:

(a) predict the flow routes of tracers (or pollutants),
(b) judiciously select dye injection sites,
(c) minimize the number of traces necessary for evaluation of a ground-water basin or a facility,
(d) interpolate flow routes in the areas between dye traces, and
(e) detect the possible influence of shale beds and other poorly permeable rocks on perching and confinement of water within a karst aquifer.

Water levels for mapping the potentiometric surface can be obtained with a steel tape, an electric tape, or an acoustic well probe [1,41,42], but such mapping must include a QA/QC program using a vertically hanging steel tape as a standard before, after, and preferably during periods of use—especially if water levels more than 30 m (100 ft) are being measured. Permanent tape stretching of as much as 2% is known to occur in electric tapes employing copper twin-lead wire similar to those used for TV antenna lead-in. Water-level measurements generally should be made during low-flow and base-flow conditions, rather than during the rainy season or after storms, when water levels can locally be significantly higher and the surface can have a different configuration.

Implicit Assumptions of Recommended Monitoring Strategy, with Examples of Exceptions

Ground water has been recognized to circulate in three different types of flow systems: local, intermediate, and regional [49,50], as reviewed in Ref 47 (pages 217–258). Many axiomatic, implicit flow-system assumptions are made by those who may use the traced spring-, cave-stream-, and well-monitoring strategy advocated here and in Refs 1 and 25, but one should realize what they are. Seven of these major implicit assumptions follow:

1. Discharge is concentrated at a point (a spring or group of springs) rather than diffused over a broad area or concentrated along a line (such as a stream).
2. Most flow systems to be monitored in karst are characterized by local flow, in the sense of Toth [49,50].
3. Ground-water velocities in karsts characterized by local flow have the high values already cited herein. The apparent velocities in karsts characterized by intermediate-flow systems and by regional-flow systems tend to be several orders of magnitude slower than those in local-flow systems.
4. A ground-water basin is a discrete entity having a specific, well-defined boundary.
5. Ground-water basins are contiguous.
6. Storm-related diversion of ground water out of a basin, if it occurs, is by means of intermediate- and high-level overflow routes (conduits) leading to adjacent ground-water basins. To restate: storm-related piracy of ground water, if it occurs, is a temporary diversion—from a pirator ground-water basin to the original piratee ground-water basin.
7. Temporary, storm-related diversion of surface water within a karst ground-water basin is not a significant problem because all the water will remain within the basin.

On the basis of more than 30 years of experience the author believes that each of these implicit, axiomatic assumptions is correct about 95% of the time—often enough to be fairly assumed until or unless data imply otherwise, but not so often as to be a certainty. One must always ask, "What are my paradigms and their assumptions?" and critically review the validity of each. Systematic analysis and review of examples of probable exceptions to the above implicit assumptions and others could be the subject of another paper [25], but here only one or two exceptions to some of them are cited and briefly discussed, and the potential relevance of these exceptions to a monitoring program is stated.

Exception 1: Nonpoint Discharge

Although some discharge in the karsted dolomite of the Door Peninsula, Wisconsin, is to springs [51], most of the discharge is through sediment and over a broad area beneath Green Bay and Lake Michigan [52,53]. Discharge from a karst aquifer through sediment over a broad area and along a line occurs in the Cano Tiburones area north and west of Barceloneta, Puerto Rico, and along a line in the valley of the Rio Grande de Manati, south of Barceloneta. These terranes are in Tertiary limestones of the alluvium and paludal sediment mantled shallow aquifer on the north coast; the aquifer seems to be characterized by diffuse flow, but springs and conduit flow are locally important [54]. Monitoring in terranes characterized by areal discharge can best be done by using randomly located wells, perhaps along fracture traces, but these wells might not intercept the relevant flow lines. Monitoring in terranes characterized by seepage along a line is best done by sampling at intervals along the line and upgradient from it. In each hydrologic setting, however, although flow velocities may be very low, tracing must be done if confidence in a monitoring effort there is what is desired. If a well proposed for monitoring in a karst terrane does not have a positive trace to it (or to its site), it isn't a monitoring well.

An example of discharge along a line occurs in the eastern Snake River Plains aquifer, in Idaho [55]. Admittedly, this basalt aquifer is not a karst aquifer, but it was because of the hydrology of such rocks that the term *pseudokarst* was first proposed more than 80 years ago (Ref 56, pages 182–183). Many of the monitoring principles advocated herein are applicable to monitoring in this basalt aquifer and in other highly fractured rocks.

Exception 2: Nonlocal Flow

There are many examples of nonlocal flow that must be monitored in karst aquifers. Some of the better known examples of such karsts are the Edwards aquifer [57,58] and the Floridan aquifer (Ref 47, pages 237–243 and 359–361) [59,60]. Flow velocities in them are likely to be much slower than in most conduit-flow aquifers and more like the low velocities characteristic of most diffuse-flow aquifers. Flow paths are likely to be braided and dispersive rather than convergent; they still, however, flow to springs.

Exception 3: Slow Movement in a Local-Flow System

Immediately beneath the soil profile of most karsts is the *epikarst,* a 3 to 10-m (10 to 30-ft) zone of karstification in which horizontal flow is dominant and storage is significant (Ref 9, pages 28–35) [1,2,61–64]. (Some authors prefer to use the synonymous term *subcutaneous karst.*) Usually the epikarst is separated from the phreatic zone by a dry, inactive waterless interval of bedrock that is locally breached by vertical percolation, but it may extend to the phreatic zone. Although tracer studies in a British epikarst have shown that vertical flow velocities (to the caves below) may exceed 100 m/h (300 ft/h), dye was still detectable 13 months later [62–64]. In spite of the fact that most dye (and, presumably, most pollutants) travels through an aquifer at high velocities, some may become "hung up" in epikarstic storage. This flow dichotomy is actually a continuum, but it suggests that the common adage, that ground-water pollution in karst areas is not a long-term problem because the aquifer is rapidly self-cleaned, is wrong—or, at least, unreliable.

Exception 4: Fuzzy and Overlapping Basin Boundaries

It is quite logical to assume that a ground-water basin has a specific, well-defined boundary. Sometimes the boundary is breached and, during response to moderate- to flood-flow conditions, some water is diverted to adjacent ground-water basins, as already discussed. This assumption is reasonable and generally correct; it can be extended to allow for temporary, slight shifts in the boundary between two basins, also in response to storm-related changes in flow conditions. In contrast, some diffuse-flow aquifers show significant exceptions to this implicit assumption; the basin boundaries may be nebulous and gradational during all flow conditions. For example, all dye injected near the center of a ground-water basin in the Great Oolite Limestone (in the Bath district of southwestern England) flows to only one spring. As different dyes from successive tests are injected at increasingly greater distances from the central axis of the basin, less dye goes to its spring. The balance of the dye goes to the major spring draining the adjacent ground-water basin. As the dye injection point gets successively farther from the central axis, more and more dye goes to the major spring in the adjacent basin. Such fuzzy boundaries are characteristic of basins in this aquifer. Smart has recommended that the gradational "boundary" between any two such ground-water basins be chosen to coincide with tracer injection springs [65]. Similar results occur elsewhere in England [66] and probably in other places where diffuse flow predominates. Nebulous, gradational boundaries between ground-water basins have recently been recognized in Missouri [24] and undoubtedly will be recognized in other karst terranes of the United States.

Gradational basin boundaries are suggested by the results of dye injection into a swallet (a sinkhole into which a stream empties) at the west boundary of the Turnhole Spring ground-water basin (Fig. 2, on the 520-ft contour). This swallet drains to two different ground-water basins: it flows both to the northeast (within the Turnhole Spring basin) and to the west, to a second major drainage basin (Ref 67, pages 48–49). This Kentucky boundary has not been studied, but I believe that the extent of overlap there is less than half a kilometre.

One should be prepared to encounter fuzzy and overlapping boundaries between ground-water basins. The possible existence of such boundaries makes it necessary to be extremely thorough in the design of dye tracing investigations and confirms the already recognized need for monitoring for dye at springs in ground-water basins adjacent to a proposed facility [27].

Overlap of ground-water basins is more convincingly illustrated by the Bear Wallow basin in Kentucky [12]. It occupies 500 km² (190 mile²) and its three subbasins, Hidden River, Three Springs, and Uno, resemble a Venn diagram (a diagram employed in symbolic logic; it uses circles and their relative position to represent sets and their relationships [68,69]). These subbasins occupy 65, 15, and 20%, respectively, of the total basin, but the Uno subbasin comprises part of the headwaters of the other two and is common to each of them. This overlap is significant because the 99 km² (38 mile²) of the Uno basin is the size of the area alluded to earlier in this paper as the terrane in Fig. 2, from which pollutants could flow to a total of 52 springs in 19 isolated segments along a 19-km (12-mile) reach of Green River. The probable consequences of leakage from a facility would be significantly fewer if it were located somewhere other than in a ground-water basin in which the headwaters are analogous to a Venn diagram and discharge is via distributary flow.

Exception 5: Noncontiguous Ground-Water Basins

Thrailkill and his students have shown that there are two physically distinct spring types in karst of the Inner Bluegrass of Kentucky: local high-level springs discharging from shallow flow paths and major low-level springs discharging from a deep, integrated conduit system [70–72]. The major springs are characterized by larger catchment areas [>10 km² (4 mile²)] and higher discharges [10 to 2700 L/s (2.6 to 700 gal/s)], in comparison with the smaller catchment areas of the local high-level springs [<2 km (0.8 mile²)] and their lower discharges [0.1 to 0.8 L/s (0.037 to 0.2 gal/s)]. The lack of integration of the local high-level spring catchments into the major low-level spring catchments can be explained in terms of the impermeability of numerous interbedded shales and the lack of fractures passing through them. The catchment of each major low-level ground-water basin is near-elliptical, unrelated to surface drainage, isolated from the nearby major basins, and commonly separated from its nearest similar neighbor by 1 to 4 km (0.6 to 2.5 mile). The reasons for the noncontiguity of these major basins are not yet understood but may be related to the nature of the epikarst between conduits and the inhibition of hydraulic integration by clay and shale [73].

Such noncontiguous basins within a karst dipping uniformly at a low angle are not known to occur elsewhere, but they probably exist. The possibility of their occurrence in various settings should be anticipated and can be detected when traces are run from each side of a suspected boundary of a ground-water basin.

It is possible that traditional, randomly located wells may provide effective monitoring in the interbasin areas between noncontiguous basins. To use these wells reliably in such a karst, however, the tracing procedures advocated herein must first be employed and competently shown to yield negative results at springs and positive results at the wells.

Exception 6: Diversion of Ground-Water to the Surface

The Poorhouse Spring ground-water basin, in Kentucky, nicely illustrates another idiosyncracy of flow in karst [12]. The basin drains 70 km^2 (27 mile^2) and most flow is to the southeast, through the 3 by 6-m (10 by 20-ft) trunk stream passage that is Steele's Cave. About once a year, after a very heavy rain, some of the trunk's flow rises about 21 m (70 ft) above its normal level, flows out of the sinkhole entrance, and is discharged onto the surface. From there it flows in the opposite direction, to the northwest and west, where it augments the flow of a surface stream which drains an area outside of the Poorhouse Spring ground-water basin. Such flow, if contaminated, could give spurious values at surface sites used to monitor pollutants or background in adjacent ground-water and surface-water basins.

Exception 7: Diversion of Surface Water to a Different Ground-Water Basin

Sometimes surface waters are diverted during storms from one ground-water basin to another. This is illustrated by the behavior of Cayton Branch, in Kentucky (Refs 12 and 67, page 22). Little Sinking Creek (the southwesternmost surface stream in Fig. 2) drains north to the Green River, as shown, and contributes to the discharge from the Turnhole Spring ground-water basin. At a point about 2 km (1.2 miles) south of the creek's northernmost swallet, where the south fork of the stream bends north at the 600-ft potentiometric contour, the creek goes out of its banks during floods and diverts some of its discharge westward. The diverted surface water flows about 600 m (2000 ft) west to a swallet that comprises part of a larger ground-water basin in which the headwaters have been captured by the Turnhole Spring ground-water basin. Although the northward piracy of the northwest-flowing south fork of Little Sinking Creek seems to be a surface piracy, it is probably related to subsurface piracy of the adjacent western ground-water basin (and others) by the Turnhole Spring ground-water basin [74]. The present-day hydraulic gradient from its swallet, northward to Mill Hole (midway between it and Turnhole Spring), is about 9 m/km (47 ft/mile), approximately four times steeper than the present-day gradient westward to the spring along Barren River to which the south fork formerly flowed and to which the storm-diverted surface water now flows [12].

This diversion of surface water from the Turnhole Spring basin to another ground-water basin could be relevant to a monitoring effort in the second basin if there were significant quantities of pollutants in the surface waters of Little Sinking Creek and if they were "exported" to the second basin, where they could adversely affect the reliability of data from cave streams used for monitoring of background.

The above exceptions to the seven implicit assumptions stated as hydrologic axioms are rare, but important. Their existence justifies a thoroughness in the design of tracer tests and in the interpretation of tracer results that, to some people, might seem almost paranoid. Their existence emphasizes the need for facility-related fieldwork. Additional exceptions are discussed in Ref 27.

Summary

In order to be relevant to monitoring for pollutants, water-quality data from karst terranes must be from springs, cave streams, and wells which have been shown by tracing to drain from the site to be monitored. Tracing should typically be done at least three times: when first convenient (during moderate-flow conditions, to quickly provide a preliminary, tentative understanding of local movement of ground water), and later, during base flow and flood flow. Sites for monitoring background should be selected on the basis of negative results of these tracing tests and should be in settings in which the rocks and waters are geochemically similar to those of the locations where the tracer tests were positive. The general prudence of tracing during all three types of flow conditions can not be overemphasized.

A map of the potentiometric surface, if it is based on enough data from an aquifer not complicated by aquicludes and aquitards, will greatly enhance one's ability to design the necessary tracing tests efficiently and to interpret them with greater confidence. One well-designed tracing test, properly done and correctly interpreted, is worth 1000 expert opinions—or 100 computer simulations of ground-water flow.

Sampling for water quality must be frequent and done before, during, and after storm and meltwater events. Base flow should be sampled between such events.

Ground-water monitoring in karst terranes can be done reliably, but the analytical costs are likely to be significantly higher than those for other terranes. It could be far less expensive to locate a proposed facility in a nonkarst terrane.

A plethora of ways exists to design the placement and sampling frequency of a ground-water monitoring network in a karst terrane so that it yields falsely negative results for the chemical compound being sought. Accordingly, environmental consultants and regulators must be ever vigilant to be sure that negative results are not falsely negative—either accidentally or intentionally.

Numerous plausible, axiomatic rules can be stated about ground-water flow in karst terranes, but they are not absolutes. Exceptions are known for each of the seven rules cited; more exceptions will be discovered.

Epilogomenon

A question can be raised of why professional geologists and engineers—who would not think of venturing into the design of a building foundation, a landfill, or a well field without first having obtained some initial experience with such matters or without a review of its design by a competent peer—all too often assume that anyone can do professional-quality dye tracing the first time it is attempted. If one's tracing experience is limited or nonexistent, the most astute, most ethical, and least expensive ways to minimize the risks of costly litigation for tracing-related malpractices are to accomplish the following tasks:

(a) obtain experience with one of the few individuals in North America who are adept at dye tracing in the field;

(b) hire one of these individuals to do the tracer investigation—with the understanding that he will provide some rudimentary training;

(c) hire one of them to design the tracer investigation; or,

(d) at an absolute minimum, hire one of them to review the design of the proposed tracer investigation.

An experienced, adept tracing consultant potentially saves his employer many times his fee—and greatly enhances the reliability of both the tracer results and the proposed monitoring system.

Acknowledgments

The ideas expressed here have been reviewed by E. Calvin Alexander, Jr., Thomas J. Aley, Ralph O. Ewers, and Arthur L. Lange. Each of them has critically reviewed successive drafts of the manuscript and made numerous valuable suggestions and improvements. Much of the fieldwork in Kentucky that initially stimulated many of these conclusions was meticulously done by Joseph A. Ray and other National Park Service assistants. Thomas J. Aley graciously allowed publication of his work in progress at Pindall, Arkansas—work in which the interest, dedication, and reliability of Jeff Henthorne greatly assisted. Some of this author's conclusions are a result of work done as a consultant for various clients. A. Richard Smith's redactional talents repeatedly improved clarity. Some of the work for this paper was supported by research funds provided by the U.S. Environmental Protection Agency, Environmental Monitoring Systems Laboratory, Las Vegas, Nevada, but this paper does not necessarily reflect the views of the EPA; neither official endorsement nor disagreement should be inferred.

References

[1] Quinlan, J. F. and Ewers, R. O., "Ground Water Flow in Limestone Terranes: Strategy Rationale and Procedure for Reliable, Efficient Monitoring of Ground Water Quality in Karst Areas," *Proceedings,* Fifth National Symposium and Exposition on Aquifer Restoration and Ground Water Monitoring, Columbus, OH, National Water Well Association, Worthington, OH, 1985, pp. 197–234.

[2] Smart, P. L. and Hobbs, S. L., "Characterization of Carbonate Aquifers: A Conceptual Base," *Proceedings,* Conference on Environmental Problems in Karst Terranes and Their Solutions, Bowling Green, KY, National Water Well Association, Dublin, OH, 1987, pp. 1–14.

[3] Quinlan, J. F. and Alexander, E. C., Jr., "How Often Should Samples Be Taken at Relevant Locations for Reliable Monitoring of Pollutants from an Agricultural, Waste Disposal, or Spill Site in a Karst Terrane? A First Approximation," *Proceedings,* Second Multidisciplinary Conference on Sinkholes and the Environmental Impacts of Karst, Orlando, FL, Balkema, Rotterdam, 1987, pp. 277–286.

[4] Crawford, N. C., "Sinkhole Flooding Associated with Urban Development upon Karstic Terrain: Bowling Green, Kentucky," *Proceedings,* Multidisciplinary Conference on Sinkholes, Orlando, FL, Balkema, Rotterdam, 1984, pp. 283–292.

[5] Aley, T., "A Model for Relating Land Use and Groundwater Quality in Southern Missouri," *Hydrologic Problems in Karst Areas,* R. R. Dilamarter and S. C. Csallany, Eds., Western Kentucky University, Bowling Green, KY, 1977, pp. 323–332.

[6] Aley, T. and Thompson, K. C., "Septic Fields and the Protection of Groundwater Quality in Greene County, Missouri, Final Report," consultant's report, Ozark Underground Laboratory, Protem, MO, 1984.

[7] Quinlan, J. F., "Recommended Procedure for Responding to Spills of Hazardous Materials in Karst Terranes," *Proceedings,* Conference on Environmental Problems in Karst Terranes and Their Solutions, Bowling Green, KY, National Water Well Association, Dublin, OH, 1987, pp. 183–196.

[8] Smoot, J. L., Mull, D. S., and Lieberman, T. D., "Quantitative Dye Tracing Techniques for Describing the Solute Transport Characteristics of Ground-Water Flow in Karst Terrane," *Pro-*

ceedings, Second Multidisciplinary Conference on Sinkholes and the Environmental Impacts of Karst, Orlando, FL, Balkema, Rotterdam, 1987, pp. 269–275.

[9] Bonacci, O., *Karst Hydrogeology, with Special Reference to the Dinaric Karst,* Springer-Verlag, New York, 1987.

[10] Milanović, P. T., *Karst Hydrogeology,* Water Resources Publications, Littleton, CO, 1979.

[11] White, W. B., *Geomorphology and Hydrology of Karst Terranes,* Oxford Press, New York, 1988.

[12] Quinlan, J. F. and Ray, J. A., "Groundwater Basins in the Mammoth Cave Region, Kentucky," *Friends of the Karst Occasional Publication,* No. 1, 1981.

[13] Palmer, A. N., "Prediction of Contaminant Paths in Karst Aquifers," *Proceedings,* Conference on Environmental Problems in Karst Terranes and Their Solutions, Bowling Green, KY, National Water Well Association, Dublin, OH, 1987, pp. 32–53.

[14] Lange, A. L., "Detection and Mapping of Karst Conduits from the Surface by Acoustic and Natural Potential Methods," research report prepared for the National Park Service and U.S. Environmental Protection Agency, The Geophysics Group, Wheat Ridge, CO, 1988.

[15] Stokowski, S. J., Jr., "Locating Groundwater Conduits in Carbonate Rocks," *Proceedings,* Multidisciplinary Conference on Sinkholes and the Environmental Impacts of Karst, Orlando, FL, Balkema, Rotterdam, 1987, pp. 185–196.

[16] Werner, E., "Photolineaments and Karst Conduit Springs in the Greenbriar Limestones of West Virginia," Abstract, *Geological Society of America Abstracts with Programs,* Vol. 20, 1988, p. 322.

[17] Parizek, R. R., "On the Nature and Significance of Fracture Traces and Lineaments in Carbonate and Other Terranes," *Karst Hydrology and Water Resources,* Vol. 1, V. Yevjevich, Ed., Water Resources Publications, Fort Collins, CO, 1976, pp. 47–100.

[18] Quinlan, J. F., Ewers, R. O., and Field, M. S., "How to Use Ground-Water Tracing to "Prove" that Leakage of Harmful Materials from a Site in a Karst Terrane Will Not Occur," *Proceedings,* Second Conference on Environmental Problems in Karst Terranes and Their Solutions, Nashville, TN, National Water Well Association, Dublin, OH, 1988, pp. 289–301.

[19] Smart, C. C. and Ford, D. C., "Structure and Function of a Conduit Aquifer," *Canadian Journal of Earth Sciences,* Vol. 23, 1986, pp. 919–929.

[20] Smart, C. C. "Quantitative Tracing of the Maligne Karst System, Alberta, Canada," *Journal of Hydrology,* Vol. 98, 1988, pp. 185–204.

[21] Smart, C. C. "Quantitative Tracer Tests for the Determination of the Structure of Conduit Aquifers," *Ground Water,* Vol. 26, 1988, pp. 445–453.

[22] Esposito, A., *Fluid Power with Applications,* Prentice-Hall, Englewood Cliffs, NJ, 1980.

[23] Quinlan, J. F. and Rowe, D. R., "Hydrology and Water Quality in the Central Kentucky Karst: Phase I," *University of Kentucky Water Resources Research Institute, Research Reports,* No. 101, 1976; reprinted 1977, with corrections, as Uplands Field Research Laboratory (National Park Service) Management Report No. 12.

[24] Aley, T., "Complex Radial Flow of Ground Water in Flat-Lying Residuum-Mantled Limestone of the Arkansas Ozarks,"*Proceedings,* Second Conference on Environmental Problems of Karst Terranes and Their Solutions, Nashville, TN, National Water Well Association, Dublin, OH, pp. 159–170.

[25] Quinlan, J. F., "Ground-Water Monitoring in Karst Terranes: Recommended Protocols and Implicit Assumptions," EPA/600/X-89/050, U.S. Environmental Protection Agency, Environmental Monitoring Systems Laboratory, Las Vegas, NV, 1989.

[26] Beck, B. F., Quinlan, J. F., and Ewers, R. O., "Presentation of the E. B. Burwell, Jr., Memorial Award to James F. Quinlan and Ralph O. Ewers," *Geological Society of America Bulletin,* Vol. 97, 1987, pp. 141–143.

[27] Aley, T. J., Quinlan, J. F., and Behrens, H., *The Joy of Dyeing: A Compendium of Practical Techniques for Tracing Groundwater, Especially in Karst Terranes,* National Water Well Association, Dublin, OH, in press.

[28] Libra, R. D., Hallberg, G. R., Hoyer, B. E., and Johnson, L. G., "Agricultural Impacts on Ground Water Quality: The Big Spring Basin Study, Iowa," *Proceedings,* Agricultural Impacts on Ground Water, Omaha, NE, National Water Well Association, Dublin, OH, 1986, pp. 253–273.

[29] Hallberg, G. R., Libra, R. D., and Hoyer, B. E., "Nonpoint Source Contamination of Ground Water in Karst-Carbonate Aquifers in Iowa," *Perspectives in Nonpoint Source Pollution,* EPA 440/5-85-001, U.S. Environmental Protection Agency, Washington, DC, 1985, pp. 109–114.

[30] Hoyer, B. E., Iowa Geological Survey, personal communication, December 1987.

[31] "Atrazine Health Advisory" (draft), *Basic Documents, National Pesticide Survey,* U.S. Environmental Protection Agency, Office of Drinking Water, Washington, DC, 1987.

[32] Sanders, T. G., Ward, R. C., Loftis, J. S., Steele, T. D., Adrian, D. D., and Yevjevich, V., *Design of Networks for Monitoring Water Quality,* Water Resources Publications, Littleton, CO, 1983.

[33] Gilbert, R. O., *Statistical Methods for Environmental Pollution Monitoring,* Van Nostrand Reinhold, New York, 1987.

[34] Makridakis, S., Wheelwright, S. C., and McGee, V. E., *Forecasting: Methods and Applications,* 2nd ed., Wiley, New York, 1983.

[35] Chatfield, C., *The Analysis of Time Series: An Introduction,* 3rd ed., Chapman and Hall, London, 1984.

[36] Bendat, J. S. and Piersol, A. G., *Random Data: Analysis and Measurement Procedures,* 2nd ed., Wiley, New York, 1986.

[37] Gibbons, R. D., "Statistical Prediction Intervals for the Evaluation of Ground-Water Quality," *Ground Water,* Vol. 25, 1987, pp. 455–465.

[38] Montgomery, R. H., Loftis, J. C., and Harris, J. "Statistical Characteristics of Ground-Water Quality Variables," *Ground Water,* Vol. 25, 1987, pp. 176–184.

[39] Beck, M. B. and van Straten, G., Eds., *Uncertainty and Forecasting of Water Quality,* Springer-Verlag, Berlin, 1983.

[40] Quinlan, J. F. and Aley, T., "Discussion of 'A New Approach to the Disposal of Solid Waste on Land,'" *Ground Water,* Vol. 25, 1987, pp. 615–616.

[41] Quinlan, J. F., "Hydrologic Research Techniques and Instrumentation Used in the Mammoth Cave Region, Kentucky," *1981 GSA Cincinnati '81 Field Trip Guidebooks,* Vol. 3, T. G. Roberts, Ed., American Geological Institute, Washington, DC, 1981, pp. 502–504.

[42] Quinlan, J. F., "Groundwater Basin Delineation with Dye-Tracing, Potentiometric Surface Mapping, and Cave Mapping, Mammoth Cave Region, Kentucky, U.S.A.," *Beitrage zur Geologie der Schweiz—Hydrologie,* Vol. 28, 1982, pp. 177–189.

[43] Smart, P. L. "A Review of the Toxicity of Twelve Fluorescent Dyes Used in Water Tracing," *National Speleological Society Bulletin,* Vol. 46, pp. 21–33.

[44] Thrailkill, J., Byrd, P. E., Sullivan, S. B., Spangler, L. E., Taylor, C. J., Nelson, G. K., and Pogue, K. R., "Studies in Dye-Tracing Techniques and Karst Hydrogeology," *University of Kentucky, Water Resources Research Institute, Research Report,* No. 140, 1983.

[45] Duley, J. W., "Water Tracing Using a Scanning Spectrofluorometer for Detection of Fluorescent Dyes," *Proceedings,* Conference on Environmental Problems in Karst Terranes and Their Solutions, Bowling Green, KY, National Water Well Association, Dublin, OH, 1987, pp. 389–406.

[46] Molz, F. J., Guven, O., Melville, J. G., Crocker, R. D., and Matteson, K. T., "Performance, Analysis, and Simulation of a Two-Well Tracer Test," *Water Resources Research,* Vol. 22, 1986, pp. 1031–1037.

[47] Fetter, C. W., *Applied Hydrogeology,* 2nd ed., Merrill, New York, 1988.

[48] Winters, S. L. and Lee, D. R., "In Situ Retardation of Trace Organics in Groundwater Discharge to a Sandy Stream Bed," *Environmental Science and Technology,* Vol. 21, 1987, pp. 1182–1186.

[49] Tóth, J., "A Theory of Ground-Water Motion in Small Drainage Basins in Central Alberta, Canada," *Journal of Geophysical Research,* Vol. 67, 1962, pp. 4375–4387.

[50] Freeze, R. A. and Witherspoon, P. A., "Theoretical Analysis of Regional Groundwater Flow," *Water Resources Research,* Vol. 3, 1967, pp. 623–634.

[51] Wiersman, J. H., Stieglitz, R. D., Cecil, D. L., and Metzler, G. M., "Characterization of the Shallow Groundwater System in an Area with Thin Soils and Sinkholes," *Proceedings,* First Multidisciplinary Conference on Sinkholes, Orlando, FL, Balkema, Rotterdam, 1984, pp. 305–310.

[52] Bradbury, K. R., "Hydrogeologic Relationships Between Green Bay of Lake Michigan and Onshore Aquifers in Door County, Wisconsin," Ph.D. thesis, University of Wisconsin–Madison, Madison, WI, 1982.

[53] Cherkauer, D. S., Taylor, R. W., and Bradbury, K. R., "Relation of Lake Bed Leakage to Geoelectrical Properties," *Ground Water,* Vol. 25, 1987, pp. 135–140.

[54] Keagy, D., Ewers, R. O., and Quinlan, J. F., studies in progress, 1988.

[55] Lindholm, G. F., "Snake River Plain Regional Aquifer-System Study," *U.S. Geological Survey Circular,* No. 1002, 1986, pp. 88–106.

[56] Von Knebel, W., *Hohlenkunde mit Berucksichtigung der Karst-phanomene,* Viewig, Braunschweig, Germany, 1906.

[57] Maclay, R. W. and Small, T. A., "Carbonate Geology and Hydrology of the Edwards Aquifer in the San Antonio Area, Texas," *U.S. Geological Survey Open-File Reports,* No. 83-537, 1984.

[58] Campana, M. E. and Mahin, D. A., "Model-Derived Estimates of Groundwater Mean Ages, Recharge Rates, Effective Porosities and Storage in a Limestone Aquifer," *Journal of Hydrology,* Vol. 76, 1985, pp. 247–264.

[59] Miller, J. A., "Hydrogeologic Framework of the Floridan Aquifer System in Florida and in Parts of Georgia, South Carolina, and Alabama," *U.S. Geological Survey Professional Papers,* No. 1403-B, 1984.

[60] Beck, B. F., "Ground Water Monitoring Considerations in Karst on Young Limestones," *Proceedings*, Conference on Environmental Problems in Karst Terranes and Their Solutions, Bowling Green, KY, National Water Well Association, Dublin, OH, 1987, pp. 229–248.

[61] Williams, P. W., "Subcutaneous Hydrology and the Development of Doline and Cockpit Karst," *Zeitschrift für Geomorphologie*, Vol. 29, 1985, pp. 463–482.

[62] Friederich, H. and Smart, P. L., "Dye Tracer Studies of the Unsaturated-Zone Recharge of the Carboniferous Limestone Aquifer of the Mendip Hills, England," *Proceedings*, Eighth International Speleological Congress, Vol. 1, Bowling Green, KY, 1981, pp. 283–286.

[63] Friederich, H. and Smart, P. L. "The Classification of Autogenic Percolation Waters in Karst Aquifers: A Study in GB Cave, Mendip Hills, England," *University of Bristol Speleological Society Proceedings*, Vol. 16, 1982, pp. 143–159.

[64] Smart, P. L. and Friederich, H., "Water Movement in the Unsaturated Zone of a Maturely Karstified Carbonate Aquifer, Mendip Hills, England," *Proceedings*, Conference on Environmental Problems in Karst Terranes and Their Solutions, Bowling Green, KY, National Water Well Association, Dublin, OH, 1987, pp. 59–87.

[65] Smart, P. L., "Catchment Delimitation in Karst Areas by Use of Quantitative Tracer Methods," *Proceedings*, Vol. 2, Third International Symposium on Underground Water Tracing, Ljubljana-Bled, Yugoslavia, 1976, pp. 291–298.

[66] Atkinson, T. C. and Smart, P. L. "Artificial Tracers in Hydrogeology," *A Survey of British Hydrogeology*, The Royal Society, London, 1981, pp. 173–190.

[67] Quinlan, J. F., Ewers, R. O., Ray, J. A., Powell, R. L., and Krothe, N. C., "Groundwater Hydrology and Geomorphology of the Mammoth Cave Region, Kentucky, and of the Mitchell Plain, Indiana," *Field Trips in Midwestern Geology*, Vol. 2, R. H. Shaver and J. A. Sunderman, Eds., Geological Society of America and Indiana Geological Survey, Bloomington, IN, 1983, pp. 1–85.

[68] Gardner, M., *Logic Machines and Diagrams*, 2nd ed., University of Chicago Press, Chicago, 1982, pp. 32, 39–54.

[69] Miller, R. W., *Study Guide, "Introduction to Logic,"* 7th ed., Macmillan, New York, 1986, pp. 75–76, 79, 83–96.

[70] Scanlon, B. R. and Thrailkill, J., "Chemical Similarities Among Physically Distinct Spring Types in a Karst Terrain," *Journal of Hydrology*, Vol. 89, 1987, pp. 259–279.

[71] Thrailkill, J., "Flow in a Limestone Aquifer as Determined from Water Tracing and Water Levels in Wells," *Journal of Hydrology*, Vol. 78, 1985, pp. 123–136.

[72] Thrailkill, J., "Hydrogeology and Environmental Geology of the Inner Bluegrass Karst Region, Kentucky," (Field Guide), Annual Meeting, Geological Society of America, Southeastern and North-Central Sections, Lexington, KY, 1984.

[73] Ewers, R. O., Department of Geology, Eastern Kentucky University, personal communication, January 1988.

[74] Quinlan, J. F. and Ewers, R. O., "Preliminary Speculations on the Evolution of Groundwater Basins in the Mammoth Cave Region, Kentucky," *GSA Cincinnati '81 Field Trip Guidebooks*, Vol. 3, T. G. Roberts, Ed., American Geological Institute, Washington, DC, 1981, pp. 496–501.

Indexes

Author Index

Subject Index